INTERNATIONAL SERIES OF MONOGRAPHS ON PHYSICS

SERIES EDITORS

J. BIRMAN	CITY UNIVERSITY OF NEW YORK
S. F. EDWARDS	UNIVERSITY OF CAMBRIDGE
R. FRIEND	UNIVERSITY OF CAMBRIDGE
M. REES	UNIVERSITY OF CAMBRIDGE
D. SHERRINGTON	UNIVERSITY OF OXFORD
G. VENEZIANO	CERN, GENEVA

INTERNATIONAL SERIES OF MONOGRAPHS ON PHYSICS

153. R. A. Klemm: *Layered superconductors, Volume 1*
152. E.L. Wolf: *Principles of electron tunneling spectroscopy, Second edition*
151. R. Blinc: *Advanced ferroelectricity*
150. L. Berthier, G. Biroli, J.-P. Bouchaud, W. van Saarloos, L. Cipelletti: *Dynamical heterogeneities in glasses, colloids and granular media*
149. J. Wesson: *Tokamaks, Fourth edition*
148. H. Asada, T. Futamase, P. Hogan: *Equations of motion in general relativity*
147. A. Yaouanc, P. Dalmas de Réotier: *Muon spin rotation, relaxation, and resonance*
146. B. McCoy: *Advanced statistical mechanics*
145. M. Bordag, G.L. Klimchitskaya, U. Mohideen, V.M. Mostepanenko: *Advances in the Casimir effect*
144. T.R. Field: *Electromagnetic scattering from random media*
143. W. Götze: *Complex dynamics of glass-forming liquids—a mode-coupling theory*
142. V.M. Agranovich: *Excitations in organic solids*
141. W.T. Grandy: *Entropy and the time evolution of macroscopic systems*
140. M. Alcubierre: *Introduction to 3+1 numerical relativity*
139. A. L. Ivanov, S. G. Tikhodeev: *Problems of condensed matter physics—quantum coherence phenomena in electron-hole and coupled matter-light systems*
138. I. M. Vardavas, F. W. Taylor: *Radiation and climate*
137. A. F. Borghesani: *Ions and electrons in liquid helium*
136. C. Kiefer: *Quantum gravity, Second edition*
135. V. Fortov, I. Iakubov, A. Khrapak: *Physics of strongly coupled plasma*
134. G. Fredrickson: *The equilibrium theory of inhomogeneous polymers*
133. H. Suhl: *Relaxation processes in micromagnetics*
132. J. Terning: *Modern supersymmetry*
131. M. Mariño: *Chern-Simons theory, matrix models, and topological strings*
130. V. Gantmakher: *Electrons and disorder in solids*
129. W. Barford: *Electronic and optical properties of conjugated polymers*
128. R. E. Raab, O. L. de Lange: *Multipole theory in electromagnetism*
127. A. Larkin, A. Varlamov: *Theory of fluctuations in superconductors*
126. P. Goldbart, N. Goldenfeld, D. Sherrington: *Stealing the gold*
125. S. Atzeni, J. Meyer-ter-Vehn: *The physics of inertial fusion*
123. T. Fujimoto: *Plasma spectroscopy*
122. K. Fujikawa, H. Suzuki: *Path integrals and quantum anomalies*
121. T. Giamarchi: *Quantum physics in one dimension*
120. M. Warner, E. Terentjev: *Liquid crystal elastomers*
119. L. Jacak, P. Sitko, K. Wieczorek, A. Wojs: *Quantum Hall systems*
118. J. Wesson: *Tokamaks, Third edition*
117. G. Volovik: *The Universe in a helium droplet*
116. L. Pitaevskii, S. Stringari: *Bose-Einstein condensation*
115. G. Dissertori, I.G. Knowles, M. Schmelling: *Quantum chromodynamics*
114. B. DeWitt: *The global approach to quantum field theory*
113. J. Zinn-Justin: *Quantum field theory and critical phenomena, Fourth edition*
112. R.M. Mazo: *Brownian motion—fluctuations, dynamics, and applications*
111. H. Nishimori: *Statistical physics of spin glasses and information processing—an introduction*
110. N.B. Kopnin: *Theory of nonequilibrium superconductivity*
109. A. Aharoni: *Introduction to the theory of ferromagnetism, Second edition*
108. R. Dobbs: *Helium three*
107. R. Wigmans: *Calorimetry*
106. J. Kübler: *Theory of itinerant electron magnetism*
105. Y. Kuramoto, Y. Kitaoka: *Dynamics of heavy electrons*
104. D. Bardin, G. Passarino: *The Standard Model in the making*
103. G.C. Branco, L. Lavoura, J.P. Silva: *CP Violation*
102. T.C. Choy: *Effective medium theory*
101. H. Araki: *Mathematical theory of quantum fields*
100. L. M. Pismen: *Vortices in nonlinear fields*
99. L. Mestel: *Stellar magnetism*
98. K. H. Bennemann: *Nonlinear optics in metals*
94. S. Chikazumi: *Physics of ferromagnetism*
91. R. A. Bertlmann: *Anomalies in quantum field theory*
90. P. K. Gosh: *Ion traps*
87. P. S. Joshi: *Global aspects in gravitation and cosmology*
86. E. R. Pike, S. Sarkar: *The quantum theory of radiation*
83. P. G. de Gennes, J. Prost: *The physics of liquid crystals*
73. M. Doi, S. F. Edwards: *The theory of polymer dynamics*
69. S. Chandrasekhar: *The mathematical theory of black holes*
51. C. Møller: *The theory of relativity*
46. H. E. Stanley: *Introduction to phase transitions and critical phenomena*
32. A. Abragam: *Principles of nuclear magnetism*
27. P. A. M. Dirac: *Principles of quantum mechanics*
23. R. E. Peierls: *Quantum theory of solids*

Advanced Ferroelectricity

Robert Blinc
Jozef Stefan Institute, Ljubljana, Slovenia

OXFORD
UNIVERSITY PRESS

OXFORD
UNIVERSITY PRESS

Great Clarendon Street, Oxford OX2 6DP

Oxford University Press is a department of the University of Oxford.
It furthers the University's objective of excellence in research, scholarship,
and education by publishing worldwide in

Oxford New York

Auckland Cape Town Dar es Salaam Hong Kong Karachi
Kuala Lumpur Madrid Melbourne Mexico City Nairobi
New Delhi Shanghai Taipei Toronto

With offices in

Argentina Austria Brazil Chile Czech Republic France Greece
Guatemala Hungary Italy Japan Poland Portugal Singapore
South Korea Switzerland Thailand Turkey Ukraine Vietnam

Oxford is a registered trade mark of Oxford University Press
in the UK and in certain other countries

Published in the United States
by Oxford University Press Inc., New York

© R. Blinc 2011

The moral rights of the author have been asserted
Database right Oxford University Press (maker)

First published 2011

All rights reserved. No part of this publication may be reproduced,
stored in a retrieval system, or transmitted, in any form or by any means,
without the prior permission in writing of Oxford University Press,
or as expressly permitted by law, or under terms agreed with the appropriate
reprographics rights organization. Enquiries concerning reproduction
outside the scope of the above should be sent to the Rights Department,
Oxford University Press, at the address above

You must not circulate this book in any other binding or cover
and you must impose the same condition on any acquirer

British Library Cataloguing in Publication Data

Data available

Library of Congress Cataloging in Publication Data

Data available

Typeset by SPI Publisher Services, Pondicherry, India
Printed in Great Britain
on acid-free paper by
CPI Antony Rowe, Chippenham, Wiltshire

ISBN 978–0–19–957094–2

1 3 5 7 9 10 8 6 4 2

Preface

The field of ferroelectricity has greatly expanded and changed recently. In addition to classical organic and inorganic ferroelectrics new fields and materials have appeared, important for both basic science and application and showing technological promise for novel multifunctional devices. Most of these fields were unknown or inactive 20 to 40 years ago. Publications and research activities in some of these fields have been increasing exponentially. Let us first mention the field of multiferroic magnetoelectric systems where time- and space-reversal symmetries are broken simultaneously and where the spontaneous polarization $\vec{P}$ and the spontaneous magnetization $\vec{M}$ are allowed to coexist. It is well known that this is strictly forbidden in conventional ferroelectrics where the presence of the electric-order parameter $\vec{P}$ excludes the existence of the magnetic-order parameter $\vec{M}$ and vice versa. The combination of electric and magnetic degrees of freedom allows for new memory devices, new magnetic field sensors and transducers as well as heterogeneous read/write devices. Spintronic applications are also possible. Another new field are 1D, 2D and 3D incommensurate ferroelectrics where the periodicity of the electric-order parameter wave is incommensurate with the periodicity of the underlying basic crystal lattice. The elementary excitations here are phasons and amplitudons. It is also interesting to note that solitons can be observed here in the ground state. Still another interesting field is ferroelectric liquid crystals (FELC) that flow like liquids but show optical and electrical properties like solid ferroelectrics. Amplitudons and phasons can here be directly observed by light scattering. One can also observe the gapless Goldstone modes that are the result of the breaking of continuous symmetry. FELC have significant applications in electro-optic devices. A specially interesting field are relaxor ferroelectrics and dipolar glasses. Relaxors are charge and site disordered crystals or solid solutions, characterized by nanosized microdomains, i.e. polar nanoclusters that determine the weakly temperature-dependent dielectric response as well as the broad frequency response. In the presence of a strong enough electric field they become ferroelectric. In the vicinity of the critical line they show a giant electromechanical and electrocaloric response that is important for applications in acoustics, robotics, micro and nanoelectronics telecommunications and medicine. In dipolar glasses on the other hand the basic unit is a single dipole and not a polar nanocluster. As a result of that an electric field here can not induce ferroelectricity in contrast to relaxors.

A lot of attention has been recently also paid to nanoferroelectrics, ferroelectric thin films and integrated ferroelectrics. Ferroelectricity is a collective phenomenon and the basic question, how many unit cells must be involved to support ferroelectricity, has reappeared. A few-nm-thick films on a metal substrate have been now shown to be still ferroelectric. Whereas 40 years ago ferroelectric memories were not considered feasible because of too high voltages necessary to switch the polarization in bulk samples, the situation has changed with the use of thin films and nanoferroelectrics. Here, the voltages necessary to reach the critical coercive fields are much lower and are of the order of a few volts in comparison with tenths of kV in bulk samples. Work on non-volatile ferroelectric random access memories (FERAMs) is now going on in different countries. Non-volatile memories are permanently keeping the stored information and make the periodic rejuvenation of the information used in the dynamic random access memories (DRAMs) unnecessary. One should mention the integration of ferroelectric (FE) films on semiconductor integrated circuits ($Ba_xSr_{1-x}TiO_3$ on GaAs) for use in mobile digital telephones. These BST/Ga chips are 50 times smaller than their predecessors. Ferroelectric RAMs are used in microprocessors. The field is thus extremely active and various new applications are appearing steadily.

The book represents a broad review of recent developments in the field of ferroelectricity after 1975, i.e. after the discovery of the soft mode. It is primarily intended for material scientists and physicists working in research or industry. It is also intended for graduate and doctoral students and can be used as a textbook in graduate courses. Finally, it should be useful for everybody following the development of modern solid-state physics.

Acknowledgement: The author would like to acknowledge the help of the Ljubljana ferroelectricity group: Special thanks go to Raša Pirc, Zdravko Kutnjak, Boštjan Zalar, Adrijan Levstik, Cene Filipič, Vid Bobnar, Denis Arčon, Anna Morozovska and M. Glynchuk. Thanks also to Gojmir Lahajnar and Tadeja Samec for their technical help in preparing this manuscript.

Contents

1 Organic, inorganic and composite ferroelectrics   1
    1.1 Introduction   1
    1.2 Organic and inorganic ferroelectrics   2
    1.3 Composite ferroelectrics   7
    1.4 Displacive and order–disorder-type phase transitions   9
    1.5 Quantum paraelectrics and incipient displacive ferroelectrics   12
    1.6 Disorder in displacive ferroelectrics   15
    1.7 New developments in KH_2PO_4-type order–disorder ferroelectrics   19

2 Incommensurate systems   23
    2.1 One-dimensionally modulated incommensurate systems   23
       2.1.1 Commensurate (C) and incommensurate (I) systems   23
       2.1.2 Phase transitions to incommensurate phases   24
       2.1.3 The observation of structurally incommensurate systems   27
       2.1.4 The theoretical side: Solitons, the Devil's staircase and phasons   28
       2.1.5 Dielectric properties   32
          2.1.5.1 Polarization in incommensurate structures   32
          2.1.5.2 The paraphase and the commensurate phase   33
          2.1.5.3 Proper ferroelectrics   33
          2.1.5.4 Improper ferroelectrics   34
          2.1.5.5 The incommensurate phase   35
          2.1.5.6 Proper ferroelectrics   35
          2.1.5.7 Improper ferroelectrics   36
          2.1.5.8 The static dielectric constant   36
       2.1.6 Neutron and X-ray scattering   38
          2.1.6.1 Probing the displacements in incommensurate structures   38
          2.1.6.2 Elastic scattering   39
       2.1.7 Magnetic resonance lineshapes in incommensurate systems   43
          2.1.7.1 Number of resonance lines and frequency distribution   43
       2.1.8 The 'plane-wave' limit: One-dimensional modulation ($m = 1$)   45
          2.1.8.1 Linear case   46
          2.1.8.2 Quadratic case   46
          2.1.8.3 Linear and quadratic terms   47
       2.1.9 The 'phase soliton' limit   47
          2.1.9.1 Soliton density and Landau theory (Fig. 2.15)   47
          2.1.9.2 The NMR lineshape in the multisoliton limit   51
       2.1.10 Phason and amplitudon excitations   52
          2.1.10.1 Dispersion relations   52

2.2	Multidimensionally modulated incommensurate systems		55
	2.2.1	NMR and multidimensional modulation	57
	2.2.2	2-q modulation	59
	2.2.3	Dispersion relations in incommensurate systems and T_1^{-1}	61
	2.2.4	3-q modulation	62
	2.2.5	The determination of the relative phases of the modulation waves	63
	2.2.6	Local and non-local case	64
	2.2.7	Multisoliton lattice limit	67
	2.2.8	Systems with a 6-component order parameter	70
		2.2.8.1 ^{75}As NQR in proustite: The non-planar 3-q case	70
2.3	Conclusions		78

3 Ferroelectric liquid crystals — 79

3.1	Modulated ferroelectric liquid crystals	81
3.2	Excitations in ferroelectric liquid crystals: The Goldstone mode, the amplitudon mode and the soft mode	84
3.3	The magnetic-field-induced Lifshitz point	89
3.4	The electric-field-induced Lifshitz point	91
3.5	Freely suspended ferroelectric smectic thin films	93

4 Dipolar glasses — 95

4.1	Introduction		95
4.2	Local structure determination and local polarization distribution function		96
4.3	Dielectric properties		99
4.4	NMR in homogeneous ferroelectrics and anti-ferroelectrics in the fast-motion regime		100
	4.4.1	Deuteron NMR and relaxation	101
	4.4.2	Oxygen-17–proton nuclear quadrupole double resonance	103
	4.4.3	Proton chemical-shift tensors	108
	4.4.4	Phosphorus-31 chemical-shift tensors	108
	4.4.5	Arsenic-75 quadrupolar coupling	110
4.5	NMR in proton and deuteron glasses		112
	4.5.1	Determination of the Edwards–Anderson order parameter q_{EA} in the fast-motion limit	112
	4.5.2	Determination of the Edwards–Anderson glass order parameter q in the slow-motion limit	115
4.6	NMR determination of order parameters in inhomogeneous ferroelectrics		117
4.7	Theory of dipolar glasses: The random-bond–random-field Ising model		118
4.8	Conclusions		120

5 Magnetoelectric ferroelectrics — 121

5.1	Introduction	121
5.2	The quadratic ME effect in Pb (Fe$_{1/2}$Nb$_{1/2}$)O$_3$	126

	5.3	Ferroelectric polarization reversal by electric and magnetic fields	128
	5.4	The modified Vogel–Fulcher relation in external fields and the polar nanocluster size	129
	5.5	Theory of bi-relaxors	131
		5.5.1 Spherical model of bi-relaxors	133
		5.5.2 Static dielectric properties under constant magnetic field	135
		5.5.3 Dynamic dielectric response	137
		5.5.4 Relaxation of dielectric polarization in magnetic field	139
6	**Relaxor ferroelectrics**		144
	6.1	Introduction	144
	6.2	Specific heat of relaxors	148
	6.3	The rigid spherical random-bond–random-field (SRBRF) model	149
	6.4	Pseudospin phonon coupling	150
	6.5	The SRBRF phase diagram	151
	6.6	Linear and non-linear dielectric response	153
		6.6.1 The difference between relaxors and ferroelectrics	155
	6.7	Ferroelectrics in random fields	159
	6.8	$PbMg_{1/3}Nb_{2/3}O_3$(PMN) and related perovskite relaxors: Phase diagrams, neutron scattering, Raman spectra and heat conductivity	159
	6.9	Effect of pressure	163
	6.10	NMR lineshapes and relaxation times in relaxor PMN: Evidence of polar clusters	165
	6.11	Electric-field-induced critical end points in PMN–PT relaxors and giant electrostriction	170
		6.11.1 Landau theory	173
		6.11.2 Experimental data	176
	6.12	Critical end points up to 8th-order terms	183
7	**Ferroelectric polymers**		187
	7.1	2D ferroelectricity	191
	7.2	Spherical model of relaxor polymers	191
		7.2.1 Polar nanoregions	191
		7.2.2 Free energy	193
		7.2.3 Order parameters	194
	7.3	Dielectric susceptibility	195
		7.3.1 Longitudinal and transverse susceptibilities	195
		7.3.2 Spontaneous polarization	198
		7.3.3 Non-linear susceptibility	199
	7.4	Electrostriction	201
8	**Electrocaloric effect in ferroelectrics and ferroelectric thin films**		203
9	**Ferroelectric thin films**		215
	9.1	The Tilley–Žekš model	219

		9.1.1 The positive–positive case	224
		9.1.2 The negative–negative case	225
		9.1.3 The mixed case	225
	9.2	Misfit-strain-induced magnetoelectric coupling in thin films	226

10 Nanoferroelectrics   230

    10.1 Surface piezoelectric, piezomagnetic and ME tensors   230
    10.2 Size-induced ferroelectricity in non-ferroelectric insulators   235
    10.3 Spontaneous flexoelectric effect in nanoferroics   239
        10.3.1 Basic equations for the flexoeffect in ferroic nanoparticles   240
        10.3.2 Thin pills   241
        10.3.3 Nanowires   241
    10.4 Ferroelectric vortex states: Phase transitions in zero-dimensional nanoferroelectrics   243

Appendix   247

References   251

Subject index   269

1
Organic, inorganic and composite ferroelectrics

1.1 Introduction

Ferroelectrics have been traditionally divided into four different classes:

1. Displacive ferroelectrics where a discrete symmetry group is broken at T_C and the ferroelectric transition can be described as the result of an instability of the anharmonic crystal lattice against a soft polar lattice vibration (e.g., $BaTiO_3$).
2. Order–disorder-type ferroelectrics where a discrete symmetry group is broken due to the ordering of ions in a strongly anharmonic rigid multisite lattice potential (e.g., KH_2PO_4).
3. Ferroelectric liquid crystals where a continuous symmetry group is broken at T_C and the doubly degenerate relaxational soft mode of the high-temperature phase splits below T_C into an 'amplitudon'-type soft mode and a symmetry-restoring Goldstone (i.e. 'phason')-mode (e.g., p-decyloxybenzylidene p'-amino-2-methylbutylcinnamate (DOBAMBC)).
4. Relaxors where there is no macroscopic symmetry breaking and where, in view of 'site' and 'charge' disorder, there is an extremely broad distribution of correlation times. The longest correlation time diverges at the freezing transition, whereas other correlation times are still finite (e.g., $Pb(Mg_{1/3}Nb_{2/3})O_3$). Here, we deal with glassy order and an Edwards–Anderson order parameter rather than with classical ferroelectric long-range order. Instead of a spontaneous polarization we have here a polarization distribution function with zero mean value.

Magnetoelectric systems have been found where the spontaneous polarization, i.e. a polar vector, and the spontaneous magnetization, i.e. an axial vector, coexist and are in some cases linearly coupled. Here, magnetic symmetry has to be taken into account in addition to spatial symmetry (Fig. 1.1).

The recent advances in the field of ferroelectricity are reflected in the number of publications and scientific meetings in this field. They have been increasing exponentially in the last 30 years. Breakthroughs have been achieved in the general understanding of this field and new physics, new subfields and new applications have emerged. Many new systems have been discovered. In addition to the classical fields of phase transitions, soft modes and critical phenomena, research has focused on the new fields of magnetoelectric systems, ferroelectric polymers, ferroelectric liquid crystals,

2 Organic, inorganic and composite ferroelectrics

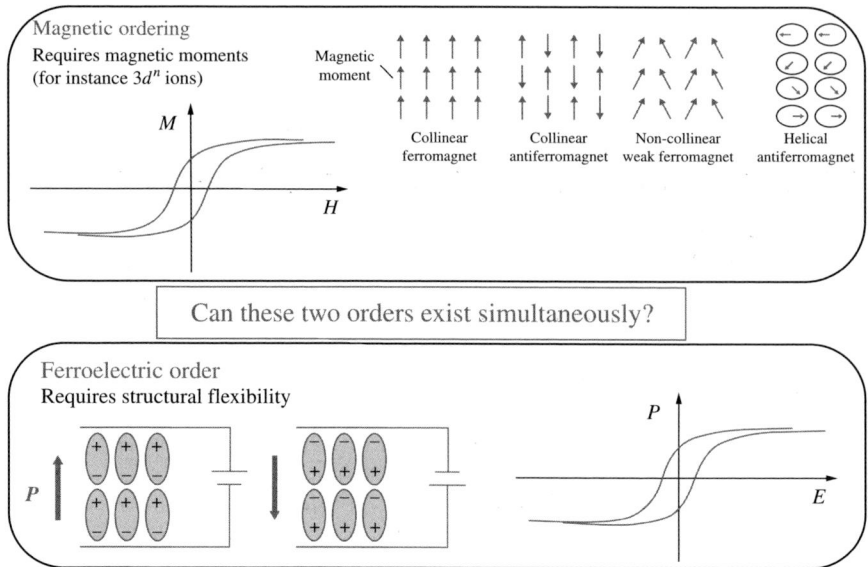

Fig. 1.1 Magnetic ordering and ferroelectric ordering in magnetoelectrics (courtesy of D. Arčon).

relaxors, dipolar glasses, incommensurate systems and thin films as well as integrated ferroelectrics. Quantum effects have been observed as well.

According to the database of the Institute of Scientific Information in Philadelphia, USA, the number of publications in the field of ferroelectrics has increased from 200 on file in 1970 to about 20 000 in 2001 (Fig. 1.2). Whereas the International Meeting on Ferroelectricity in Prague in 1966 was the first conference of its kind, there are at present at least seven kinds of conferences devoted to various aspects of ferroelectric research (Table 1.1): IMF meetings from 1966, European Meetings on Ferroelectricity (EMF) from 1996, International Symposia on Applied Ferroelectrics (ISAF) from 1971, European Conference on Applications of Polar Dielectrics (ECAPD) from 1988, International Symposia on Ferroelectric Liquid Crystals (FLC) from 1987, International Symposia on Integrated Ferroelectrics (ISIF) from 1989 and Asian Conferences on Ferroelectricity.

A partial list of meetings on ferroelectricity is shown in Table 1.1. The increase in the number of publications in the field of ferroelectricity between 1979 and 2009 is illustrated in Figs. 1.2(a) and (b).

1.2 Organic and inorganic ferroelectrics

Ferroelectrics are polar substances [1.1–1.32] that occur in either solid or polymeric or liquid-crystalline form and where the spontaneously generated electric polarization can be reversed by inverting the external electric field. The number of known

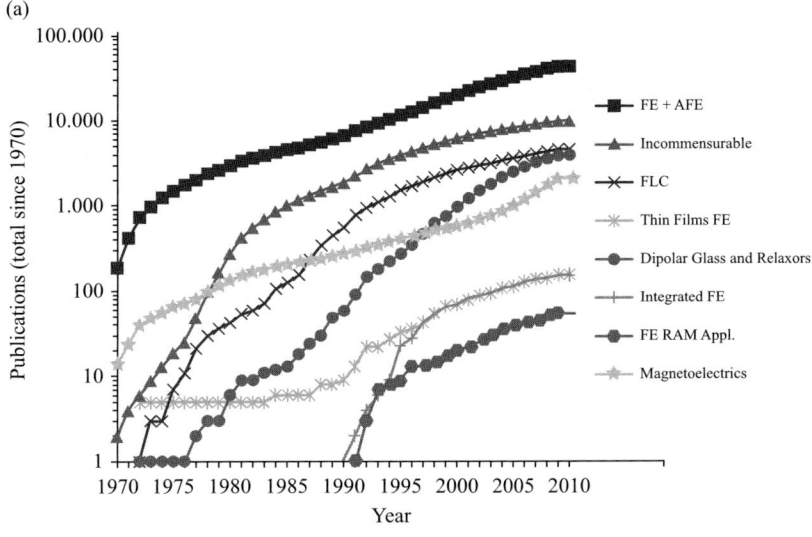

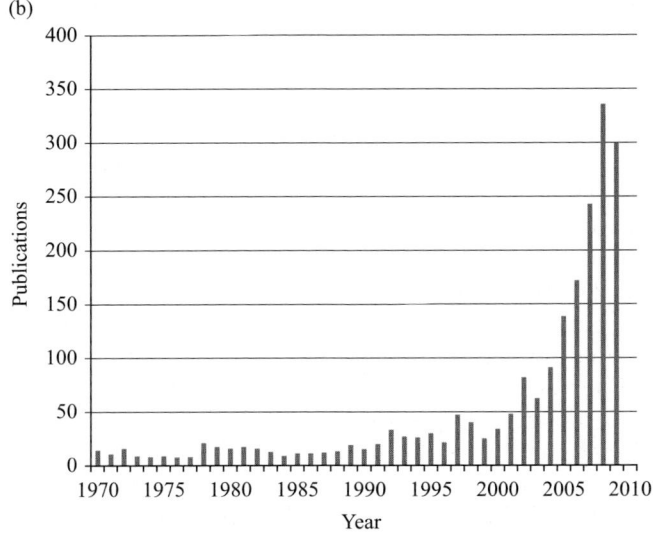

Fig. 1.2 (a) Publications in the field of ferroelectricity between 1970 and 2009* (* = until 29.10.2009). (b) Publications in the field of magnetoelectrics between 1970 and 2009* (* = until 29.10.2009).
Source: ISI-Web of Science.

organic and inorganic ferroelectrics has been greatly enhanced recently. The first known ferroelectric is Rochelle salt [1.33]. This is an inorganic-organic compound suitable for piezoelectric applications. There are more than 700 ferroelectrics known so far. Most ferroelectrics are inorganic oxides, like $BaTiO_3$ and $PbTiO_3$ [1.1]. Fluoride ferroelectrics have been discovered, too [1.34]. Because of that only some

Table 1.1 Meetings on ferroelectricity.

ISIF (International Symposium on Integrated Ferroelectrics):		ISAF (International Symposium on Applied Ferroelectrics):	
1st	1989 Colorado, USA;	1st	1971 New York, USA;
2nd	1990 Monterey, USA;	2nd	1975 New Mexico, USA;
3rd	1991 Colorado, USA;	3rd	1979 Minnesota, USA;
4th	1992 Monterey, USA;	4th	1983 Maryland, USA;
5th	1993 Colorado, USA;	5th	1986 Pennsylvania, USA;
6th	1994 Monterey, USA;	6th	1988 Zürich, Switzerland;
7th	1995 Colorado, USA;	7th	1990 Illinois, USA;
8th	1996 Tempe, USA;	8th	1992 South Carolina, USA;
9th	1997 New Mexico, USA;	9th	1994 Pennsylvania, USA;
10th	1998 Monterey, USA;	10th	1996 Brunswick, NJ, USA;
11th	1999 Colorado, USA;	11th	1998 Montreux, Switzerland;
12th	2000 Aachen, Germany;	12th	2000 Honolulu, USA;
13th	2001 Colorado, USA;	13th	2002 Nara, Japan;
14th	2002 Nara, Japan;	14th	2004 Montreal, Canada;
15th	2003 Colorado, USA;	15th	2006 North Carolina, USA;
16th	2004 Gyeongju, South Korea;	16th	2007 Nara, Japan;
17th	2005 Shanghai, China;	17th	2008 New Mexico, USA;
18th	2006 Honolulu, USA;	18th	2009 Xi'an, China.
19th	2007 Bordeaux, France;		
20th	2008 Singapore;		
21th	2009 Colorado, USA.		

IMF (International Meeting on Ferroelectricity):		EMF (European Meeting on Ferroelectricity):	
1st	1966 Prague, Czechoslovakia;	1st	1969 Saarbrücken, Germany;
2nd	1969 Kyoto, Japan;	2nd	1971 Dijon, France;
3rd	1973 Edinburgh, UK;	3rd	1975 Zürich, Switzerland;
4th	1977 Leningrad, USSR;	4th	1979 Portorož, Slovenia;
5th	1981 Pennsylvania, USA;	5th	1983 Málaga, Spain;
6th	1985 Kobe, Japan;	6th	1987 Poznan, Poland;
7th	1989 Saarbrücken, Germany;	7th	1991 Dijon, France;
8th	1993 Maryland, USA;	8th	1995 Nijmegen, Netherlands;
9th	1997 Seoul, Korea;	9th	1999 Prague, Czech Republic;
10th	2001 Madrid, Spain;	10th	2003 Cambridge, UK;
11th	2005 Foz do Iguacu, Brazil;	11th	2007 Bled, Slovenia.
12th	2009 Xi'an, China.		

FLC (International Symposium on Ferroelectric Liquid Crystals):

1st	1987	Bordeaux, France;
2nd	1989	Göteborg, Sweden;
3rd	1991	Colorado, USA;
4th	1993	Tokyo, Japan;
5th	1995	Cambridge, UK;
6th	1997	Brest, France;
7th	1999	Darmstadt, Germany;
8th	2001	Washington, USA;
9th	2003	Dublin, Ireland;
10th	2005	Warsaw, Poland;
11th	2007	Sapporo, Japan;
12th	2009	Zaragoza, Spain.

ECAPD (European Conference on Applications of Polar Dielectrics):

1st	1988	Zürich, Switzerland;
2nd	1992	London, UK;
3rd	1996	Bled, Slovenia;
4th	1998	Montreux, Switzerland;
5th	2000	Jurmala, Latvia;
6th	2002	Aveiro, Portugal;
7th	2004	Liberec, Czech Republic;
8th	2006	Metz, France;
9th	2008	Rome, Italy.

representative examples of organic and inorganic ferroelectrics are listed in Table 1.2. The critical electric field for reversing the polarization is called the coercive field. The electric displacement versus electric field ($D - E$ loop) relation can be used for non-volatile memory elements such as FERAM (ferroelectric random access memories) and ferroelectric field effect transistors (FET). As the temperature approaches the paraelectric–ferroelectric transition temperature T_c, the dielectric constant, obeying the Curie–Weiss law, becomes very large. This effect can be exploited for high dielectric constant (high ε) condensers or capacitors. Ferroelectrics are also piezoelectrics. This can be exploited in piezoelectric transducers and sensors.

Another important feature is pyroelectricity. The temperature dependence of the spontaneous polarization produces an electric current on heating or cooling. This can be used as an infrared detector or thermal imaging sensor. The polar crystal structure also allows for second-order harmonic generation and a linear electro-optic effect. Both effects are important in electro-optics.

Piezoelectric, dielectric and electrostrictive responses are particularly large near the morphotropic phase boundary (MPB) between the tetragonal (T) and rhombohedral (R) phases in some perovskite ferroelectrics [1.35]. The corresponding Landau-type potential is

$$f = \frac{\alpha}{2}\left(p_1^2 + p_2^2 + p_3^2\right) + \frac{\beta_1'}{4}\left(p_1^4 + p_2^4 + p_3^4\right) + \frac{\beta_2'}{2}\left(p_1^2 p_2^2 + p_2^2 p_3^2 + p_3^2 p_1^2\right) \quad (1)$$

Here, p_1, p_2 and p_3 are components of the spontaneous polarization, $\alpha = a(T - T_C)$ and β_1', β_2' are constants. The condition for the MPB is $\beta_1' = \beta_2'$. The Landau potential now becomes:

$$f = \frac{\alpha}{2}\left(p_1^2 + p_2^2 + p_3^2\right) + \frac{\beta_1'}{4}\left(p_1^2 + p_2^2 + p_3^2\right)^2 \quad (2)$$

Table 1.2 Some examples of (A) Organic ferroelectrics and (B) Inorganic ferroelectrics.

Materials	References	Transition temperature (K)		C ($\times 10^3$ K)	P_s (μC cm^{-2}) temperature	E_c (kV cm^{-1})
		T_c	T_c^D			
A. Organic compounds						
Thiourea	1.7	169	185	3.7	3.2; 120 K	0.2
TEMPO	1.8	287	288	-	0.5	-
CDA	1.9	397		-	-	-
TCAA	1.10, 1.11	355		0.0076	0.2; RT	4
Benzil	1.12	84	88		3.6 × 10^{-3}; 70 K*	-
DNP	1.13, 1.14	46		0.026	0.24; 10 K*	-
TCHM	1.15	104		2.7	6 × 10^{-3}; 96 K*	-
VDF oligomer	1.16	-			13; RT	1200
Croconic acid ($H_2C_5O_5$)	1.17	> 450			21	11
CT complexes						
TTF-CA	1.18	81	84	5.7	-	-
TTF-BA	1.19	50		-	-	-
Phz–H_2ca	1.20	253	304†	5.0	1.8; 160 K*	0.8
Phz–H_2ba	1.20	138	204†	4.0	0.8; 105 K*	0.5
[H-55dmbp][Hia]	1.22	269	338†	14	4.2*; 110 K*	2
Clathrate						
β-Quinol–methanol	1.23	63.7		-	6 × 10^{-3}; 25 K*	-
Polymers						
VDF$_{0.65}$-TrFE$_{0.35}$	1.24	363		-	8; RT	500
Nylon-11	1.25	-		-	5; RT	600
B. Inorganic compounds						
BiFeO$_3$	1.23	1100			55; RT	
NaNO$_2$	1.25	437		4.7	10; 140 K	5
BaTiO$_3$	1.25	381		150	26; RT	10
PbTiO$_3$	1.25	763		410	75; RT	7
SbSI	1.25	293		233	20; 270 K	-

KH_2PO_4(KDP)	1.25	123	213	2.9	5.0	0.1
$HdabcoReO_4$	1.26	374		-	16; RT*	> 30
TGS	1.27, 1.25	323	333	3.2	3.8; 220 K	0.9
TSCC	1.25	127		-	0.27; 80 K	3
Rochelle salt	1.25	297	308	2.24	0.25; 276 K	0.2
$K_3Fe_5F_{15}$	1.4	490				

β-Quinol, hydroquinone; CDA, cyclohexan-1, 1′-diacetic acid; dabco, diazabicyclo[2.2.2]octane; DNP, 1,6-bis(2,4-dinitrophenoxy)-2,4-hexadiyne; TCAA, trichloroacetamide; TCHM, tricyclohexylmethanol; TEMPO, 2,2,6,6-tetramethyl-1-piperidinyloxy (tanane); TGS, triglycine sulphate; TSCC, Tris-sarcosine calcium chloride; VDF oligomer, $CF_3(CH_2CF_2)_{17}I$ (thin film); $VDF_{0.65}$-$TrFE_{0.35}$, vinylidene fluoride–trifluoroethylene copolymer $((CH_2CF_2)_{0.65}(CHF-CF_2)_{0.35})_n$; H-55dmbp, H_2ca with 5.5′-dimethyl-2.2′-bipyridine; T_c^D, transition point for deuterated compound, C, paraelectric Curie constant; P_s, spontaneous polarization; E_c, coercive field; RT, room temperature.
* Pyroelectric charge measurements.
† Deuterated only in the hydrogen bond.

At the MPB the equilibrium values of $\overline{P} = (p_1, p_2, p_3)$ are $p_1 = p_2 = 0$ and $p_3 \neq 0$. The Landau potential becomes isotropic. The essential point for the giant response near the MPB is the transversal instability (Fig. 1.3) [1.35], allowing for large fluctuations in P without a significant change in f.

Significant breakthroughs have also been made in organic ferroelectrics [1.3] like croconic acid, $H_2C_5O_5$, as well as inorganic ones like $BaTiO_3$- and $SrTiO_3$- [1.4] and KH_2PO_4- type systems [1.5, 1.6]. This will be discussed later.

It should be mentioned that croconic acid has, in spite of its small size, the largest polarization among all low molecular mass organic ferroelectric crystals. The nominal Curie temperature is also the highest. The crystal decomposes before the paraelectric phase is reached. The polarity reversal is achieved by a simultaneous intermolecular proton transfer from an O—H...O bond to an O...H—O bond without any molecular reorientation (Fig. 1.4). Such a proton tautomerism requires much less energy to overcome the steric hindrance for polarity reversal than bulk molecular rotation. This is rather different from conventional organic ferroelectrics where molecular dipoles can be reversed only by a bulk molecular rotation as for instance in squaric acid and leads to operations at small voltages. Thus, new electronic and opto-electronic applications should be possible in view of the large optical non-linearities involved [1.17].

1.3 Composite ferroelectrics

Composite ferroelectrics are mainly man made. Until recently, research of ferroelectrics and multiferroics was focused on single-phase materials. The recent revival of magnetoelectric research [1.29] has also revived the research on composite ferroelectrics and other composite materials and multiphase systems, where electric polarization and spontaneous magnetization coexist.

8 Organic, inorganic and composite ferroelectrics

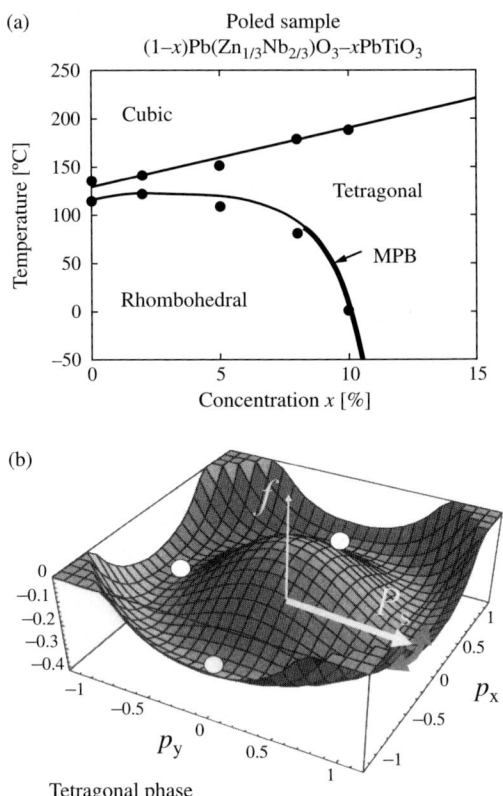

Fig. 1.3 (a) Phase diagram in PZN-PT and (b) thermodynamic potential in the tetragonal phase near the MPB [1.35].

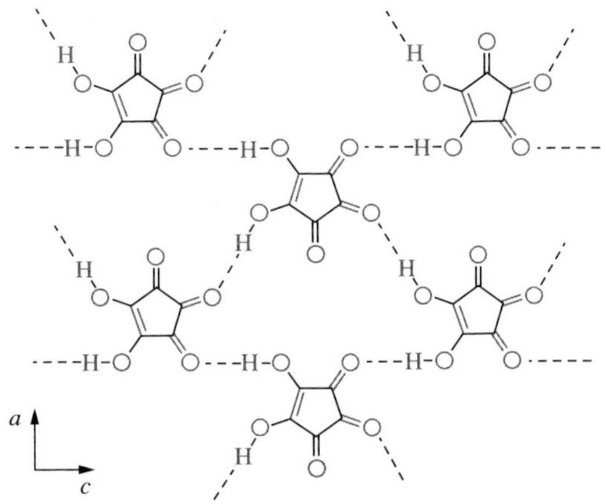

Fig. 1.4 Structure and positions of the protons in the hydrogen bonds of croconic acid. The polarization points along the c-axis.

Table 1.3 Magnetoelectric coupling constants in two-phase systems (following Scott [1.29]).

	Material	Coupling Constant [mV cm^{-1} Oe^{-1}]	Reference
Composite	BaTiO$_3$ and CoFe$_2$O$_4$	50	1.3
Composite	Terfenol-D and PZT in polymer matrix	42	1.4
Laminated Composite	Terfenol-D in polymer matrix/PZT in polymer matrix	3000	1.5

Here, Terfenol-D stands for ($Tb_x Dy_{1-x} Fe_2$).

Magnetoelectric coupling [1.29] may arise indirectly via strain or directly as a linear coupling between the two order parameters, e.g., the spontaneous polarization and the spontaneous magnetization (see Chapter on Magnetoelectrics).

The most important class of composites are those exhibiting product properties. Here, novel effects are present at the composites but not at the ends of the concentration scale of the mixed compounds. This is, e.g., achieved by combining piezoelectric and magnetostrictive compounds. A prominent example is ferroelectric and piezoelectric BaTiO$_3$ and ferromagnetic and piezomagnetic CoFe$_2$O$_4$ [1.1]. A magnetic field applied to the composite will induce a strain in the piezoelectric component and therefore an electric polarization.

In addition to the pioneering work in BaTiO$_3$/CoFe$_2$O$_4$ a number of other titanate/ferrite systems has been investigated.

Some examples of composite magnetoelectric systems are presented in Table 1.3.

Ferroelectrics exhibit a spontaneous polarization due to long range dipolar interactions. Relaxors, on the other hand, show no macroscopic polarization. Short-range interactions at the mesoscopic scale here lead to broad anomalies of the dielectric properties and to a large dielectric dispersion. It is usually very difficult to find these two behaviors to coexist in the same material if no external electric field is applied. Recently, such a coexistence was found in a novel solid solution Ba$_2$Pr$_x$Nd$_{1-x}$FeNb$_4$O$_{15}$ (Fig. 1.5) [1.36]. By increasing the temperature a ferroelectric to relaxor cross-over is found for the Pr content between 20% and 80%. The cross-over from long-range order by varying the temperature is here ascribed to the tetragonal tungsten bronze network, which is much more open and flexible than the perovskite structure.

1.4 Displacive and order–disorder-type phase transitions

In 1959 Cochran [1.37] showed that in weakly anharmonic systems, ferroelectric phase transitions are the result of an instability of the lattice against a soft polar phonon at the centre of the Brillouin zone

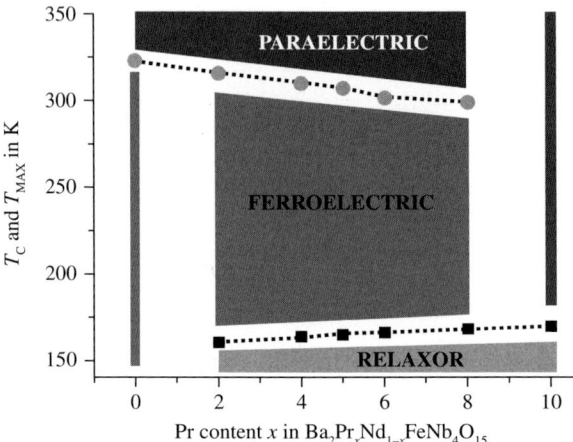

Fig. 1.5 Qualitative phase diagram of $Ba_2Pr_xNd_{1-x}FeNb_4O_{15}$ showing coexistence between ferroelectricity and the relaxor state in the intermediate composition range [1.36].

$$\omega^2(q) = K(T - T_0) + A^2(\vec{q} - \vec{q}_0)^2 + \ldots \qquad (3)$$

As $T \to T_0, \omega \to 0$ for $\vec{q} = \vec{q}_0 = 0$. The structure of the new phase is determined by the eigenvector of the soft mode and the structure of the high-temperature phase. As for ferroelectrics $\vec{q}_0 = 0$, the displacements are identical for all unit cells and there is no change in the number of atoms per unit cell. For anti-ferroelectrics, $\vec{q}_0 \neq 0$, and the magnitude and direction of $\vec{q}_0$, which lies at the Brillouin zone boundary, determines the size of the new unit cell. The static part of the soft-mode eigenvector is in both cases the order parameter of the transition.

The nature of the soft mode is best seen from the early theory of Slater [1.38]. The equation of motion of a Ti ion in $BaTiO_3$ along the ferroelectric x-axis can be written as

$$m\ddot{x} + \gamma\dot{x} + (K_S - K_L + B \cdot T)x = eE_0 e^{i\omega t} \qquad (4)$$

Here, K_S is the short range restoring force constant, K_L is the long-range electrostatic force which tends to drag the ion away from its equilibrium position and $B \cdot T$ is the effective anharmonic restoring force constant that arises from the inclusion of anharmonic terms in lowest order. Expressing the local polarization P as

$$P = N \cdot e \cdot x = \chi E_0 e^{i\omega t} \qquad (5)$$

where N is the number of ions per unit volume, e the effective charge and χ the dielectric susceptibility, we find by solving Eq. (4)

$$\chi = \frac{Ne^2/m}{\overline{\omega}^2 - \omega^2 + i\omega\gamma/m} \qquad (6)$$

where the soft-mode frequency $\bar{\omega}$ is given by

$$\bar{\omega}^2 = (BT - K_L + K_S)/m = \frac{B}{m}\left(T - \frac{K_L - K_S}{B}\right) \propto (T - T_0) \tag{7}$$

and

$$T_0 = (K_L - K_S)/B \tag{8}$$

Obviously here, for $\omega = 0$ a Curie–Weiss law is found:

$$\chi(0) = \frac{Ne^2/m}{\bar{\omega}^2} = \frac{C}{T - T_0} \tag{9}$$

Whereas in displacive soft-mode systems, we deal with a single-well, weakly anharmonic potential, this is not the case in order–disorder systems. Here, we have a strongly anharmonic double- or multiple-well potential. In case of a symmetric double-well potential the system can be described by an Ising model in a transverse-field Hamiltonian [1.39]

$$\mathcal{H} = -\Omega \sum_i S_i^x - \frac{1}{2}\sum_{i,j} J_{ij} S_i^z S_j^z \tag{10}$$

Ω is the tunnelling integral measuring the overlap of the wavefunctions in the 'left' and 'right' potential well, whereas J_{ij} represents the coupling between different double-well potentials i and j. The pseudospins $\vec{S}_i = (S_i^x, S_i^y, S_i^z)$ are vectors, the components of which are the three spin-$1/2$ Pauli matrices

$$S^x = \begin{vmatrix} 0 & 1/2 \\ 1/2 & 0 \end{vmatrix}, \quad S^y = \begin{vmatrix} 0 & -i/2 \\ i/2 & 0 \end{vmatrix}, \quad S^z = \begin{vmatrix} 1/2 & 0 \\ 0 & -1/2 \end{vmatrix} \tag{11}$$

The expectation value of S^z measures the difference in the occupation of the 'left' and 'right' equilibrium site in the double-well potential. Within the molecular-field approximation (MFA) the Hamiltonian (10) is replaced by an effective one

$$\mathcal{H}_i^{MFA} = -\vec{H}_i \vec{S}_i \tag{12}$$

where $\vec{H}_i = \left(\Omega, 0, \sum_i J_{ij}\langle S_i^z \rangle\right)$ is the molecular field and the brackets designate the thermal expectation value.

The expectation value of the pseudospin at site i is now obtained as

$$\langle \vec{S}_i \rangle = \frac{\text{Tr}\, \vec{S}_i\, e^{-\beta \mathcal{H}_i^{MFA}}}{\text{Tr}\, e^{-\beta \mathcal{H}_i^{MFA}}} = \frac{1}{2}\frac{\vec{H}_i}{H_i}\text{tgh}\left(\frac{1}{2}\beta H_i\right) \tag{13}$$

where $\beta = 1/(kT)$.

12 Organic, inorganic and composite ferroelectrics

In the paraelectric phase $\langle S^z \rangle = 0$. Below T_C, $\langle S^x \rangle$ and H_z are different from zero and the molecular field points along a general direction in the x-z plane. $\langle S^z \rangle$ is now obtained as

$$\langle S_i^z \rangle = \frac{1}{2} \frac{\sum_i J_{ij} \langle S_i^z \rangle}{H_i} \operatorname{tgh}\left(\frac{1}{2}\beta H_i\right) \tag{14}$$

where $H_i = |\vec{H}_i| = \left[\Omega^2 + \left(\sum_j J_{ij} \langle S_i^z \rangle\right)^2\right]^{1/2}$. The transition temperature is found from

$$\frac{2\Omega}{J_0} = \operatorname{tgh}\left(\frac{1}{2}\beta_C \Omega\right), \quad \beta_C = 1/(kT_C) \tag{15}$$

where $J_0 = \sum_j J_{ij}$.

If 2Ω is larger than J_0, no ordering can occur and there is no real solution for T_C. Expression (15) predicts a significant isotope effect if the mass of the ion m increases – e.g., on deuteration – or if Ω decreases when the distance between the two equilibrium sites increases.

The Heisenberg equations of motion predict a free precession of the pseudospins around the molecular field. There is a soft pseudospin wave that condenses at T_C similarly as a soft phonon in displacive systems.

It should be mentioned that recent work by Dalal and Bussmann-Holder [1.5] has shown that in H-bonded order–disorder systems like squaric acid a small displacive component exists in addition to the order–disorder one.

1.5 Quantum paraelectrics and incipient displacive ferroelectrics

Quantum fluctuations of the electrical polarization close to a continuous zero-temperature phase transition (quantum critical point) produce rather unconventional forms of the temperature dependence of the dielectric constant $\varepsilon^{-1} \propto T^2$ and other quantities such as the thermal expansion coefficient and soft-mode frequencies. Quantum zero-point fluctuations may in such a case even suppress the onset of ferroelectricity. Such materials are called 'quantum paraelectrics' after Müller *et al.* [1.40].

The situation is illustrated in Fig. 1.6, where the schematic phase diagram of a ferroelectric quantum critical point is shown. The paraelectric and ferroelectric regions are separated by a second-order phase transition line (solid) that terminates at the quantum critical point (QCP) at zero temperature. χ^{-1} designates the inverse dielectric susceptibility and the dashed line the transition between the paraelectric and quantum paraelectric, respectively quantum critical, regime. The preferred quantum tuning parameter is hydrostatic pressure, which can be used to vary the soft-mode frequency and T_c to zero. In some cases pure materials fall into this region at ambient pressure, as seems to be the case with SrTiO$_3$ and KTaO$_3$. Above 50 K we generally observe a cross-over to classical behaviour.

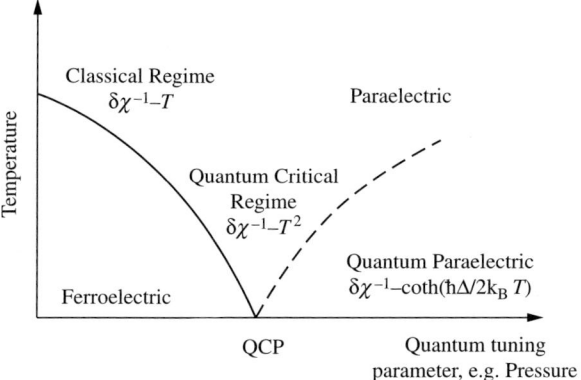

Fig. 1.6 Schematic phase diagram of the ferroelectric quantum critical point. The quantum tuning parameter used to suppress the transition towards zero temperature could be pressure or alternatively chemical or isotopic substitution [1.41].

The simplest systems of this kind are displacive [1.39] ferroelectrics where the lowest transverse optical phonon mode becomes soft at the zone centre and eventually freezes out with zero frequency at a critical temperature T_c. Below this temperature the inversion lattice symmetry is broken and spontaneous polarization due to ion displacements appears.

The results at low temperatures for $SrTiO_3$ and $KTaO_3$ are shown in Figs. 1.7, 1.8 and 1.9 [1.41, 1.42]. Both are close to the quantum critical point but on the paraelectric side (Fig. 1.6). Elemental substitution and strain [1.43, 1.44] lead to finite-temperature ferroelectricity. Substituting the ^{16}O isotope in $SrTiO_3$ with ^{18}O, for example, results in a ferroelectric transition at 25 K [1.45]. $SrTi^{18}O_3$ is thus at ambient pressure at the left of the quantum critical point (Fig. 1.6). It should be noted that at high

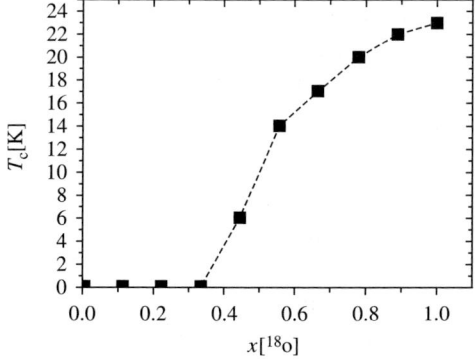

Fig. 1.7 Theoretical concentration (x) dependence of the ferroelectric phase transition temperature T_c in ^{18}O substituted $SrTiO_3$. The points for $x < 0.35$ do not refer to $T_c = 0$ K, but indicate that quantum fluctuations suppress the ferroelectric instability [1.42].

14 Organic, inorganic and composite ferroelectrics

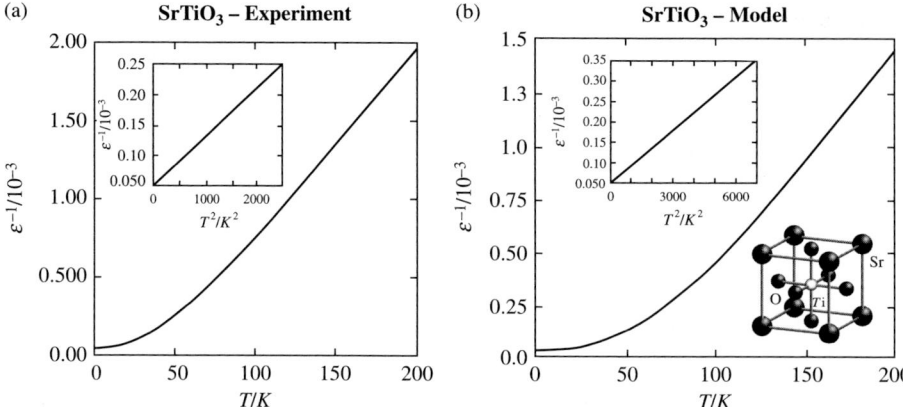

Fig. 1.8 Experimental (a) and self-consistent phonon model results (b) for the inverse dielectric constant in SrTiO$_3$. The upper left insets show the inverse dielectric constant plotted against temperature squared. The lower right inset in (b) shows the crystal structure of SrTiO$_3$ [1.41].

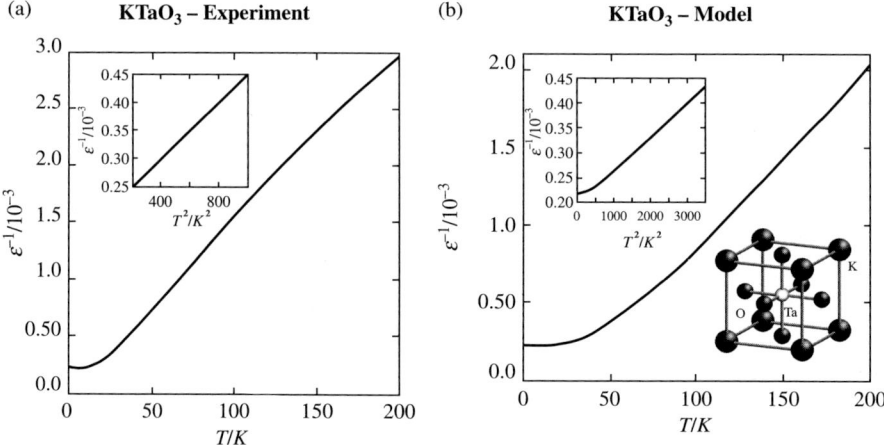

Fig. 1.9 Experimental (a) and self-consistent phonon model results (b) for the inverse dielectric constant in KTaO$_3$. The upper left insets show the inverse dielectric constant plotted against temperature squared. The lower right inset in (b) shows the crystal structure of KTaO$_3$ [1.41].

temperatures the inverse dielectric constant of SrTiO$_3$ (Fig. 1.7) varies according to the Curie–Weiss law. Extrapolating the classical χ^{-1} fit indicates an expected T_c of 35 K. However, at temperatures below 50 K the inverse dielectric constant starts to flatten out. Below this cross-over temperature, SrTiO$_3$ enters a region dominated by quantum critical fluctuations and $\varepsilon^{-1} \propto T^2$ (Fig. 1.8).

The results for KTaO$_3$ (Fig. 1.9) are similar to those of SrTiO$_3$. However, KTaO$_3$ is further from the quantum critical point and deeper in the paraelectric regime than SrTiO$_3$. The cross-over from classical to quantum behaviour here is found around 30 K.

For theoretical work on quantum effects in ferroelectrics see references [1.41, 1.46, 1.47, 1.48, 1.49]. It should be pointed out that a classical to quantum cross-over has also been observed in proton glasses as described at the end of this chapter and in the chapter on dipolar glasses.

1.6 Disorder in displacive ferroelectrics

Phase transitions in BaTiO$_3$ and SrTiO$_3$ are generally considered to be classical examples of displacive soft-mode-type transitions [1.41, 1.42] describable by anharmonic lattice dynamics [1.43, 1.50]. The cubic (Pm$\bar{3}$m) to tetragonal (P4mm) phase transition in BaTiO$_3$ around 415 K occurs at the Brillouin-zone centre. The cubic (Pm$\bar{3}$m) to tetragonal (I4/mcm) transition at 150 K in SrTiO$_3$, on the other hand, occurs at the Brillouin-zone boundary. It is anti-ferrodistortive and accompanied by a doubling of the unit cell size. Inelastic neutron-scattering data [1.51] have clearly revealed phonon softening both in cubic BaTiO$_3$ and in cubic SrTiO$_3$ [1.52, 1.53]. Itoh et al. [1.45] discovered a ferroelectric state in ^{18}O enriched SrTiO$_3$ below $T = 25$ K.

The validity of the transverse optical mode softening scenario of the paraelectric to ferroelectric transition has been first challenged by electron paramagnetic resonance (EPR) measurements performed on crystals doped with paramagnetic centres [1.54]. These studies revealed the existence of strong anharmonicity of the local potential of Ti ions. In support of EPR results diffuse X-ray scattering experiments in BaTiO$_3$ [1.55] have been interpreted in terms of pretransitional correlated clusters of quasi-static off-centre Ti ions as expected for an order–disorder transition. It has, however, been shown [1.50] that this scattering may also result from a strong anisotropy of the soft phonon-dispersion relation. The EPR results could be similarly due to doping and not representative of pure crystals.

The problem has been recently solved by quadrupole perturbed ^{47}Ti($I = 5/2$) and ^{49}Ti($I = 7/2$) NMR [1.4] of BaTiO$_3$, SrTiO$_3$ and ^{18}O enriched SrTiO$_3$ single crystals (Fig. 1.10).

The first-order satellites ($\pm 1/2 \rightarrow \pm 3/2$, $\pm 3/2 \rightarrow \pm 5/2$, $\pm 5/2 \rightarrow \pm 7/2$) have been observed as a broad background in the high-temperature cubic paraelectric phase and as sharp lines in the low-temperature phase demonstrating the presence of non-zero quadrupole coupling at the Ti sites in all these systems. This is incompatible with the central position of the Ti ions in the oxygen octahedron and requires the presence of off-centre Ti sites and dynamic disorder even in the paraelectric cubic phase [1.55]. It should be mentioned that EXAFS studies [1.56] have also shown the existence of off-centre positions of Ti ions in tetragonal – but not cubic – BaTiO$_3$ and some other related perovskites.

NMR studies [1.4] have further shown that the Ti potential surface does not change significantly at the cubic–tetragonal transition. It does not exhibit a minimum at the centre of the oxygen cage but has eight (Fig. 1.11) off-centre minima [1.57] between

16 Organic, inorganic and composite ferroelectrics

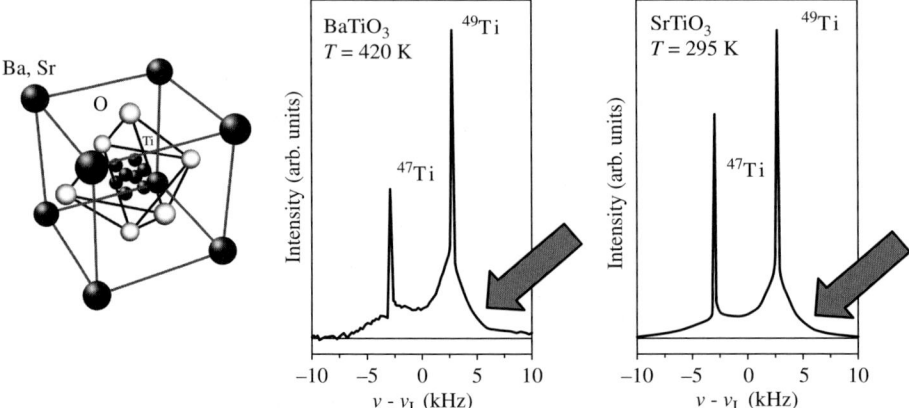

Fig. 1.10 Quadrupole perturbed ^{47}Ti and ^{49}Ti NMR spectra of BaTiO$_3$ and SrTiO$_3$ above T_c. Broken cubic symmetry in the paraelectric phase high above T_C and corresponding dynamic tetragonal distortions result in a non-zero quadrupole coupling at the Ti sites [1.4].

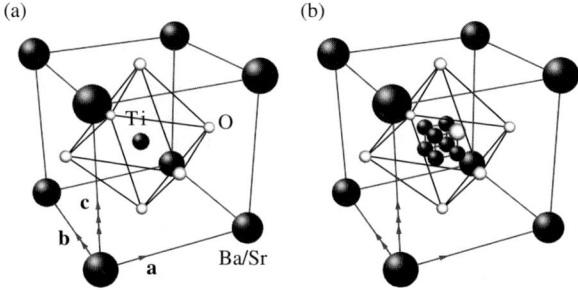

Fig. 1.11 Possible Ti ion positions, compatible with the O_h symmetry in the paraelectric phase: (a) static Ti ion at the centre of the oxygen octahedron; (b) Ti disordered between eight off-centre displaced sites along the [111] directions. Ti shifts are greatly exaggerated.

which the Ti ion is dynamically disordered. The agreement between the theoretical and experimental ratio of the satellite vs. central ($1/2 \rightarrow -1/2$) line intensities shows that all Ti ions are off-centre.

The pure 8-site 'order–disorder' model, however, does not agree with the NMR experiment. A good agreement with the experimental data is obtained if in addition to the dynamic Ti disorder, the 'displacive' soft-mode-induced deformation of the unit-cell shape and the existence of dynamic tetragonal nanodomains, present already in the cubic phase [1.4], is taken into account (Fig. 1.12).

The angular dependencies of the second moments of the ^{49}Ti quadrupole perturbed NMR spectra (Fig. 1.13) of cubic BaTiO$_3$ at 450 K for the [1-10] $\perp \vec{B}_0$ crystal rotation and of cubic SrTiO$_3$ at 294 K for the [100] $\perp \vec{B}_0$ crystal rotation are shown in Fig. 1.13. The solid line is the theoretical displacement scenario.

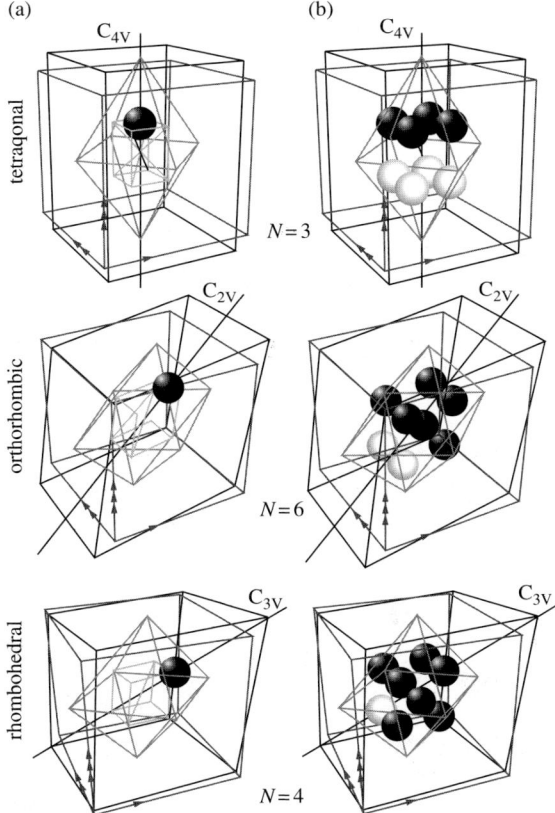

Fig. 1.12 (a) Displacive soft-mode-induced distortions of the perovskite unit cell with off-centre Ti ions. The undistorted cubic frame is shown in light gray with arrows denoting the [100] cubic axes. (b) Order–disorder-type configurations with Ti ions exhibiting biased dynamical exchange between 8 off-centre sites. Different shades of gray designate different site occupancy. It is assumed that the biased Ti potential exhibits the same symmetry as the soft-mode-induced distortion of the shape of the unit cell. N denotes the number of physically distinguishable Ti off-centre configurations in the non-cubic environment. $N = 3$ for tetragonal (C_{4v}) symmetry with displacements along the six [100] directions, $N = 6$ for orthorhombic (C_{2v}) symmetry with displacements along the twelve [110] directions, and $N = 4$ for rhombohedral (C_{3v}) symmetry with displacements along the eight [111] directions.

The fact that tetragonal local symmetry breaking of the high-temperature symmetry occurs not only for $BaTiO_3$ but also for $SrTiO_3$ and $SrTi^{18}O_3$ though the data for $BaTiO_3$ are taken several tens of K above T_C, whereas they are taken 190 K above the cubic-tetragonal transition in $SrTiO_3$, demonstrates that we do not deal with critical fluctuations. Rather, we have genuine off-centre sites in the Ti potential surface in the cubic phase.

18 Organic, inorganic and composite ferroelectrics

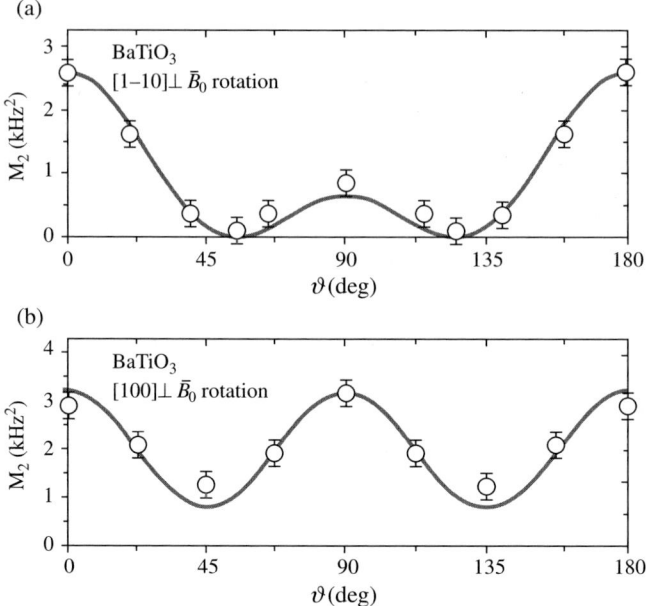

Fig. 1.13 Angular dependencies of the second moments of the ^{49}Ti quadrupole perturbed NMR spectra for (a) cubic BaTiO$_3$ at 450 K and (b) cubic SrTiO$_3$ at 294 K. The solid line designates the theoretical fit for the tetragonal displacement scenario. The central Ti position with a cubic environment would result in no angular dependence [1.4].

The randomly oriented dynamic tetragonal nanodomains of the cubic phase freeze out below T_C and form the static domains of the tetragonal ferroelectric phase. It should be noted that the dynamic exchange of orientation of the tetragonal nanodomains in the cubic phase, i.e. 90° 'flipping' between tetragonal distortions along the a-, b- and c-axes, is much slower than the Ti 'hopping' between the 8 off-centre sites (Fig. 1.14).

The NMR T_2 data allow for a determination of the 'two-time scale' dynamics (Fig. 1.14) [1.4].

The Ti ion hopping in the paraelectric phase is of the order of 10^{-6} s and is about two orders of magnitude faster than the flipping between the six tetragonal 90° domains, which is in the ms range. Both of this motions are much slower than the dielectric mode 10^{-8}–10^{-9} s. The difference between the EXAFS and NMR data in the cubic phase seems to be due to the different time scales involved.

The simultaneous presence of Ti ion disorder and of the soft mode thus leads to a special type of phase transition with both 'displacive' and 'order–disorder' character. It thus seems that soft-mode-type displacive systems, which are inherently weakly anharmonic, also have some disorder present. This can be seen if the technique used is local and sufficiently sensitive.

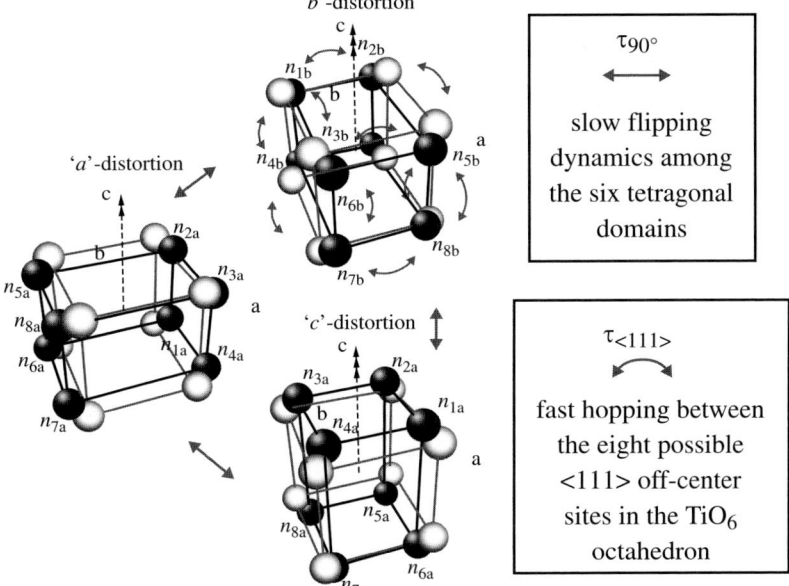

Fig. 1.14 Two-time scale dynamics [1.4].

1.7 New developments in KH$_2$PO$_4$-type order–disorder ferroelectrics

In contrast to the soft-mode-type 'displacive' systems, 'order–disorder' systems are strongly anharmonic and it is the ordering of the ions that induces the paraelectric to ferroelectric phase transition. Chemical shift NMR studies [1.5, 1.6] by Dalal et al. in squaric acid and several KH$_2$PO$_4$ type systems have, however, shown that a small displacive component appears close to the paraelectric–ferroelectric transition. 'Order–disorder' type systems are thus also of a mixed nature since a soft-mode-type component is superimposed on the ion-ordering mechanism.

Another important breakthrough has been the direct determination of the H-bond dynamics in KH$_2$PO$_4$-type systems [1.58] and the confirmation of the validity of the Slater ice rules. Since the pioneering work of Slater in 1941 on the ferroelectric transition in KH$_2$PO$_4$ (abbreviated as KDP) it is known that due to the so-called Pauling ice rules [1.59] only two out of the four protons linking adjacent PO$_4$ groups via double-minimum-type O—H...O bonds are close to each PO$_4$ ion. A whole variety of such 'Slater' lattices can be built with Slater H$_2$PO$_4$ groups, leading to a macroscopic degeneracy of possible ground states. Reiter et al. [1.60] have found by Compton high-energy neutron scattering that on the time scale of this experiment the O—H...O double minimum potential in KD$_2$PO$_4$ and KH$_2$PO$_4$ is symmetric [1.39]. The symmetric form of the double-minimum potential is due to the fact that in view of the short time scale involved the protons are essentially decoupled from the lattice and we deal with the bare proton potential [1.39]. On the other hand, we know that the instantaneous double-well potential of a proton in an individual O—H...O bond that

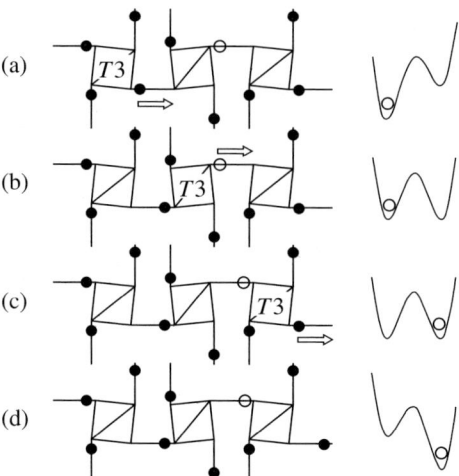

Fig. 1.15 Schematic illustration of the H-bond bias inversion by the passage of an H_3PO_4 Takagi group (T3) in a random Slater lattice. The left side shows the configuration evolution (a)–(d), whereas the corresponding double-well potential of the proton (open circle) is shown on the right side [1.58].

is not decoupled from the lattice is asymmetric. This is so since an isolated intrabond proton transfer from one well to another creates a Takagi pair [1.61] $HPO_4-H_3PO_4$ which has a higher energy than a Slater pair $H_2PO_4-H_2PO_4$. In contrast to that an $O-H\ldots O$ bond linking an unpaired Takagi defect with a Slater group has a symmetric double-well potential (Fig. 1.15). The switching of the Slater configurations is driven by two processes: the creation and annihilation of Takagi pairs and the diffusion of unpaired Takagi defects over the lattice. The Takagi pairs diffuse in a nearly symmetric double-well potential. The visits of Takagi groups to a given Slater group invert the bias of the H bonds in a stochastic way, thus symmetrizing the double-well H-bond potential on the time average.

At high temperatures the concentration of Takagi groups is high and the H-bond dynamics is determined by Takagi-group creation and annihilation. The intrabond deuteron jump time here is $10^{-11}-10^{-12}$ s [1.61]. At low temperatures, on the other hand, the H-bond dynamics is determined by the diffusion of unpaired Takagi groups. These groups move around for long distances in a nearly symmetric double-well $O-H\ldots O$ potential and rearrange many Slater groups before recombination. The time scale of the exchange between different Slater configuration, i.e. the deuteron intrabond jump time, is here of the order of 15 s at 45 K [1.58].

As shown by Feng et al. [1.58] a classical to quantum cross-over takes place at low temperatures in RADP-72 and RADP-35 where tunnelling of Takagi groups in a symmetric double-well potential becomes rate determining as $T \to 0$ (Fig. 1.16). These data have been observed by dielectric dispersion measurements as well as neutron Compton scattering data [1.58].

New developments in KH$_2$PO$_4$-type order–disorder ferroelectrics 21

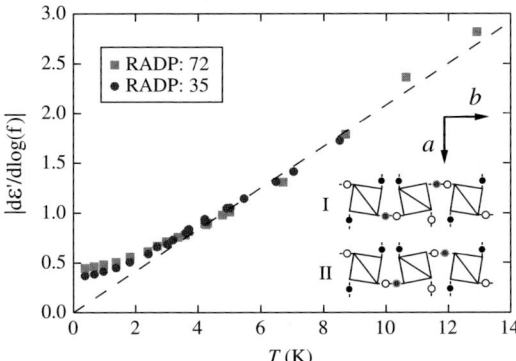

Fig. 1.16 Classical to quantum cross-over in proton glasses [1.58]. f stands for frequency.

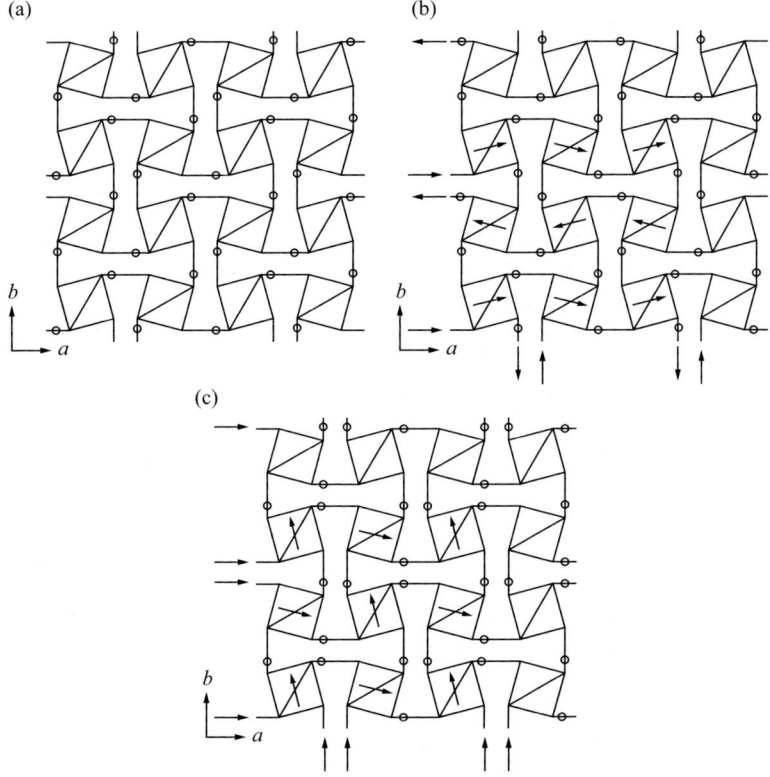

Fig. 1.17 Proton configurations allowed by the ice condition at $T = 0$: (a) KH$_2$PO$_4$ (b) antiferroelectric NH$_4$H$_2$PO$_4$ (c) monoclinic ferroelectric phase [1.64].

KH_2PO_4 shows a large isotope effect in the ferroelectric transition temperature on deuteration [1.39]. T_C shifts from 123 K in KH_2PO_4 to 224 K in KD_2PO_4. It was found that the T_C enhancement is due to proton tunnelling (which is reduced by the larger mass of the deuteron), but as a feedback effect on the O—H...O potential geometry [1.62, 1.63].

Finally, one should mention that significant progress has been made in the understanding of the stability of the various Slater lattices that are degenerate if only the ice condition is taken into account (Fig. 1.17). To single out a specific structure, dipole–dipole interactions have to be introduced. If this Ishibashi interaction [1.64] is repulsive the anti-ferroelectric state found in $NH_4H_2PO_4$ is stabilized.

2
Incommensurate systems

2.1 One-dimensionally modulated incommensurate systems

Translational lattice periodicity is usually considered as one of the most basic characterizing features of a crystal. Identical unit cells where atoms occupy certain regularly spaced positions are indefinitely repeating, thus forming a crystal lattice. Depending on the arrangement of atoms inside the unit cell and the symmetry of the crystal lattice, each crystal has been traditionally assigned to one of the three-dimensional crystallographic space groups. The concepts of symmetry and translational periodicity of crystals have had a central role in the development of solid-state physics since they lead to a great deal of simplifications. By and large, these concepts have come to be accepted as a completely universal property of the crystalline state.

2.1.1 Commensurate (C) and incommensurate (I) systems

In the past decade, however, a great variety of materials has been found that should certainly be called crystals but that show a long-range correlated deviation from three dimensional translational lattice periodicity. In these systems a local atomic property is modulated with a period that is incommensurate [2.1] with the underlying lattice, i.e. the wavelength λ of the modulation is not an integral multiple of the unit-cell edge $a(\lambda \neq n \cdot a, n = 1, 2, 3, \ldots)$. This means that $\frac{\lambda}{a}$ can not be expressed as the ratio of two rational numbers:

$$\frac{\lambda}{a} \neq \frac{M}{N}; M, N = 1, 2, 3 \ldots \tag{1}$$

Such a property can be an atomic position in structurally incommensurate systems, the spin magnetization in helicoidally ordered anti-ferromagnets [2.2], the electron charge density in organic charge-transfer salts (TTF–TCNQ), layered transition-metal dichalcogenides, linear-chain platinum complexes [2.3] and other charge-density wave (CDW) systems or the composition in certain alloys [2.4]. Here, we shall be concerned with structurally incommensurate (I) insulating systems that are characterized by a basic lattice structure and a superimposed incommensurate mass density (i.e. lattice deformation) wave. In such systems at least one atomic position does not exactly repeat from cell to cell and the translational symmetry of the crystal is lost.

The atomic arrangement in a classical crystal and in the incommensurate phase is schematically shown in Figs. 2.1(a) and 2.1(b). In the incommesurate phase no two atoms along the distorted chain are displaced by the same distance from their positions in the undistorted phase. It is therefore impossible to find any lattice translation that

24 Incommensurate systems

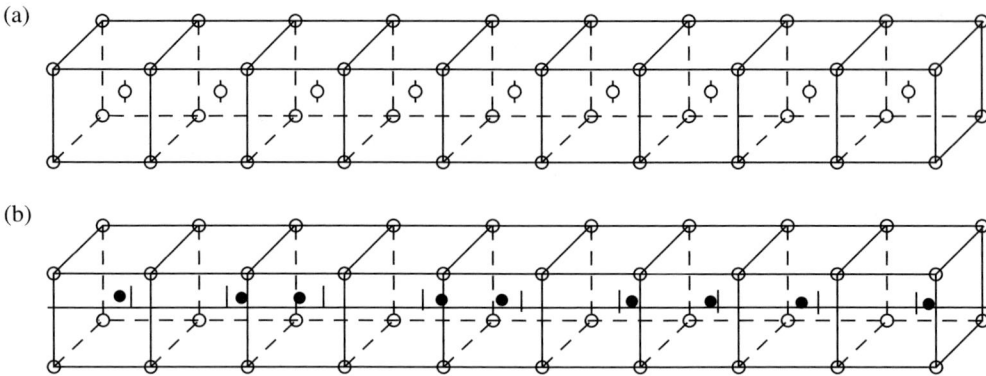

Fig. 2.1 Atomic arrangements in (a) commensurate and (b) incommensurate crystals.

maps the crystal into itself. The crystal cannot be described within the framework of the 230 three-dimensional crystallographic space groups. The conventional description of crystals in terms of three-dimensional repeating unit cells is clearly no longer adequate as the whole crystal represents the unit cell.

Translational lattice periodicity is generally restored at lower temperatures at a 'lock-in' phase transformation where the incommensurate modulation ($\lambda \neq n \cdot a$) changes into a commensurate one ($\lambda = n \cdot a$). The new unit-cell edge will now be an integral multiple of the high-temperature unit cell ($a' = n \cdot a$).

Incommensurate crystals are interesting from the point of view of basic physics since the breaking of the three-dimensional lattice periodicity destroys one of the most fundamental assumptions of solid-state theory. They show a number of new phenomena that are not found in translationally periodic crystals. Since they possess three-dimensional long-range order but lack translational periodicity, incommensurate systems are intermediate between classical periodic crystals and aperiodic biological systems. The study of incommensurate systems may thus – in addition to discovering qualitatively new physical phenomena – lead to an improved understanding of aperiodic materials and perhaps even living matter.

2.1.2 Phase transitions to incommensurate phases

Phase transitions to incommensurate phases can be well described by a generalization of the soft-mode theory of structural phase transitions [2.5].

A crystal with N atoms is said to be stable as long as all $3N-6$ vibrational normal mode frequencies, ω_i, are non-zero and positive. Displaced atoms will always return to their equilibrium positions as long as the normal-mode frequencies and the corresponding effective restoring force constants are > 0. If this condition is not fulfilled for a particular lattice vibration – the so-called soft mode – a structural phase transition occurs. This happens if there are at least two contributions to the effective interatomic force constants that are of opposite sign and that tend to cancel each other. If, because of such a cancellation, the effective restoring force constants vanish in the harmonic approximation, anharmonic interactions – which are proportional to temperature – will stabilize the system above a certain temperature. The observable soft-mode frequency will be temperature dependent as the anharmonic contribution decreases with decreasing temperature. The soft-mode frequency will decrease to zero as the temperature is lowered to a critical temperature T_0. At T_0 the soft mode has condensed out, $\omega_0^2 = 0$, and the frozen in soft-mode displacements will determine the structure of the low-temperature phase.

Structural phase transitions are thus the result of the instability of the crystal against a vibrational soft mode of wavelength λ the frequency of which vanishes for $\lambda \to \lambda_{\text{crit}}$ at T_0 as

$$\omega_0^2 \approx K_1(T - T_0) + K_2(2\pi/\lambda - 2\pi/\lambda_{\text{crit}})^2 \qquad (2)$$

At T_0, the displaced atoms do not return to their old equilibrium positions and the amplitude of the soft mode develops a static component. If the wavelength λ of the condensed out soft mode is infinite ($\lambda_{\text{crit}} = \infty$), the transition occurs at the Brillouin-zone centre and the atomic arrangements inside all unit cells are changed in the same way. The unit-cell size a does not change at $T_0 (a' = a)$. Such a transition is called a ferrodistortive one. If the wavelength of the soft mode is an integral multiple of the high-temperature unit-cell edge ($\lambda_{\text{crit}} = n \cdot a$), the size of the low-temperature unit cell will be an integral multiple of the high-temperature unit-cell size ($a' = n \cdot a$). Such transitions are called anti-ferrodistortive ones. The resulting modulation is commensurate with the underlying crystal lattice. If, however, the wavelength of the condensed out soft mode λ_{crit} is neither an integral multiple of the high-temperature unit cell ($\lambda_{\text{crit}} \neq n \cdot a$) nor infinite, the transition occurs at a general point of the Brillouin zone (Fig. 2.2) and the resulting low-temperature structure will be incommensurate with the basic crystal lattice. The translational periodicity will be lost.

The physical effects driving the transition to structurally incommensurate phases are competing interatomic forces of different ranges and similar magnitudes. For interactions that act only between nearest neighbours such a transition cannot occur.

The order parameter Q of one-dimensionally modulated I systems is generally two-dimensional:

$$Q = A e^{i\Phi(x)}. \qquad (3)$$

Here, A is assumed to be a constant.

The possibility of phase transitions leading to a modulated incommensurate phase is implicitly contained in the theory of structural phase transitions developed long ago by Landau and Lifshitz. Any system, for which a so-called Lifshitz invariant (composed of terms bilinear in the order-parameter components and their spatial first derivatives) is not forbidden by symmetry, can become incommensurate.

To be more specific, let us consider the case of $(NH_4)_2BeF_4$. The soft mode that induces the transition to the commensurate ferroelectric phase $Pn2_1a$ from the paraelectric prototype phase $Pnam$ in this crystal belongs to a two-dimensional representation at the X-point $[k_0 = (\pi/a, 0, 0)]$ of the Brillouin-zone boundary. From the basis functions p and q of this irreducible representation a Lifshitz invariant of the form $\delta(p \cdot dq/dx - q \cdot dp/dx)$ can be constructed where x represents a vector component.

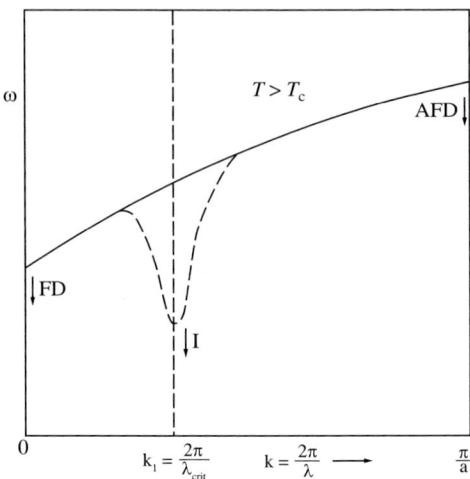

Fig. 2.2 Frequency ω versus wavevector k relation for a typical optical lattice vibration in a crystal. Structural phase transitions are the result of the instability of the crystal lattice against a vibrational soft mode ($\omega_i \to 0$). If the instability occurs at $k = 0$, i.e. at infinite wavelength the transition is a *ferrodistortive* (FD) one and the unit-cell size does not change at the transition ($a' = a$). If the instability occurs at a wavelength that is an integral multiple of the high-temperature unit cell the transition is connected with an increase in the unit-cell size ($a' = n \cdot a; n = 2, 3, 4, \ldots$) and the transition is called an anti-*ferrodistortive* (AFD) one. If, however, the instability occurs at a general point of the ω versus wavevector curve the transition is *incommensurate* (I) and the wavelength of the condensed out soft mode is not a simple multiple of the high-temperature unit-cell dimensions.

The elastic term is of the form $\frac{\kappa}{L}[(\frac{dp}{dx})^2 + (\frac{dq}{dx})^2]$, where $\kappa > 0$. The high-temperature phase with $p = 0, q = 0$ becomes below a certain temperature $T_I, T_I > T_0$, unstable with respect to a modulation wave with $k_I = |\delta|/\kappa$.

Since the coefficients of the Lifshitz and the elastic terms δ and κ are not related to the lattice period it is clear that the wavelength of the modulation wave $\lambda = 2\pi/k_I$ is generally not a simple multiple of the lattice period of the paraelectric phase.

The presence of the Lifshitz term thus prevents the direct transition from the paraelectric into the commensurate phase since the frequency of the soft mode becomes zero for the wavevector k_I that is displaced from $k_0 = (\pi/a, 0, 0)$, as shown in Figs. 2.2 and 2.3.

With decreasing temperature well below T_I the amplitudes of p and q will become large and the contributions of higher-order terms become more and more important. The inclusion of higher harmonics induces the occurrence of regions of nearly commensurate structure and confines the occurrence of modulation to narrow regions, thus producing a domain-like structure.

At still lower temperatures a transition to the commensurate phase (i.e. the transition $k_I \to k_0$) will occur.

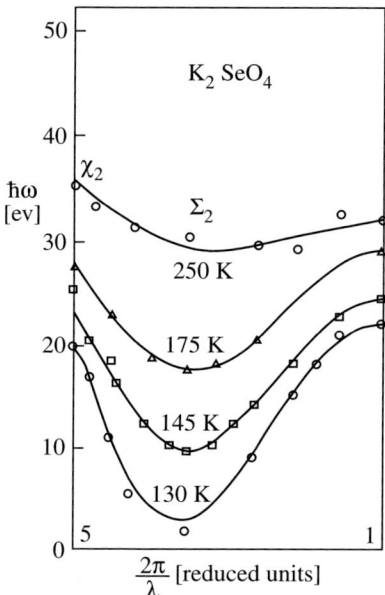

Fig. 2.3 Condensing incommensurate soft mode observed in the high-temperature paraelectric phase of K$_2$SeO$_4$ by inelastic neutron scattering [2.7].

Crystals where the Lifshitz invariant is allowed by symmetry represent type I incommensurate systems. Incommensurate phases may, under certain circumstances, occur also in systems where the Lifshitz invariant is forbidden by symmetry (i.e. in type II incommensurate systems such as NaNO$_2$, SC(NH$_2$)$_2$ or Cs$_2$HgBr$_4$). In these systems the minimum of the soft branch occurs at the Γ point (i.e. Brillouin zone centre) and the irreducible representation of the soft mode is one-dimensional so that no Lifshitz invariant can be constructed and the dispersion curve has zero slope at the Γ-point.

2.1.3 The observation of structurally incommensurate systems

The great majority of structural phase transitions observed so far represents transitions to commensurate phases: the unit-cell size either remains the same or changes by an integral multiple at the transition.

The first system where a structurally incommensurate phase was discovered was NaNO$_2$ [2.6]. The incommensurate phase exists in this system in a narrow temperature range between the paraelectric $T > T_I = 166°$C and the commensurate ferroelectric phase $T < T_c = 164.8°$C. It represents a sinusoidally modulated anti-ferroelectric phase where the wavelength λ of the electric polarization

$$P = P_0 \cdot \cos\left(\frac{2\pi}{\lambda}x\right) \tag{4}$$

is incommensurate to the basic lattice and amounts to about 8.4 unit-cell edges ($\lambda \approx 8.4a$). Below T_c the crystal is homogeneously polarized ($\lambda = \infty, P = P_0$) and the system becomes commensurate. The incommensurate structure was identified from the X-ray diffraction pattern by the presence of extra reflections, the so-called satellite reflections, which are separated from the principal reflections by multiples of $2\pi/\lambda$.

The progress in this field was relatively slow until some years ago when several important discoveries were made nearly simultaneously.

Iizumi, Axe, Shirane and Shimaoka from Brookhaven National Laboratory [2.7] were the first to discover a well-defined condensing incommensurate soft mode, $\omega_0 \to 0$, by inelastic neutron scattering (Fig. 2.3). The crystal they investigated was K_2SeO_4. The incommensurate phase exists here between $T_I = 130$ K and $T_c = 93$ K. At T_c a 'lock-in' transition to a commensurate ferroelectric phase takes place that is connected with a triplication of the unit-cell size. In this system, force constants between first- and second-neighbour layers of atoms decrease, whereas third interlayer force constants increase with decreasing temperature above T_I. It is believed that the competition between these forces is responsible for the occurrence of the incommensurate phase in K_2SeO_4.

Nearly identical effects were independently discovered at about the same time in Rb_2ZnBr_4 by de Pater [2.8] from Delft. Since then, structurally incommensurate phases and 'lock-in' transitions have been observed in Rb_2ZnCl_4, $(NH_4)_2BeF_4$, $RbH_3(SeO_3)_2$, ammonium Rochelle salt, thiourea and many other crystals. Incommensurate phases have also been found in chiral ferroelectric liquid crystals, monolayers of rare gas atoms or D_2 adsorbed on a graphite substrate, pseudo-one-dimensional solids such as $Hg_{3-\delta}AsF_6$, and organic systems such as biphenyl.

2.1.4 The theoretical side: Solitons, the Devil's staircase and phasons

The development of the theoretical understanding of incommensurate systems has been even more striking than the progress in the experimental field.

Symmetry classifications of structurally incommensurate systems have been developed that explain the systematic satellite extinctions observed in the X-ray diffraction pattern of these systems [2.9, 2.10]. The incommensurate modulation wave is mapped into itself by an operation that translates the wave by the distance $n \cdot a (n = 1, 2, 3, \ldots)$ and at the same time changes the phase of the wave by $n \cdot a \frac{2\pi}{\lambda}$. The translational symmetry group of the crystal thus consists of combined operations that translate the crystal and slide the modulation wave with respect to the crystal. The symmetry operations thus act in a superspace whose dimensionality is greater than the dimensionality of the crystal. The additional dimensions can be understood as internal degrees of freedom describing the position of the modulation waves with respect to the crystal lattice. For a three-dimensional crystal with a single one-dimensional modulation wave this superspace is four-dimensional and one needs 4 integer indices to label the diffraction peaks. In the general case of a 3D crystal with an m-dimensional modulation the superspace has $(3 + m)$ dimensions.

In $NaNO_2$ the incommensurate modulation is characterized by a plane wave

$$u = A\cos\phi(x), \phi(x) = \frac{2\pi}{\lambda} \cdot x + \phi_0, \phi_0 = \text{const.} \tag{5}$$

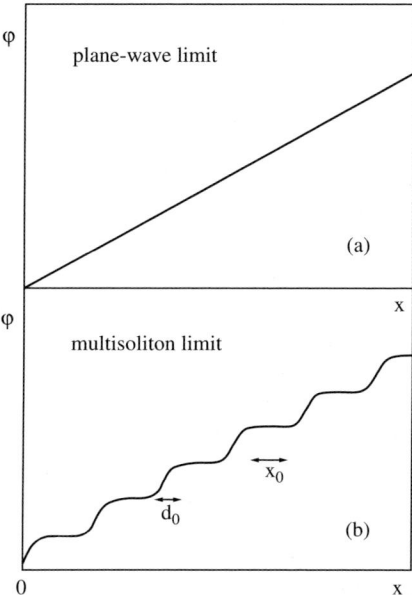

Fig. 2.4 Variation of the phase $\phi(x)$ of the incommensurate modulation wave $u = \cos\phi(x)$ with distance x in the plane-wave (a) and multisoliton (b) limits. In the plane-wave limit ϕ linearly increases with x, whereas in the multisoliton limit the incommensurate phase consists of commensurate regions – where the phase is constant – which are separated by a regular array of domain walls – or phase solitons – where the phase changes rapidly.

so that the phase $\phi(x)$ linearly increases with x (Fig. 2.4(a)). In other systems higher harmonics of the incommensurate modulation wave were observed [2.11]. To explain these results McMillan [2.12, 2.13] suggested that the incommensurate phase can be best described as consisting of commensurate regions – where the phase is constant – separated by a regular array of domain walls – or phase solitons – where the phase changes rapidly (Fig. 2.4(b)). The distortion of the phase minimizes the free energy of the system. The non-linear differential equation that results from this minimization in the continuum limit – where the discrete atomic positions are described by a continuous variable x – is known as the sine-Gordon equation. The phase steps in Fig. 2.4(b) are examples of solitary waves or solitons that are solutions to many non-linear wave equations. The fact that solitons form a part of the incommensurate ground state makes incommensurate systems a particularly suitable field for the study of non-linear phenomena. Soliton-like domain walls are part of the excitation spectrum in a commensurate structural phase transition, whereas they are part of the ground state – and thus easier to demonstrate – in incommensurate systems.

The difference between the commensurably distorted low-temperature phase and the incommensurably distorted phase (Fig. 2.4(b)) is that the latter contains a finite number of phase solitons. The intersoliton distance increases and the soliton density $n_s = \frac{d_0}{x_0}$, where d_0 is the width of the domain wall and x_0 the intersoliton distance

(Fig. 2.4(b)), decreases with decreasing temperature. The soliton density n_s measures the volume fraction of the crystal in the I phase. At the lock-in transition T_c the number of solitons becomes zero and the periodicity of the distortion 'locks-in' to that of the underlying lattice. The soliton density is thus the order parameter of the incommensurate–commensurate 'lock-in' phase transition similarly as the spontaneous magnetization is the order parameter of the ferromagnetic to paramagnetic transition. The analogous quantity for the incommensurate–high-temperature paraelectric phase transition is the amplitude of the modulation wave.

If the solitons are broad the multisoliton lattice will be generally incommensurate with the underlying lattice. If, however, the solitons are narrow – so that the soliton thickness is only a few lattice constants – it will be energetically favourable for solitons to be centred at a particular site within the unit cell. The intersoliton distance will then be an integral number of lattice spacings and the structure will be commensurate. With decreasing temperature the soliton density will not change continuously but in discrete steps. The system will undergo a series of first-order phase transitions between different – commensurate – soliton spacings until the commensurate ground state is reached. This theoretically predicted sequence of first-order phase transitions – which is known as the Devil's staircase [2.14, 2.15] – is somewhat similar to the successive phase transitions observed in $(N(CH_3)_4)_2 ZnCl_4$.

One may say that when the discreteness of the crystal lattice is taken into account – and the continuum model is dropped – the soliton picture remains valid even when the soliton width is only a few unit cells. The main effect of the discreteness of the crystal lattice is to produce a pinning energy that will lock the solitons into a commensurate superlattice giving rise to a 'Devil's-staircase'-type behaviour. When the pinning energy is low enough it may be overcome by thermal fluctuations and the solitons become unpinned. For high enough temperatures solitons are thus able to move freely through the lattice and the intersoliton spacing is not restricted to an integral multiple of the unit-cell dimension. The phase is truly incommensurate and the soliton density varies continuously with temperature.

Whereas there is ample evidence for the existence of solitons in the low-temperature part of the incommensurate phase close to the lock-in transition, quantitative information on the nature of the solitons and the variation of the soliton density with temperature has been lacking until recently.

Still another interesting open problem is the nature of the elementary excitations in the incommensurate phase.

The incommensurate modulation is characterized by the amplitude of the wave and the phase of the wave. The excitation spectrum thus consists of two modes: one – the so-called amplitudon mode – corresponds to space and time variations or, more precisely oscillations of the amplitude, whereas the other – the phason mode – as named by Overhauser [2.16] – corresponds to oscillations of the phase of the displacement wave (Fig. 2.5). In the soliton limit the phason corresponds to oscillations in the soliton positions. The amplitudon mode is optic-like and its frequency varies with temperature, as expected for a classical soft mode (Fig. 2.6). In contrast the phason mode is in the continuum limit acoustic-like and gapless (Fig. 2.6) as it takes zero energy to slide the whole condensed modulation wave throughout the crystal.

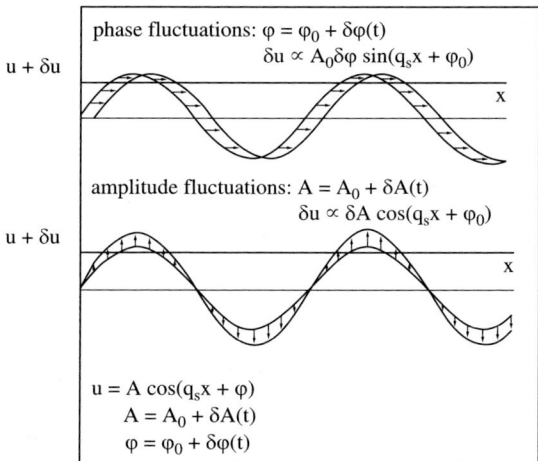

Fig. 2.5 'Amplitudon' and 'phason' fluctuation modes in plane-wave-like incommensurate systems.

The incommensurate equilibrium state is continuously degenerated with respect to the choice of the initial phase of the incommensurate modulation wave. The zero-frequency sliding mode – i.e. the phason – represents the Goldstone boson recovering the broken continuous phase symmetry in the distorted incommensurate phase that corresponds to a particular choice of the phase. A crystal exhibiting an m-dimensional incommensurate modulation thus has m such symmetry recovering Goldstone phasons.

The fact that the energy of the system is unchanged if the modulation is translated with respect to the lattice is unique and true only for incommensurate systems. In these systems the phase is different at each lattice point and there is no optimum phase that could minimize the free energy. For commensurably modulated systems, on the other hand, there is always an optimum phase for which the free energy is a minimum. The arguments establishing the existence of a gapless phason mode thus depend crucially upon the assumption of the strict incommensurability of the frozen-out modulation wave. The pinning of the solitons – due to the discreteness of the crystal lattice – will thus introduce a gap into the phason spectrum that could make the experimental detection of phasons difficult or impossible (Fig. 2.6).

The unpinning of the solitons by thermal fluctuations will, on the other hand, produce a truly incommensurate floating soliton phase where a gapless sliding mode, i.e. a phason, is expected.

The experimental evidence concerning the dynamics of the incommensurate phase has been rather scarce until recently. Whereas the amplitudon was observed to follow the soft-mode frequency versus temperature relation over the whole incommensurate phase well into the commensurate region, high-frequency optic-like phasons have been seen by light and neutron scattering only in the commensurate phase. The attempts to observe gapless or nearly gapless phasons in the incommensurate phase by scattering

32 Incommensurate systems

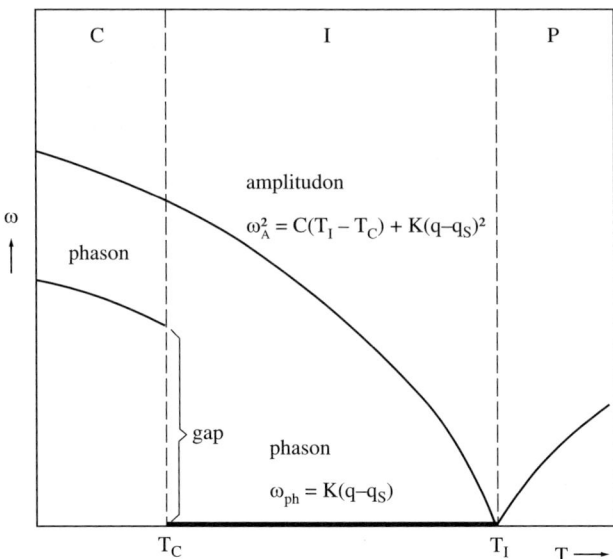

Fig. 2.6 Temperature dependence of the frequency of the incommensurate soft mode in the high-temperature paraelectric (P) phase and the amplitudon (A) and phason (Ph) frequencies in the incommensurate (I) and low-temperature commensurate (C) phase.

techniques have all been unsuccessful until recently. The breakthrough was achieved in 1978–1980 by nuclear magnetic resonance and relaxation.

2.1.5 Dielectric properties

2.1.5.1 Polarization in incommensurate structures

In incommensurate (I) dielectric structures the modulation of the lattice distortion is generally connected with the appearance of a spatially varying polarization. In their low temperature C phases, if they exist, most dielectric substances investigated so far behave as ferroelectrics, i.e. they exhibit a finite spontaneous polarization. Similarly to normal ferroelectrics it is useful to classify incommensurate dielectric structures according to the properties of their C phases:

(a) In proper ferroelectrics the spontaneous polarization P_s can be considered as a primary order parameter. The C ferroelectric phase is induced by the instability of the mode having a wavevector $q = 0$, corresponding to the centre of the Brillouin zone. The I phase, which in fact appears between the high-temperature paraelectric phase and the ferroelectric phase, can be regarded as a modulated C structure. The representatives of the proper ferroelectric class that exhibit intermediate I phases are $NaNO_2$ and thiourea $SC(NH_2)_2$.

(b) The wide majority of substances that are of interest for dielectric studies fall within the class of improper ferroelectrics [2.17]. The ordered C phase is generated by the instability of the lattice distortion mode having a non-zero wavevector

$q = G/p$. Here, G is one of the reciprocal wavevectors and p is just the multiplication of the unit cell when cooling from the P phase to the C phase. The polarization appears as a second-order parameter induced by the lattice distortion. The intermediate I phase can be considered as a superposition of modulated C structures. A well-known example of I improper ferroelectrics is the A_2BX_4 family. The latter includes K_2SeO_4, Rb_2ZnCl_4, Rb_2ZnBr_4 and K_2ZnCl_4 having $p = 3$, $(NH_4)_2BeF_4$ with $p = 2$ and $[N(CH_4)_3]_2ZnCl_4$ with $p = 10/2 = 5$.

2.1.5.2 The paraphase and the commensurate phase

Outside the temperature range of the I phase, i.e. within the P phase and the C phase, the dielectric properties of the investigated substances are expected to be essentially equivalent to the properties of ordinary ferroelectric structures [2.18]. The theory has so far not predicted substantial deviations. Thus, the properties crucially depend merely on the class to which the particular substance belongs. The similarity to ordinary ferroelectrics has also been the reason why in several substances the narrow intermediate I phases have been overlooked in the earlier investigations.

2.1.5.3 Proper ferroelectrics

The representatives of the proper ferroelectrics, i.e. $NaNO_2$ [2.19] and thiourea [2.20] are characterized by the typical Curie–Weiss behaviour of the static dielectric constant (Figs. 2.7 and 2.8):

$$\varepsilon - \varepsilon_\infty = C/(T - T_0), \quad T > T_1 \tag{6a}$$

$$\varepsilon - \varepsilon_\infty = C'/(T_0' - T), \quad T > T_C \tag{6b}$$

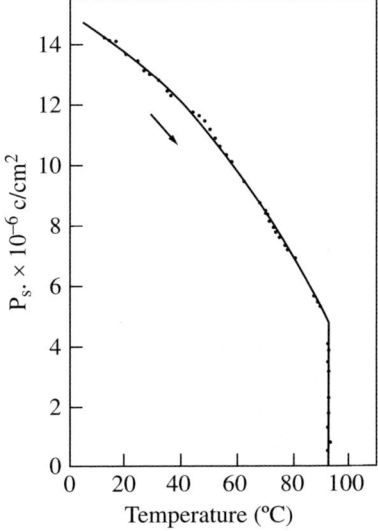

Fig. 2.7 Spontaneous polarization P_S vs. temperature in K_2SeO_4 [2.25].

34 Incommensurate systems

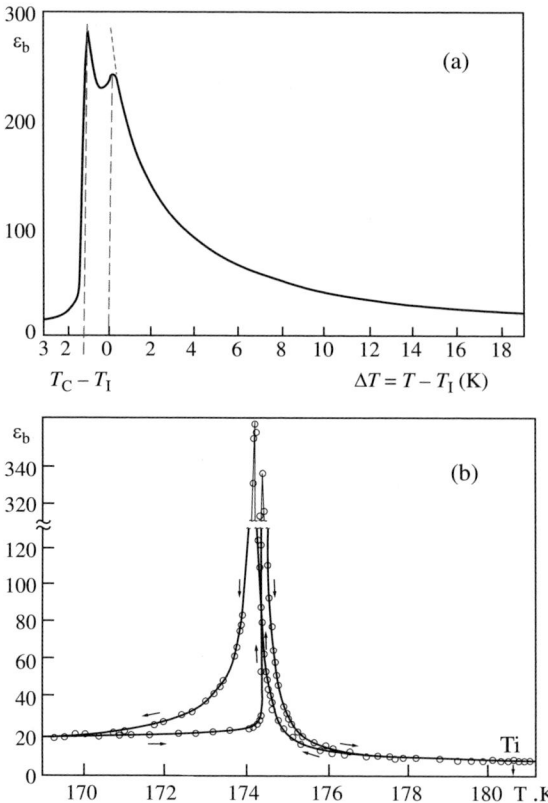

Fig. 2.8 (a) Dielectric constant along the b-direction, ε_b, in $NaNO_2$ at 1 MHz vs. the temperature difference T–$T_I(T_I = 163.09°C)$ as obtained on heating [2.19]. (b) Dielectric constant along the b-direction, ε_b, in $(NH_4)_2BeF_4$ vs. temperature, as obtained on heating and cooling runs [2.27].

which is obeyed both above and below the I phase. Here, T_I is paraelectric-incommensurate and T_C the incommensurate-commensurate transition temperature. The temperatures T_0 and $T_0' \approx T_0$ lie, however, within the I phase, $T_C < T_0 < T_I$, so that ε remains finite at T_c as well as T_I. Also, the temperature variation of the spontaneous polarization in the C phase and the hysteresis loop is in no respect different from ordinary proper ferroelectrics.

2.1.5.4 Improper ferroelectrics

In improper ferroelectrics the polarization arises from the higher-order (non-linear) coupling to the primary order parameter [2.17]. As a consequence, the static dielectric response, i.e. the dielectric constant should not diverge at the P–I transition. Moreover, ε is expected to be nearly constant through the P phase and the C phase. In improper ferroelectrics without the intermediate I phase, rather weak anomalies in ε are observed in the very vicinity of T_c. The intermediate I phase is in most cases broader, so that such anomalies are not expected to be visible in I ferroelectric structures. As a rule, the spontaneous polarization in improper ferroelectrics is smaller by an order of magnitude than in the proper ones.

2.1.5.5 The incommensurate phase

The mechanism that drives the proper ferroelectric structure to prefer an intermediate I phase is substantially different from the driving mechanism for an improper I ferroelectric. In addition to first- also second-nearest-neighbor interactions are important. Hence, there are also differences in the qualitative behaviour of the polarization.

2.1.5.6 Proper ferroelectrics

At the P–I transition temperature the paraphase becomes unstable against the formation of a modulated polarization wave. For the case of the one-dimensional modulation (valid for $NaNO_2$ and $SC(NH_2)_2$) it can be represented as

$$P(x) = P_0 \sin q_0 x \tag{7}$$

where q_0 is the wavevector lying in the vicinity of the Brillouin-zone centre. P_0 is the amplitude of the polarization modulation that vanishes at T_I. It follows from Eq. (7) that the average polarization $\overline{P}$ vanishes in the I phase. However, the existence of an additional homogeneous polarization component P_{sl} is not prohibited by symmetry [2.21]. Indeed, there seem to exist a few examples of polar (ferroelectric) I phases, namely the phase III in thiourea [2.20] and the I phase in ammonium Rochelle salt [2.22].

Approaching the C phase the deviations from the plane-wave variation (7) could become noticeable. Still there are several arguments that they are not very important, in contrast to the situation in the improper I ferroelectrics.

The external electric field perturbs the simple plane-wave modulation (7). A meaningful discussion of the polarization response and the dielectric results can be performed with the help of the underlying Landau theory. The free energy density in proper I ferroelectrics can be given in terms of the spatially varying polarization $P(x)$ [2.23]:

$$f(x) = \frac{\alpha}{2}P^2 + \frac{\beta}{4}P^4 + \frac{\kappa}{2}(\frac{dP}{dx})^2 + \frac{\lambda}{2}(\frac{d^2P}{dx^2})^2 - PE \tag{8}$$

where E is the external field, $\alpha = \alpha_0(T - T_0)$ and $\beta, \lambda > 0$, while $\kappa < 0$ induces the P–I transition. The corresponding Euler–Lagrange equation

$$\lambda\frac{d^4P}{dx^4} - \kappa\frac{d^2P}{dx^2} + \alpha P + \beta P^3 = E \tag{9}$$

can be used to generalize the modulation (7). Just below $T_1, E > 0$ leads to an additional homogeneous component $\overline{P}$ given by

$$\overline{P} = \frac{1}{\alpha + 3\beta P_0^2}E \tag{10}$$

2.1.5.7 Improper ferroelectrics

The polarization in improper ferroelectrics arises from the coupling to the complex order parameter $Q(x)$ that represents the slow modulation of the C distorted structure

$$P = \chi_0 E - \zeta\chi_0(Q^p + Q^{*p}), p > 1 \tag{11}$$

The simplest Landau free-energy functional for the A_2BX_4 class can be written [2.24] as

$$f = \frac{\alpha}{2}|Q|^2 + \frac{\beta}{4}|Q|^4 + i\frac{\delta}{2}(Q\frac{dQ^*}{dx} - Q^*\frac{dQ}{dx}) + \frac{\kappa}{2}\left|\frac{dQ}{dx}\right|^2$$
$$- \frac{\Gamma}{2}(Q^n + Q^{*n}) + \zeta(Q^p + Q^{*p})P + \frac{P^2}{2\chi_0} - PE \tag{12}$$

2.1.5.8 The static dielectric constant

The measurements of the static dielectric constant ε in I structures have been very numerous. They have proved to be one of the easiest experiments to perform and seem to be also the most reproducible ones. In addition, the temperature stability and resolution allowed detailed studies of the phase transitions, so that ε measurements have in most cases preceded other techniques. Also, the anomalies in ε due to the I modulation are quite large, especially near the I–C transition.

The static dielectric constant ε probes the response of the I modulation to small external electric fields:

$$\varepsilon = 1 + \frac{\partial\overline{P}}{\varepsilon_0\partial E} \tag{13}$$

Here, ε_0 is the dielectric constant of the vacuum.

In an I structure the finite E favours the regions having the same direction of the polarization. Thus, the regions of the opposite polarization are reduced at the expense of the preferred ones. Anomalies near the onset of the I phase, i.e. at T_I are much more pronounced for the class of proper ferroelectrics. As follows from Eqs. (10) and (13) they should show at $T \sim T_I$ a cusp-like behaviour characteristic of anti-ferroelectrics, too. This has indeed been observed in $NaNO_2$ [2.26] and thiourea [2.20], as presented in Fig. 2.8(a) for the case of $NaNO_2$.

As already noted, at $T > T_I$ the Curie–Weiss law (6a) is followed with $T_0 < T_I$. The sharpness of the cusp at $T = T_I$ thus depends on the (smallness of the) difference of T_I–T_0, The effects within the class of improper ferroelectrics near T_I are much smaller. ε is nearly constant ($\varepsilon \sim \varepsilon_\infty$) in the paraelectric phase. At T_I again a cusp [2.28, 2.29, 2.30] appears that is, however, hardly visible in some substances, e.g., in $(NH_4)_2BeF_4$ [2.27, 2.31] (see Fig. 2.8(b)). The free-energy functional (12) alone is not sufficient for the explanation of the cusp and higher-order terms have to be included. An isotropic coupling $\eta|Q|^2P^2/2$ modifies the T dependence of the dielectric constant

$$\chi = 1/\lfloor\chi_0^{-1} + \eta|Q|^2\rfloor \tag{14}$$

so that below T_I, χ decreases nearly linearly due to $|Q|^2 \propto |T_I - T|^{2\beta}$. Variation as predicted by Eq. (14) seems to be consistent with experiments [2.30].

Further away from the P–I transition the deviations from the simple plane-wave modulation become important. Entering the multisoliton regime the polarization modulation becomes much more sensitive to the external field. This effect can be attributed to the weakening of the soliton interaction, which is exponentially dependent on the intersoliton distance [2.32]. Thus, a finite field easily extends the regions of preferred polarization at the expense of the opposite polarization. A strongly increasing ε is thus a direct indication of the well-defined multisoliton regime. Moreover, ε gives rather straightforward information on the nature of the I–C transition. In the case of the continuous I–C transition with the vanishing of the soliton density $n_s \to 0$ at T_c, ε should exhibit a divergency at $T \to T_c$, which is, however, in experimental situations usually limited by defects [2.33, 2.34, 2.35]. On the other hand, a discontinuous I–C transition would show up as a jump in ε that cannot be modified by improved sample preparation.

Both representatives of proper ferroelectrics, i.e. $NaNO_2$ and thiourea, exhibit above T_c a weak increase of ε. This fact is consistent with the theory that predicts a discontinuous I–C transition and only small deviations from the plane-wave behaviour above T_c. As already discussed, thiourea shows a well-defined Curie–Weiss anomaly of ε below T_c (Eq. (6b)), typical of proper ferroelectrics. The corresponding temperature T_0' is, however, larger than T_c.

More pronounced are dielectric anomalies near T_c for improper ferroelectrics. For continuous I–C transitions the linear response of the multisoliton lattice, as calculated within the Landau theory [2.36], yields a Curie–Weiss law for the susceptibility (Fig. 2.9)

$$\varepsilon - \varepsilon_\infty = \frac{C'}{T - T_c'} \qquad (15)$$

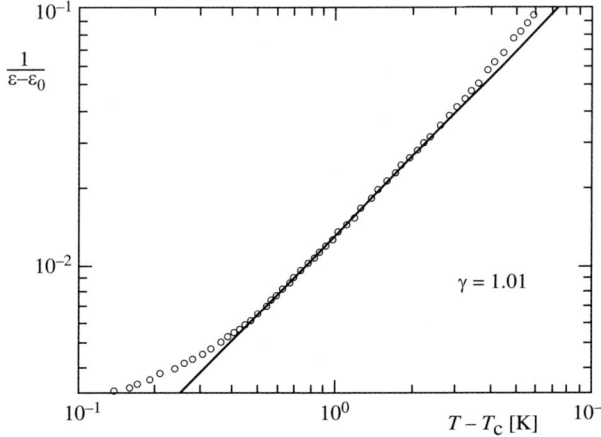

Fig. 2.9 Log–log plot of the reciprocal dielectric constant vs. T–T_c in the I phase of Rb_2ZnCl_4 [2.36]. The best fit with the power-law behaviour is obtained for a critical exponent $\gamma = 1.01 \pm 0.05$.

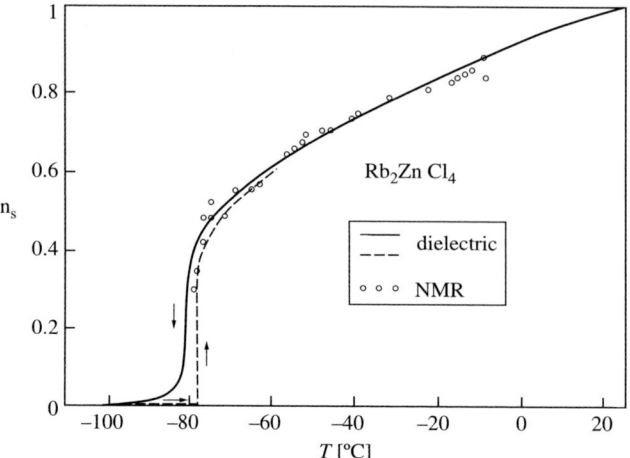

Fig. 2.10 Temperature dependence of the soliton density n_s in Rb_2ZnCl_4, as derived from the dielectric data in cooling and heating runs with the help of the Landau theory [2.36]. The circles show the soliton density, derived from the $^{87}Rb \frac{1}{2} \rightarrow -\frac{1}{2}$ NMR lineshapes.

The temperature dependence of the soliton density $n_s = \frac{d_0}{x_0}$ is shown in Fig. 2.10. Here, d_0 is the soliton width and x_0 the intersoliton distance.

2.1.6 Neutron and X-ray scattering

2.1.6.1 Probing the displacements in incommensurate structures

Neutron scattering and X-ray scattering represent the most powerful techniques for the study of incommensurate (I) structures. Both methods are appropriate for the determination of the average structure of the crystal that exhibits the sequence of paraphase, I and commensurate (C) phases. The deconvolution of I structures has proved to be quite difficult in principle. Bragg reflections and satellites have been therefore extensively studied, both in dielectric as well as in charge-density wave I structures.

Diffuse scattering and inelastic scattering provide information on the fluctuations around the average structure and are appropriate for the study of the order-parameter modes in I systems.

In general, scattering experiments yield the scattering cross-section $d^2\sigma/d\Omega d\omega$, which is connected to the dynamical structure factor $S(\vec{K}, \omega)$ [2.5]:

$$\frac{d^2\sigma}{d\Omega d\omega} = \frac{k_2}{2\pi k_1} S(\vec{K}, \omega) \tag{16}$$

where $\vec{K} = \vec{k}_1 - \vec{k}_2$ and $\vec{k}_1, \vec{k}_2$ are the wavevectors of the incident and the outgoing particle wave. The frequency ω is given by the energy transfer, i.e. $\hbar\omega = E_1 - E_2$, E_1 and E_2 being the incident and outgoing particle energies. On the other hand, the structure factor $S(\vec{K},\omega)$ can be related to the microscopical displacements via time correlation of scattering-density operators $F(\vec{K},t)$:

$$S(\vec{K},\omega) = \frac{1}{2\pi} \int_{-\infty}^{\infty} \left\langle F(\vec{K},t)F(-\vec{K},0) \right\rangle e^{i\omega t} dt \tag{17}$$

$F(\vec{K},t)$ can be, in the case of neutron and X-ray scattering, represented as a coherent sum of single-atom scattering-lengths contributions. Here, the atoms are characterized by their scattering length b_k:

$$F(\vec{K},t) = \sum_{l,k} b_k e^{i\vec{K}\cdot \vec{r}_{lk}(t)} \tag{18}$$

where the summation is performed over all l (high-temperature) unit cells with k atoms within each cell.

It is convenient to separate $\vec{r}_{lk}$ into average positions $\vec{r}_{lk}^{\,0}$ in the paraphase, the average displacements $\vec{u}_{lk}^{\,0}$ and the remaining, in general, time-dependent deviations $\Delta \vec{u}_{lk}$ [2.37]:

$$\vec{r}_{lk}(t) = \vec{r}_{lk}^{\,0} + \vec{u}_{lk}^{\,0} + \Delta \vec{u}_{lk}(t) \tag{19}$$

In order to analyze the scattering experiment results one must be able to represent the displacements $\vec{u}_{lk}^{\,0}$ and $\Delta \vec{u}_{lk}(t)$ in the I structure in the appropriate form. We consider only the structures modulated along one dimension, i.e. characterized by a single C wavevector $\vec{q}_c$. This is indeed the case for most dielectric I phases. Within the plane wave (PW) regime the displacement u_{lk}^0 can then be expressed as

$$\vec{u}_{lk}^{\,0} = Q(\vec{r}_{lk}) \vec{e}_k e^{i\vec{q}_c \cdot \vec{r}_{lk}^{\,0}} + c.c. \tag{20}$$

where the eigenvector $\vec{e}_k$ is in general complex [2.7, 2.38], i.e.

$$\vec{e}_k = \cos\vartheta\, \vec{e}_k^{\,1} - i\sin\vartheta\, \vec{e}_k^{\,2} \tag{21}$$

The complex modulation factor $Q(\vec{r}_{lk}) = Ae^{i\varphi}$ varies in the phase only, $\varphi = \vec{q}_1 \cdot \vec{r} + \varphi_0$. For $\vec{u}_{lk}^{\,0}$ it follows that

$$\vec{u}_{lk}^{\,0} = A\left[\cos\vartheta\, \vec{e}_k^{\,1} \cos(\vec{q}_c \cdot \vec{r}_{lk} + \varphi) + \sin\vartheta\, \vec{e}_k^{\,2} \sin(\vec{q}_c \cdot \vec{r}_{lk} + \varphi)\right] \tag{22}$$

For collinear eigenvectors $\vec{e}^{\,1}$ and $\vec{e}^{\,2}$, it is possible to represent $\vec{u}_{lk}^{\,0}$ as a simple modulation $\vec{u}_{lk}^{\,0} = A\zeta_k\, \vec{e}^{\,1} \cos[q_c \cdot r_{lk} + \varphi - \vartheta_k]$, where $\zeta_k = \cos\vartheta/\cos\vartheta_k$.

2.1.6.2 Elastic scattering

The elastic part of the dynamical structure factor $S_e(\vec{K},\omega) = \delta(\omega)|\langle F(\vec{K})\rangle|^2$ is given by the elastic scattering function $\langle F(\vec{K})\rangle$ [2.37]:

$$\left\langle F(\vec{K}) \right\rangle = \left\langle \sum_{lk} b_k e^{i\vec{K}\cdot\vec{r}^{\,0}_{lk}} e^{i\vec{K}\cdot\vec{u}_{lk}} \right\rangle \tag{23a}$$

This can be expressed with the help of the Debye–Waller factor W_k as

$$\left\langle F(\vec{K}) \right\rangle = \sum_{lk} b_k e^{-W_k} e^{i\vec{K}\cdot(\vec{r}^{\,0}_{lk}+\vec{u}^{\,0}_{lk})} \tag{23b}$$

where for a small modulation the Debye–Waller factor [2.39] is assumed to be independent of the cell coordinate $\vec{r}_l$, but still dependent on k, $\vec{K}$ and on the temperature, i.e. $W_k = W_k(\vec{K}, T)$. The separation performed in eq. (23b) is only approximate, since the extent of fluctuations depends in general on the position.

Plane-wave regime In the plane-wave regime where the phase varies as $\varphi = \vec{q}_I \cdot \vec{r} + \varphi_0$, the diffraction pattern can be calculated by using the identity

$$e^{iZ\cos\psi} = \sum_{n=-\infty}^{\infty} i^n J_n(Z) e^{in\psi} \tag{24}$$

The structure factor (23b) can be resolved with the help of Eq. (24) into fundamental reflections and satellites [2.16]:

$$F(\vec{K}) = \sum_G \sum_{n=-\infty}^{\infty} N \left| f(\vec{G}+n\vec{q}) \right|^2 \delta(\vec{K}-\vec{G}-n\vec{q}) \tag{25}$$

where the sum runs over all reciprocal lattice vectors $\vec{G}$ and all corresponding satellite orders n. Assuming for simplicity that the modulation is of the collinear type, one gets for the form factors [2.38]:

$$f(\vec{G}+n\vec{q}_c) = \sum_k \tilde{b}_k J_n(Z_k) i^n e^{-i\vartheta_k n} \tag{26}$$

where the renormalized scattering length $\tilde{b}_k = b_k \exp(-W_k)$ still depends on K and

$$Z_k = A\zeta_k \vec{K} \cdot \vec{e}^{\,1}_k \tag{27}$$

From expressions (25) and (26) it follows that the relative intensity of fundamental reflections, first-order and higher-order satellites crucially depends on the amplitude A. For small enough amplitudes we can use $J_n(Z) \approx Z^n/2^n n!$, so that the relative intensity of the first order satellite is

$$\frac{I_{\pm 1}(\vec{G})}{I_0(\vec{G})} = \left| \frac{f(\vec{G}\pm\vec{q})}{f(\vec{G})} \right|^2 = \frac{A^2 \left| \sum_k \tilde{b}_k \zeta_k \vec{K}\cdot\vec{e}^{\,1}_k e^{\pm i\vartheta_k} \right|^2}{\left| \sum_k \tilde{b}_k \right|^2} \tag{28}$$

Expressions (25)–(28) represent the basis for the analysis of the neutron and X-ray Bragg reflections. From the position of satellites one determines the wavevector $\vec{q}$ and in particular its incommensurate part $\vec{q}_I = \vec{q} - \vec{q}_c$

From the variation of intensity of the first order satellite I_1 one can determine the behaviour of the amplitude just below T_I. The scattering experiments can thus test the critical behaviour $A \propto (T_I - T)^\beta$ with

$$I_1 = B_1(T_1 - T)^{2\beta}, T < T_I \tag{29}$$

This critical behaviour (29) has been most carefully studied in K_2SeO_4 [2.40] yielding $2\beta = 0.75 \pm 0.05$ and $RbZnCl_4$ [2.41] yielding $2\beta = 0.69 \pm 0.01$, which is consistent with the theoretical predication $2\beta = 0.7 \pm 0.04$.

The critical behaviour of higher-order satellite intensities has been shown to be somewhat more subtle [2.40].

Distorted plane wave and the multisoliton lattice Further within the I phase the modulation becomes distorted. To measure the distortion by elastic scattering one should be able to resolve in satellites the contributions of the pure plane wave and its corrections. Within the constant-amplitude approximation the displacement can be generalized by including the phase function $\theta(\vec{r})$ describing the corrections to the plane-wave variations $\varphi = q_I x + \varphi_0$,

$$\vec{u}^0_{lk} = A\xi_k \vec{e}^1_k \cos(\vec{q} \cdot \vec{r}_{lk} + \varphi_0 - \vartheta_k + \theta(\vec{r}_{lk})) \tag{30}$$

The form factors calculated for the plane-wave regime (26) are now modified by an additional form factor $g(\vec{K})$ [2.42]:

$$f(\vec{G} \pm n\vec{q}) = \sum_k b_i Z_k^n e^{\pm in(\vartheta_k + \varphi_0)} g(\vec{K} - n\vec{q}_c) \tag{31}$$

with

$$g(\vec{K}) = \frac{1}{\lambda_I} \int_{-\lambda_I/2}^{\lambda_I/2} e^{i(K_x x - \theta(x) - q_I x)} dx \tag{32}$$

The form factor $g_n = g(\vec{K} - n\vec{q}_c)$ is simple in two limiting situations: (a) in the PW regime we have $g_n = 1$, (b) g_n can be easily calculated also in the narrow soliton limit, where $\varphi(x) = \theta(x) + q_I x$ is nearly constant in commensurate regions and changes only within narrow solitons [2.42]. Then g_n is

$$g_n = \frac{1}{n} \sin\frac{\pi}{p} e^{-i\pi/p} \tag{33}$$

As compared to some other phenomena, e.g., the dielectric constant [2.36] and NMR spectra [2.43], the elastic scattering results are less sensitive to the appearance of the soliton lattice. Only a few attempts [2.44] have therefore been made to compare the measured satellite intensities near the I–C transition to the calculated ones.

Diffuse scattering Whereas the Bragg reflections yield the information on the ordered structure, the diffuse scattering can be used for the examination of the low-frequency fluctuations of the system.

With the term diffuse scattering one usually refers to the inelastic scattering in the regime of low energy transfer. Several experimental approaches are used for the investigation of diffuse scattering:

(a) In X-ray scattering experiments the energy resolution is very small, so that the measured quantity is the diffuse scattering intensity

$$S(\vec{K}) = \int d\omega\, S(\vec{K},\omega) = S_\mathrm{B}(\vec{K}) + S_\mathrm{d}(\vec{K}) \tag{34}$$

which represents the integrated dynamical structure factor. Here, B stands for Bragg and d for diffuse scattering. $S(\vec{K})$ now incorporates the Bragg reflections due to the long-range order as well as the contributions due to fluctuations. Whereas both components are easily separated far from the transition temperature, the separation becomes less trivial [2.45] in the critical regime. Then, the diffuse scattering contribution appears to be very narrow in $\vec{K}$ space, comparable to the $\vec{K}$ resolution of Bragg peaks. This difficulty, which is common also to the neutron-scattering technique, increases the uncertainty of the determination of T_I. It also enhances the errors in the evaluation of Bragg and diffuse scattering intensities, and consequently influences the accuracy of critical exponents.

(b) Neutron-scattering experiments have a high energy resolution and the whole energy profile of the diffuse scattering can be studied. The measurement of the energy spectra allows by integration [2.37] the calculation of the diffuse scattering intensity $S(\vec{K})$ (34). It is, however, easier to follow the wavevector dependence at a chosen energy transfer, in particular the $E = 0$ case [2.7]. The $E = 0$ neutron scattering thus sees the intensity that lies within a resolved finite energy interval around $E = 0$. It should be noted however that in both methods one should subtract the background intensity that is due to incoherent scattering [2.46].

The temperature dependence of the neutron-scattering results for the amplitudon and phason modes are compared with the Raman scattering data in Fig. 2.11.

Critical scattering $(T > T_I)$ For the qualitative description of the diffuse scattering in the critical regime above T_I, the simple damped oscillator form for $S(\vec{K},\omega)$ can be used [2.37]:

$$S(\vec{K},\omega) = \frac{\Gamma}{(\omega_0^2 - \omega^2)^2 + (\Gamma\omega)^2} \tag{35}$$

where for the soft-mode frequency $\omega_0(\vec{K})$ one can assume the conventional slowing-down form

$$\omega_0^2(\vec{K}) = a\chi(\vec{K})^{-1} \tag{36}$$

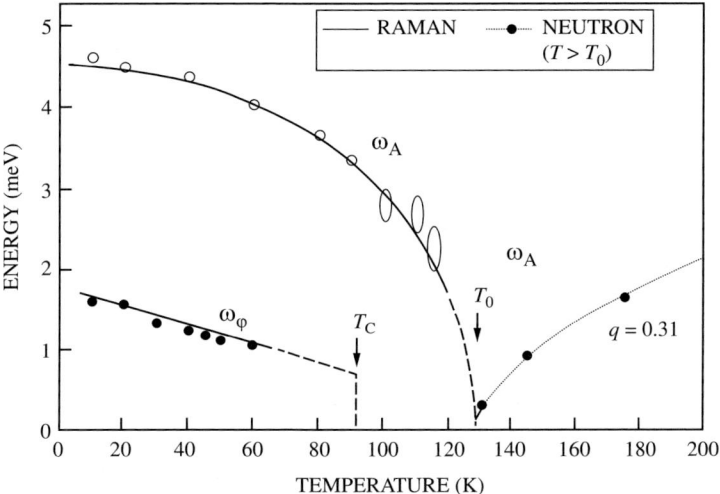

Fig. 2.11 Temperature dependence of the $\vec{q}=0$ phase-mode frequency ω_ϕ and amplitude-mode frequency ω_A in K_2SeO_4. Solid lines indicate Raman scattering results, data points are neutron results. The dashed portion of the ω_ϕ curve indicates that ω_ϕ is expected to vanish at T_C [2.47].

and $\chi(\vec{K})$ is the generalized static susceptibility [2.37, 2.44]:

$$\chi(\vec{K}) = \frac{\chi(\vec{q})}{1 + \sum_\alpha (K_\alpha - q_\alpha)^2/x_\alpha^2} \tag{37}$$

Here, $\chi(\vec{q})$ is the divergent susceptibility at the critical wavevector $\vec{q} = \vec{q}_c + \vec{q}_I$ and x_α are the components of the inverse correlation length. From Eqs. (35) and (36) it follows that the integrated diffuse scattering behaves as

$$S(\vec{K}) \propto \chi(\vec{q}). \tag{38}$$

Critical scattering above T_I has been studied in several substances and results are generally consistent with the above analysis. Rather detailed X-ray scattering studies have been performed in pure [2.44] and K-doped [2.44] Rb_2ZnCl_4. The temperature variation of the diffuse scattering profiles in Rb_2ZnCl_4 follows the expected critical behaviour $\chi(q) \propto |T - T_I|^{-\gamma}$ where the fitted value $\gamma \sim 1.26$ is in agreement with the theory. The inverse correlation length x_α also has been shown to scale as $x_\alpha = c_\alpha |T - T_I|^\nu$, where the observed critical exponent was $\nu \sim 0.693$.

2.1.7 Magnetic resonance lineshapes in incommensurate systems

2.1.7.1 Number of resonance lines and frequency distribution

The usefulness of nuclear magnetic resonance (NMR), nuclear quadrupole resonance (NQR), and electron paramagnetic resonance (EPR) for the study of incommensurate

44 Incommensurate systems

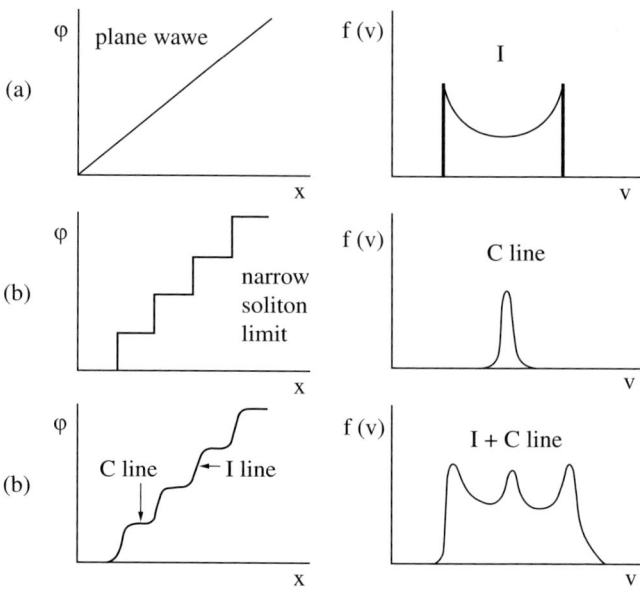

Fig. 2.12 NMR lineshapes in one-dimensionally modulated incommensurate systems: (a) plane-wave limit, (b) narrow soliton case, (c) broad soliton case.

systems is based on the fact that the resonance frequency varies in space in a way that reflects the spatial variation of the incommensurate modulation. In commensurate systems the number of magnetic resonance lines in the spectrum is equal to the – usually small – number of physically non-equivalent nuclei or paramagnetic sites per unit cell.

In incommensurate systems where the translational lattice periodicity is lost there is an essentially infinite number of non-equivalent nuclei or paramagnetic sites that contribute to the magnetic resonance spectrum. Except in the narrow soliton limit one thus expects to see a quasi-continuous distribution of NMR frequencies instead of sharp lines as in commensurate crystals.

As can be seen from Fig. 2.12 the nature of this distribution depends critically on the space variation of the phase of the incommensurate modulation, $\phi = \phi(x)$. A plane-wave modulation thus produces a lineshape that differs strongly from the one expected for a multisoliton lattice. With decreasing width of the phase soliton and increasing soliton–soliton spacing more and more nuclei will find themselves in locally commensurate domains so that new lines will appear at positions corresponding to the ones in the commensurate phase. These lines become stronger as the soliton density decreases, allowing for a quantitative determination of the variation of the soliton density with temperature.

NMR, NQR and EPR thus represent one of the most sensitive means to study the local nature of the incommensurate modulation. Let us now evaluate the resonance lineshape quantitatively.

2.1.8 The 'plane-wave' limit: One-dimensional modulation ($m = 1$)

In the $m = 1$ 'plane-wave' modulation limit the phase ϕ is a linear function of x and the incommensurate distortion is characterized by a single Fourier component of the displacement field:

$$u = A \cos \phi(x) = A \cos(\vec{k}_1 x + \phi_0), \quad \phi_0 = \text{const.} \tag{39}$$

The incommensurate wavevector $\pm \vec{k}_\text{I}$ characterizing this distortion equals the wavevector for which the high-temperature soft mode becomes unstable at the paraelectric–incommensurate transition temperature T_I.

When the wavelength of the incommensurate modulation is large compared to the radius of the region where the dominant contribution to the NMR, NQR or EPR frequency comes from, the resonance frequency ν at a given lattice site can be expanded in powers of the displacement field u

$$\nu = \nu_0 + a_1 u + \frac{1}{2} a_2 u^2 + \ldots \tag{40}$$

which is the order parameter for the paraelectric–incommensurate transition. Here, ν_0 stands for the resonance frequency in the high-temperature paraelectric phase. Inserting (39) into (40) we thus find for $m = 1$:

$$\nu = \nu_0 + \nu_1 \cos \phi(x) + \frac{1}{2} \nu_2 \cos^2 \phi(x) + \ldots \tag{41}$$

where ν_1 is proportional to A, ν_2 to A^2, etc.

The above expansion is valid in the commensurate as well as in the incommensurate phase. In the commensurate phase $\cos \phi$ takes on one or more discrete values, while in the incommensurate phase the whole crystal is a unit cell and $\cos \phi(x)$ takes on nearly continuously all values between $+1$ and -1 as x runs over lattice sites that are equivalent in the high-temperature phase. Reducing the phases $\phi = q_s x + \phi_0$ to the interval $(0, 2\pi)$ one can introduce the phase density $\rho(\phi)$ that is constant within this interval and zero outside. The density of the spectral lines $f(\nu)$ is thus obtained from [2.48, 2.49, 2.50].

$$f(\nu) d\nu = \rho(\phi) d\phi \tag{42}$$

as

$$f(\nu) = 1/(2\pi \, d\nu/d\phi) \tag{43}$$

Here, $\int f(\nu) d\nu = 1$.

The derivative appearing in Eq. (43) is given by:

$$d\nu/d\phi = -(\nu_1 + \nu_2 \cos \phi + \nu_3 \cos^2 \phi + \ldots) \cdot \sin \phi \tag{44}$$

The spectral density $f(\nu)$ – which equals the inhomogeneous lineshape – will be peaked whenever $d\nu/d\phi$ becomes small, i.e. when $\sin \phi = 0$ or when $(\nu_1 + \nu_2 \cos \phi + \nu_3 \cos^2 \phi + \ldots)$ will be zero. If a single line corresponding to a given site has the lineshape $L(\nu - \nu_c)$, the inhomogeneous lineshape of the composite spectrum is given by:

$$F(\nu) = \int L(\nu - \nu_c) f(\nu_c) d\nu_c. \tag{45}$$

In evaluating the frequency distribution $f(\nu)$ some symmetry considerations are useful. The coefficients $a_1, a_2, \ldots$ as well as $\nu_1, \nu_2, \ldots$ depend on, among other things, the symmetry of the nuclear site and the direction of the external magnetic field with respect to the symmetry elements of the crystal lattice. It can be easily shown that all odd terms in expansions (41) and (42) are identically zero if the nucleus occupies in the high-temperature phase a site that is invariant with respect to a symmetry element that vanishes at the transition to the low-temperature commensurate phase [2.51].

2.1.8.1 Linear case

Let us assume that the nucleus occupies in the high-temperature phase a general position and that the term in Eqs. (43), (44) and (45), linear in the order parameter, is dominant:

$$\nu \approx \nu_0 + \nu_1 \cos \phi(x). \tag{46}$$

The frequency distribution is in this simplest case obtained from Eqs. (44) and (45) as:

$$f(\nu) = \frac{1}{2\pi\nu_1 \cdot |\sin \phi|} = \frac{1}{2\pi\nu_1 (1 - X^2)^{1/2}} \tag{47}$$

where $X = (\nu - \nu_0)/\nu_1$.

The edge singularities corresponding to $X = \pm 1$ will occur at

$$\nu = \nu_0 \pm \nu_1 \tag{48}$$

and reflect the fact that the density of states has a maximum at the extreme displacements of the incommensurate distortion wave.

The frequency separation between the two edge singularities will be proportional to the amplitude of the incommensurate distortion wave, i.e. to the order parameter:

$$\Delta\nu = \nu_+ - \nu_- = 2\nu_1 \propto (T - T_\mathrm{I})^\beta \tag{49}$$

where β is the critical exponent.

2.1.8.2 Quadratic case

If the site symmetry of the nucleus in the high-temperature phase is such that ν is a symmetric function of the order parameter and if the quadratic term is dominant

$$\nu = \nu_0 + \frac{1}{2}\nu_2 \cos^2 \phi(x) \tag{50}$$

one finds:

$$f(v) = \left\{2\pi\nu_2 |\cos\phi \cdot \sin\phi|\right\}^{-1} = \left\{2\pi\nu_2 \sqrt{\left(\frac{\nu - \nu_0}{\nu_2/2}\right)\left(1 - \frac{\nu - \nu_0}{\nu_2/2}\right)}\right\}^{-1} \tag{51}$$

The frequency separation between the two singularities at

$$\nu_{(a)} = \nu_0 \tag{52}$$

and

$$\nu_{(b)} = \nu_0 + \nu_2/2 \tag{53}$$

is now smaller than before. The position of the singularity at $\nu_{(a)} = \nu_0$ will not depend on temperature, whereas the position of the singularity at $\nu_{(b)} = \nu_0 + \nu_2/2$ will shift proportionally to the order parameter squared, i.e.

$$\Delta \nu = \nu_{(b)} - \nu_{(a)} \propto (T - T_I)^{2\beta} \tag{54}$$

2.1.8.3 Linear and quadratic terms

If both linear and quadratic terms have to be taken into account, the positions of the two singularities will be given by:

$$\nu_{(a)} = \nu_0 + \nu_1 + \frac{1}{2}\nu_2 \tag{55}$$

and

$$\nu_{(b)} = \nu_0 - \nu_1 + \frac{1}{2}\nu_2 \tag{56}$$

The difference $\nu_{(a)}-\nu_{(b)}$ is proportional to the order parameter and does not depend on ν_2. If $|\nu_1| \leq |\nu_2|$, a third singularity appears at

$$\nu_{(c)} = \nu_0 - \nu_1^2/2\nu_2 \tag{57}$$

This singularity is a consequence of the non-linearity of the relation between the resonance frequency and the order parameter. It is shifted by a constant amount with respect to ν_0 and its position is independent of the order parameter.

The frequency distributions $f(\nu)$ corresponding to the linear and quadratic terms are schematically shown in Fig. 2.13. The experimental spectra of Rb_2ZnCl_4 are shown in Fig. 2.14 for $T > T_I$ and $T < T_I$.

2.1.9 The 'phase soliton' limit

2.1.9.1 Soliton density and Landau theory (Fig. 2.15)

For systems of interest like Rb_2ZnCl_4, $(NH_4)_2BeF_4$, etc. where the order parameter Q has two components, the Landau free-energy density can be expressed as

$$f(x) = \frac{\alpha}{2}(Q^*Q) + \frac{\beta}{4}(Q^*Q)^2 + \frac{\Gamma}{4}\left[Q^n + Q^{*n} + 2(Q^*Q)^{n/2}\right]$$
$$- i\frac{\delta}{2}\left(Q\frac{dQ^*}{dx} - Q^*\frac{dQ}{dx}\right) + \frac{\kappa}{2}\left|\frac{dQ}{dx}\right|^2 \tag{58}$$

where Q is complex.

48 Incommensurate systems

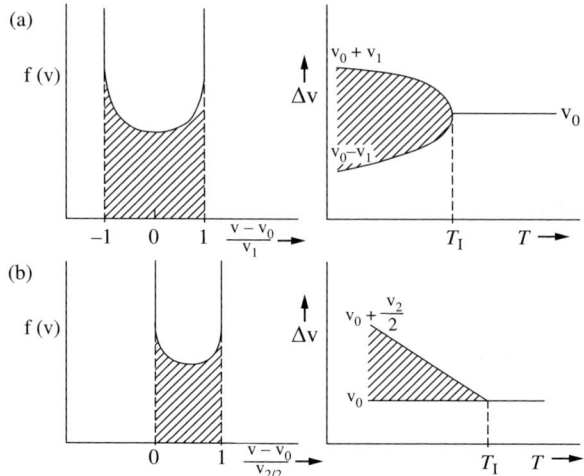

Fig. 2.13 Magnetic resonance lineshapes obtained in the 'plane-wave' limit for the case of a one-dimensional ($m = 1$) modulation: (a) linear terms, (b) quadratic terms present in the expansion of the frequency in terms of the order parameter ($\beta = \frac{1}{2}$).

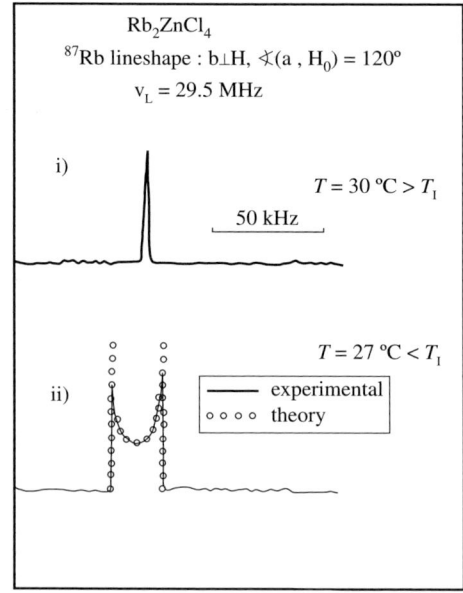

Fig. 2.14 NMR lineshape of the ^{87}Rb $\frac{1}{2} \to -\frac{1}{2}$ transition in the (i) paraelectric and the (ii) incommensurate phase of Rb_2ZnCl_4.

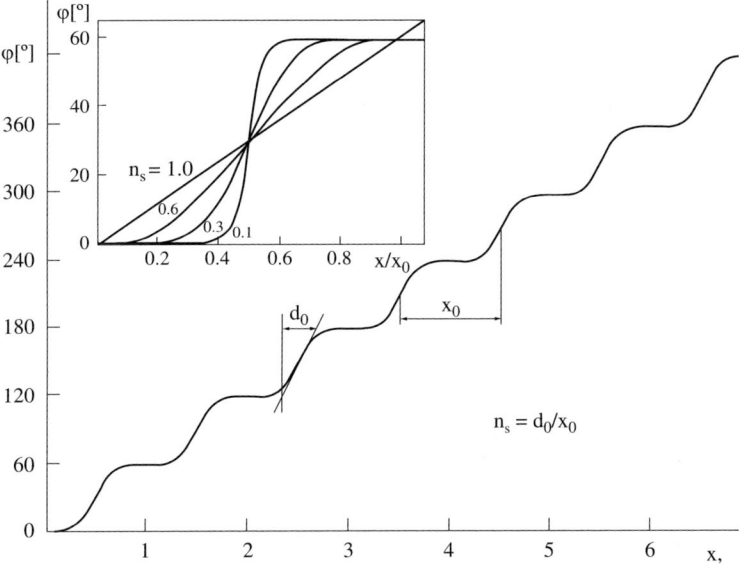

Fig. 2.15 Spatial variation of the phase of the modulation wave for different soliton densities $n_s = d_0/x_0$. Here, d_0 is the soliton width and x_0 the intersoliton spacing. The spatial variable x is measured in units of x_0.

For convenience we have retained a higher-order term $(Q^*Q)^{n/2}$ considering only even n, e.g., $n = 10$ for [N(CH$_3$)$_4$]$_2$ZnCl$_4$, $n = 6$ for Rb$_2$ZnCl$_4$, $n = 4$ for (NH$_4$)$_2$BeF$_4$ and $n = 2$ for chiral smectic liquid crystals in an external magnetic field. Here, we have $\alpha = \alpha_0(T - T_0)$ and $\beta, \kappa, \Gamma, \delta > 0$. Introducing polar coordinates $Q = A\exp(i\phi)$ Eq. (58) can be rewritten as

$$f(x) = \frac{\alpha}{2}A^2 + \frac{\beta}{4}A^4 + \Gamma A^n \cos^2\left(\frac{n\varphi}{2}\right) - \delta A^2 \varphi' + \frac{\kappa}{2}A^2 \varphi'^2 \tag{59}$$

Minimizing $F = (1/x_0)\int_0^{x_0} f(x)\,dx$ with respect to ϕ in the constant-amplitude approximation $A = A_0 \neq A(x)$ and making use of the identity $\cos^2(n\phi/2) = (1 + \cos n\phi)/2$ one finds the Euler equation of our problem (i.e. the sine-Gordon equation) as

$$\frac{\partial f}{\partial \varphi} - \frac{d}{dx}\frac{\partial f}{\partial \varphi'} = 0 \tag{60}$$

yielding

$$\kappa \varphi'' = -\frac{n}{2}\Gamma A_0^{n-2}\sin(n\varphi) \tag{61}$$

with $\phi' = d\phi/dx$ and $\phi'' = d^2\phi/dx^2$.

The first integral of this equation is

$$\frac{\kappa}{2}\varphi'^2 = \eta + \frac{\Gamma}{2}A_0^{n-2}[\cos(n\varphi) - 1] = \eta - \Gamma A_0^{n-2}\sin^2\left(\frac{n}{2}\varphi\right) \tag{62}$$

where η is an integration constant. The maximum value of the derivative of the phase $\varphi'_{\max}$ determines the soliton width d_0 (Fig. 2.15). It is given by

$$\varphi'_{\max} = \frac{2\pi}{d_0 n} = \sqrt{\frac{2\eta}{\kappa}} \tag{63}$$

The intersoliton distance is, on the other hand, obtained as

$$x_0 = \sqrt{\frac{\kappa}{2}} \int_0^{2\pi/n} \frac{d\varphi}{\sqrt{\eta - \Gamma A_0^{n-2} \cdot \sin^2\left(\frac{n}{2}\varphi\right)}} \tag{64}$$

Introducing a new variable $\phi = n/2\varphi$ we can express x_0 as

$$x_0 = \sqrt{\frac{\kappa}{2\Gamma A_0^{n-2}}} \frac{4}{n} kK(k) \tag{65}$$

where $K(k)$ is the complete elliptic integral of the first kind:

$$K(k) = \int_0^{\pi/2} \frac{d\phi}{\sqrt{1 - k^2 \sin^2 \phi}} \tag{66a}$$

and

$$k^2 = \frac{1}{1 + \Delta^2} = \frac{\Gamma A_0^{n-2}}{\eta} \tag{66b}$$

The relation between φ and x, as well as between x_0 and d_0, is illustrated in Fig. 2.15.

The soliton density (Fig. 2.15) which measures the volume fraction of the crystal in the incommensurate domain walls [2.1, 2.52] is now obtained as

$$n_s = \frac{d_0}{x_0} = \frac{\pi/2}{K(k)} \tag{67}$$

The parameter k is within the constant-amplitude approximation determined by minimizing $F = (1/x_0) \int_0^{x_0} f(x)\,dx$ with respect to η yielding

$$\frac{E(k)}{k} = \frac{\pi}{4} \cdot \delta \sqrt{\frac{2}{\kappa \Gamma A_0^{n-2}}} \tag{68}$$

The temperature T enters the above expression through the amplitude of the order parameter $A_0^2 = \alpha_0(T_{\rm I} - T)/\beta$. Here, $E(k)$ is the complete elliptic integral of the second kind

$$E(k) = \int_0^{\pi/2} \sqrt{1 - k^2 \sin^2 \phi}\, d\phi \tag{69}$$

It should be noted that for $T \to T_{\text{I}}$ (i.e. $A_0 \to 0$), $k \to 0, \Delta^2 \to \infty$ and $K(k) \to \pi/2$, so that $n_{\text{s}} \to 1$. For $T \to T_{\text{c}}$ one finds on the other hand $\Delta^2 \to 0, k^2 \to 1, K(k) \to \infty$ and $n_{\text{s}} \to 0$. It should also be pointed out that the soliton density n_{s} and the density of the commensurate domains n_{c} are related by $n_{\text{s}} + n_{\text{c}} = 1$.

2.1.9.2 The NMR lineshape in the multisoliton limit

The frequency distribution function for a static one-dimensional modulation wave in the constant amplitude approximation [2.1, 2.52] is

$$f(\nu) = \frac{\text{const.}}{d\nu/dx} = \frac{\text{const.}}{(d\nu/d\psi)(d\psi/dx)} \tag{70}$$

where $\psi(x) = \varphi(x) + \phi_0$.

The inhomogeneous resonance lineshape $F(\nu)$ is determined by the convolution of $f(\nu)$ with the lineshape function $L(\nu-\nu_{\text{c}})$ of a paraelectric line:

$$F(\nu) = \int L(\nu-\nu_{\text{c}}) f(\nu_{\text{c}}) d\nu_{\text{c}} \tag{71}$$

Singularities will appear in the spectrum when:

(i) $d\nu/d\psi \to 0$, and/or
(ii) $d\psi/dx \to 0$.

In the 'plane wave' limit $d\psi/dx = \text{const.}$ and the only singularities are associated with $d\nu/d\psi \to 0$. In the general case $d\psi/dx$ is not constant. The phase ψ will be nearly constant in the commensurate domains where $d\psi/dx \to 0$, whereas it will rapidly increase with x in the soliton-like domain walls, where $d\psi/dx \neq 0$. The appearance of commensurate domains in the multisoliton lattice modulation regime will thus result in the appearance of new commensurate lines and in a reduction in the intensity of the incommensurate background and edge singularities.

To discuss the spatial variation of the phase of the modulation wave, let us first note that in the simplest possible case

$$\psi(x) = \varphi(x) + \phi_0 \tag{72}$$

where ϕ_0 is an initial phase that depends on the position of the nucleus in the paraelectric unit cell and the relative strength of the two components of the modulation wave. $\varphi(x)$ is – within the continuum Landau theory – a solution of the sine-Gordon equation describing the formation of the multisoliton lattice.

In the constant-amplitude approximation one finds $d\varphi/dx$ as

$$\frac{d\varphi}{dx} = \text{const.} \left[\Delta^2 + \cos^2(n(\psi - \phi_0)/2)\right]^{\frac{1}{2}} \tag{73}$$

where Δ^2 is related to the soliton density n_{s} by Eqs. (66b) and (67), with $K(k) = K(1/\sqrt{1+\Delta^2})$ being defined by Eq. (66a). Here, Δ^2 is determined from the experiment by fitting the observed lineshape to Eqs. (70) and (73) [2.52].

It should be noted that $n_s = 1$ in the plane-wave limit where $\Delta \gg 1$, whereas n_s approaches zero in the multisoliton limit when $\Delta \ll 1$.

In the multisoliton limit, expression (73) yields up to n new lines for $\Delta \to 0$ in addition to the incommensurate edge singularities. The new lines appear for $\Delta \ll 1$ when $\cos^2(n(\psi - \phi_0)/2) = 0$, i.e. when

$$\psi = (2m+1)\pi/n + \phi_0, m = 0, 1, 2, \ldots, n-1 \qquad (74)$$

whereas the edge singularities will appear when $d\nu/d\psi = 0$, e.g., in the 'linear case' when $\sin \psi = 0$ and ψ is an integer multiple of π. Expression (74) together with the relation between the frequency and the nuclear displacement determines the position of the C lines in the multisoliton limit.

2.1.10 Phason and amplitudon excitations

In the 'plane-wave modulation' limit the incommensurate distortion is characterized by a single Fourier component of the displacement:

$$u(z) = u_0 \cos(q_s z + \phi_0), \phi_0 = \text{const.} \qquad (75)$$

The excitation spectrum consists of two modes: the amplitudon branch corresponds to oscillations of the amplitude of the displacement, $u = u_0 + \delta u(t)$, while the phason branch corresponds to oscillations of the phase of the displacement profile, $\phi = \phi_0 + \delta\phi(t)$ (Fig. 2.5). To simplify our treatment as much as possible we shall use a classical mean-field approximation (MFA) description of the soft-mode behaviour and isotropic dispersion relations [2.53].

2.1.10.1 Dispersion relations

The phason frequency vanishes when $\vec{k} = \vec{q} - \vec{q}_s \to 0$:

$$\omega_{\varphi k}^2 = \kappa \cdot k^2, \kappa = \text{const.}, \boldsymbol{T < T_I} \qquad (76)$$

while the amplitudon frequency is finite in this limit:

$$\omega_{Ak}^2 = 2a(T_I - T) + \kappa \cdot k^2, a = \text{const.}, \boldsymbol{T < T_I} \qquad (77)$$

Any phase-pinning perturbation due to defects will induce a gap [2.53, 2.54], $\Delta = \text{const.}$, in the phason spectrum (76).

The amplitudon frequency will be in the C phase still given by expression (77), whereas the phason will now exhibit an intrinsic non-zero gap. Any extrinsic phason gap present in the I phase will increase in the C phase. The transition from the I to the C phase will be thus always accompanied by a sudden increase in the phason frequency, while the amplitudon mode smoothly increases with decreasing temperature.

Both the phason and the amplitudon branch may be strongly damped. To describe this damping we shall introduce a constant damping factor

$$\Gamma_\beta = \text{const.}, \beta = \Gamma, A \tag{78}$$

The incommensurate soft mode is two-fold degenerate in the high-temperature paraelectric phase where $u_0 = 0$. It exhibits a classical soft-mode-like dispersion:

$$\omega_{sk}^2 = a(T - T_\mathrm{I}) + \kappa \cdot k^2, T > T_\mathrm{I} \tag{79}$$

The dispersion relations (76), (77) and (79) for the order-parameter fluctuations in the I and the C phases can be thus summarized as

$$\omega_{\beta k}^2 = \kappa \cdot k^2 + \omega_{\beta 0}^2, \beta = \varphi, A, s \tag{80}$$

where ω_{A0}^2 and ω_{s0}^2 are given by the corresponding terms in Eqs. 77 and 79. The excitation spectrum is schematically shown in Figs. 2.6 and 2.16.

The phason excitation spectrum consists in the multisoliton limit of two branches (Fig. 2.17) [2.53, 2.55, 2.56]. The phason spectrum is 'acoustic-like' for $k_z < \pi/x_0$, whereas it is 'optic-like' for $k_z > \pi/x_0$, where x_0 is the intersoliton distance (Fig. 2.17). The 'acoustic-like' branch corresponds to phase oscillations of the incommensurate multisoliton lattice, whereas the 'optic-like' branch corresponds to phase oscillations in the commensurate regions. The dispersion relation for the 'acoustic-like' branch is given by

$$\omega_{\varphi k 1}^2 = \omega_{\varphi 0 V}^2 + \kappa \cdot k_\perp^2 + \kappa_z k_z^2 \tag{81}$$

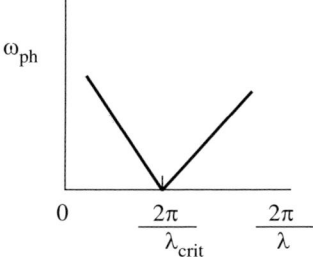

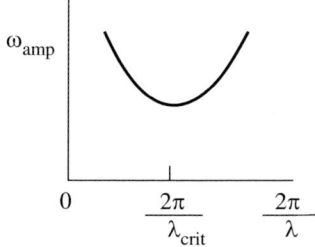

Fig. 2.16 Frequency dispersion of amplitudon and phason modes for the case that there is no phason gap.

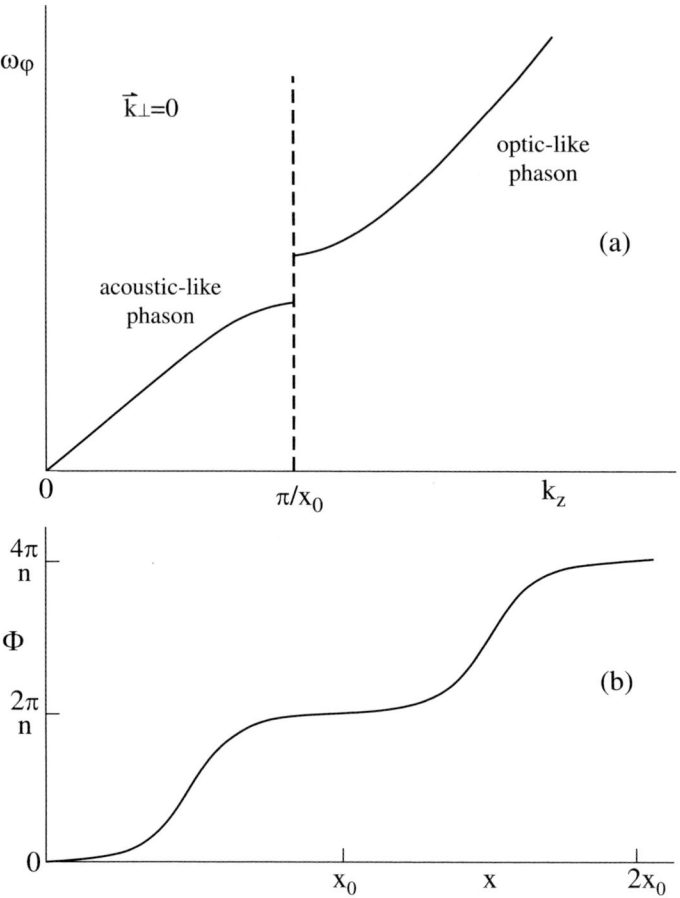

Fig. 2.17 Phason excitation spectrum (a) and spatial dependence of the phase (b) in the multisoliton limit.

κ = const., and $\omega_{\varphi 0V} \neq 0$ even in the absence of the defect-induced phase pinning perturbations [2.52]. In the ideal case $\omega_{\varphi 0V}$ non-critically increases with decreasing temperature, whereas

$$\omega_{\varphi 0V} = \text{const.} \tag{82}$$

when defects dominate the pinning.

The dispersion relation for the 'optic-like' phason branch can be, on the other hand, written as:

$$\omega_{\varphi kC}^2 = \omega_{\varphi 0C}^2 + \kappa \left[k_\perp^2 + (k_z - \pi/b)^2\right]. \tag{83}$$

where the gap $\omega_{\varphi 0C}$ is very nearly the same as in the C phase. The intersoliton distance increases with decreasing temperature as

$$x_0 \propto \ln(1 - T_c/T) \tag{84}$$

according to the Landau theory and as

$$x_0 \propto (1 - T_c/T)^{-1/2} \tag{85}$$

if fluctuations are taken into account [2.55, 2.56].

2.2 Multidimensionally modulated incommensurate systems

Phase transitions leading to conventional incommensurate phases are described by order parameters with a minimum dimensionality two, four, or six, resulting in single-q, double-q, or triple-q modulated structures. The NMR spectra and relaxation of incommensurate systems with a two-component order parameter [2.1, 2.52, 2.57–2.59]

$$Q_\pm = \rho e^{\pm i\phi} \tag{86}$$

corresponding to a one-dimensional modulation wave with a phase $\phi = \vec{q}\cdot\vec{r} + \phi_0$ and an amplitude ρ, are by now relatively well understood [2.52]. Much less is known [2.57, 2.60–2.67] about incommensurate systems where the order parameter has four or more components such as biphenyl [2.57, 2.67, 2.68], barium sodium niobate [2.68], BaMnF$_4$ [2.69], quartz [2.70, 2.71], proustite, or many charge-density-wave compounds.

In systems with a four-component order parameter [2.67–2.69]

$$Q_{1\pm} = \rho_1 e^{\pm i\phi_1} \tag{87a}$$

and

$$Q_{2\pm} = \rho_2 e^{\pm i\phi_2} \tag{87b}$$

two rather different incommensurate states may exist. The first of these two possible states $\rho_1 = \rho_2 \neq 0$ corresponds to a 'quilt-like' structure with a simultaneous freezing of all four vectors $\pm\vec{q}_1$ and $\pm\vec{q}_2$ [2.52, 2.62]. It is incommensurately modulated along two different directions. Another possible state corresponds to a 'stripe-like' domain structure with either $\rho_1 \neq 0, \rho_2 = 0$ or $\rho_1 = 0, \rho_2 \neq 0$, i.e. to a freezing of order-parameter components associated with a single pair $\pm\vec{q}_1$ or $\pm\vec{q}_2$ of opposite wavevectors. Each domain is here modulated in a single direction but the direction of the modulation wave varies from one domain to the other. Several cases are known where the angle between the two modulation directions $\vec{q}_1$ and $\vec{q}_2$ suddenly changes and, e.g., a two-q structure becomes a single-q structure [2.67, 2.68] (Table 2.1).

In systems with a six-component order parameter:

$$Q_{1\pm} = \rho_1 e^{\pm i\phi_1} \tag{88a}$$
$$Q_{2\pm} = \rho_2 e^{\pm i\phi_2} \tag{88b}$$
$$Q_{3\pm} = \rho_3 e^{\pm i\phi_3} \tag{88c}$$

there are several rather different situations:

Table 2.1 Selected examples of multiple-q modulated incommensurate systems (I OP = incommensurate order parameter).

Substance	I OP	Number of independent modulation directions
Ag_3AsS_3 (Proustite)	6	1–3
$AlPO_4$ (Berlinite)	6	2
Quartz (SiO_2)	6	2
2H-$TaSe_2$	6	2
Biphenyl ($C_{12}H_{10}$)	4	1–2
$Ba_2NaNb_5O_{15}$	4	1–2
$BaMnF_4$	4	1

(i) the case of three modulation waves with non-planar $\vec{q}$ vectors (i.e. the genuine triple-q modulated case),

(ii) the case of three coplanar modulation waves that in effect reduces to a double-q modulated structure.

In each of these two cases two rather different states may exist:

(i) a single-q (or 'stripe') multidomain state where each domain is modulated in a single direction but the direction of the modulation vector varies from one domain to another, and

(ii) an 'all-q' state where the amplitudes of all modulation waves are different from zero.

Here, we first look at the behaviour of incommensurate insulators with a multiple-q modulation with a special emphasis on systems with a double-q and triple-q modulation. For NMR in one-dimensional incommensurate systems we refer to the recent work of Petersson and Michel [2.71] and to the other older reviews [2.52–2.57, 2.72].

The main points that distinguish these multiple-q modulated incommensurate systems from 'standard' I systems like Rb_2ZnCl_4 with a 2-dimensional I order-parameter dimension and a single modulation direction are as follows.

(i) The phase diagram can exhibit several I phases differing by their number of modulation directions and their effective 'point group' symmetry.

(ii) Modulation waves and phase solitons can form complex spatial patterns.

(iii) The order-parameter dynamics is complex and contains more than just one phase and amplitude excitation mode. For a 2-q system we have two phasons and two amplitudons, whereas we have 3 phasons and 3 amplitudons for a 3-q system. Some of these modes may be degenerate.

X-ray and neutron scattering experiments can hardly distinguish between several possible I subphases (e.g., 'quilt-' and 'stripe-' type modulations) in these systems as they measure just the spatial average. NMR experiments, on the other hand, measure local properties and are very well suited for a determination of 'quilt-' or 'stripe-' type modulations and the exact nature of the I modulation waves. Similarly NMR can easily differentiate between 'plane-wave'- and 'soliton'-type modulations.

2.2.1 NMR and multidimensional modulation

Let us now examine an incommensurate system with a $2n$-component order parameter

$$Q_{n\pm} = \rho_n e^{\pm i\phi_n}, n = 1, 2, 3, \ldots \tag{89}$$

corresponding to a n-dimensional plane-wave-type modulation wave. Further, we assume that the relation between the NMR frequency shift and the nuclear displacements is purely local and linear so that

$$\nu(\phi_1, \ldots, \phi_n) = \nu_0 + \sum_{i=1}^{n} \nu_i \cos\phi_i \tag{90}$$

Here ν_i is proportional to the amplitude of the i-th component of the modulation wave. In the plane-wave limit we have

$$\phi_i = \vec{q}_i \cdot \vec{r}_i + \psi_i \tag{91}$$

where the various $\vec{q}_i$ are the modulation wavevectors that are incommensurate to the basic crystal lattice. The ν_i are proportional to the amplitudes ρ_i of the incommensurate order parameters. The phases ϕ_i take on any value in the interval $[-\infty, +\infty]$ with equal probability, but may be correlated. Instead of a single sharp NMR line we find for each physically non-equivalent site in the high-temperature unit cell in the incommensurate phase a characteristic frequency distribution [2.73]:

$$f(\nu) = \int d\phi_1 d\phi_2 \ldots d\phi_n \delta(\nu - \nu(\phi_1, \phi_2, \ldots, \phi_n)) \tag{92}$$

Here, the integration over the phase space

$$\left| \int d\phi_1 d\phi_2 \ldots d\phi_n \right|$$

must take note of possible correlations. When the modulation waves are independent the frequency distribution $f(\nu)$ is given by a multiple convolution

$$f(\nu) = \int_{-\infty}^{+\infty} f_1(\nu'_1) f_2(\nu'_2) \ldots f_{n-1}(\nu'_{n-1}) \times f_n\left(\nu - \sum_{j=1}^{n-1} \nu'_j\right) d\nu'_1 d\nu'_2 \ldots d\nu'_{n-1} \tag{93}$$

of frequency distributions $f_i(\nu)$ for purely one-dimensional modulations, $\nu = \nu_0 + \nu_i \cos\phi_i$, respectively [2.58]:

$$f_i(\nu') = \int_{-\infty}^{+\infty} \delta[\nu' - \nu(\phi_i)] d\phi_i = \frac{\text{const.}}{d\nu'/d\phi_i} = \frac{\text{const.}}{(\nu_i^2 - \nu'^2)^{1/2}} \tag{94}$$

This well-known distribution is shown in Fig. 2.18(a).

58 Incommensurate systems

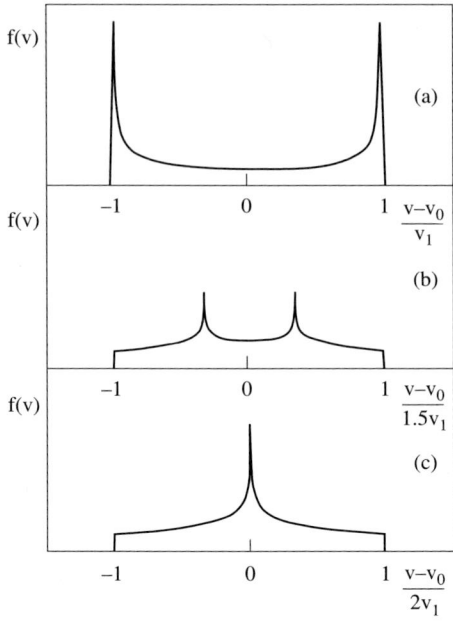

Fig. 2.18 NMR lineshape $f(\nu)$ for multiple-q plane-wave modulations with different relative amplitudes [2.73] (see Eq. (90)): (a) 1-q modulation; (b) 2-q modulation with $\nu_1 = 2\nu_2$, and (c) 2-q modulation with $\nu_1 = \nu_2$.

It should be stressed that the modulation waves are independent if the $\vec{q}$ vectors are independent or if the corresponding phases ψ_i are not correlated. If, however, the number of modulation waves is larger than the dimension of the space spanned by the corresponding wavevectors, and if their phases ψ_i are also correlated, the NMR spectrum strongly depends on the relative phases of these modulation waves. Such a situation occurs, for example, in the incommensurate phase of quartz [2.70] where the 'symmetric' triple-q structure is realized.

Even in a simple 1-q incommensurate system the spin-lattice relaxation rate varies [2.52, 2.58] over the incommensurate lineshape. We have

$$\frac{1}{T_1} = C\left[\cos^2\phi_i J_{A_i} + \sin^2\phi_i J_{\varphi_i}\right] \qquad (95)$$

where the J_{A_i} and J_{ϕ_i} are the local spectral densities of the amplitudon and phason order-parameter fluctuation modes, respectively, and the ϕ_i are given by Eq. (91).

For a system with n modulation waves linearly and locally related to the frequency shift one can write the spin-lattice relaxation rate as:

$$\frac{1}{T_1}(\phi_1, \phi_2, \ldots, \phi_n) = K \sum_{i=1}^{n} \left(\cos^2\phi_i J_{A_i} + \sin^2\phi_i J_{\phi_i}\right) \qquad (96)$$

The spin-lattice relaxation rates $\frac{1}{T_{1A}}$ and $\frac{1}{T_{1\varphi}}$ are proportional to the corresponding spectral densities J_A and J_Φ. Whereas there is a simple relation (Fig. 2.19(a))

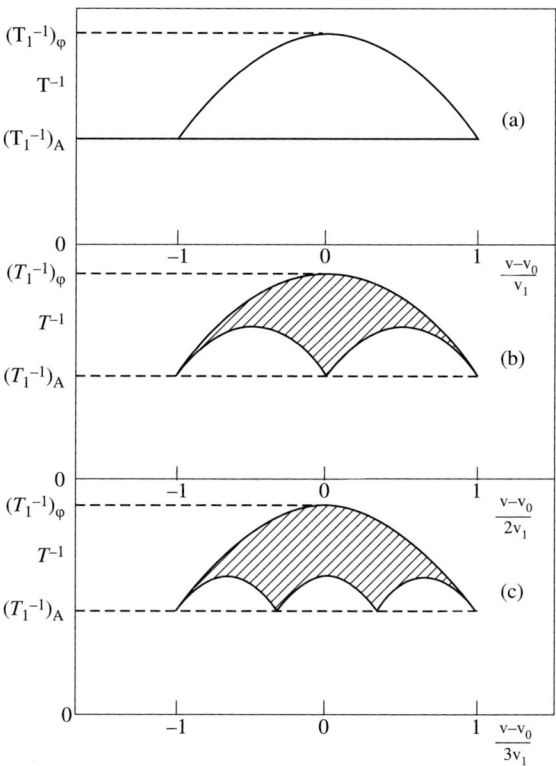

Fig. 2.19 Relation between T_1^{-1} and the frequency shift $\nu-\nu_0$ for a multiple-q modulation [2.73]. (a) 1-q modulation. Maximum and minimum values of the relaxation rate T_1^{-1} as function of the frequency shift for (b) a 2-q modulation with $\nu_1 = \nu_2$ and (c) a 3-q modulation with $\nu_1 = \nu_2 = \nu_3$ are represented by solid lines. In the multiple-q case the hatched dashed area stands for the region where a distribution of T_1^{-1} appears.

between the effective spin-lattice relaxation rate T_1^{-1} and the NMR frequency shift ν (Fig. 2.18(a)) in one-dimensionally modulated systems [2.58], this is no longer the case in multidimensionally modulated incommensurate systems. At any given frequency ν we have here a distribution of spin-lattice relaxation rates:

$$F\left(\frac{1}{T_1},\nu\right) = \int d\phi_1 d\phi_2 \ldots d\phi_n \delta(\nu - \nu(\phi_1,\phi_2,\ldots,\phi_n)) \times \delta\left[\frac{1}{T_1} - \frac{1}{T_1'}(\phi_1,\phi_2,\ldots,\phi_n)\right] \tag{97}$$

In the following, we discuss a few examples.

2.2.2   2-q modulation

Let us consider the case of a two-dimensionally modulated incommensurate structure where the angle between the two modulation wavevectors $\vec{q}_1$ and $\vec{q}_2$ equals γ [2.74]. The modulation wave is given by:

$$u = u_{01}\cos\phi_1 + u_{02}\cos\phi_2 \tag{98}$$

where ϕ_1 and ϕ_2 can be expressed in the plane-wave modulation approximation as

$$\phi_1 = q_1 x \tag{99a}$$

and
$$\phi_2 = q_2 x \cos\gamma + q_2 y \sin\gamma + \phi_0 \tag{99b}$$

Let us again stress that the relation between the NMR frequency shift and the nuclear displacement is taken to be purely local. According to Eq. (93) we find

$$f(\nu) = \int_{-\infty}^{+\infty} f_1(\nu') f_2(\nu - \nu') \, d\nu' \tag{100}$$

with f_1 and f_2 given by Eq. (92). Expression (100) can be evaluated analytically in terms of elliptic functions. It is easy to see that for $u_{01} = u_{02}$ and $\nu_1 = \nu_2$ the incommensurate frequency distribution $f(\nu)$ given by Eq. (100) exhibits an infinite logarithmic singularity at $\nu - \nu_0 \to 0$ (Fig. 2.18(c)):

$$f(\nu)_{\nu-\nu_0 \to 0} \approx -\ln[(\nu - \nu_0)/\nu_1]/\nu_1 \tag{101}$$

and two step discontinuities at $\pm 2\nu_1$ (Fig. 2.18(c)). This is in sharp contrast to the frequency distribution (Eq. 94) for a one-dimensional modulation (Fig. 2.18(a)) where we have two edge singularities at $\nu - \nu_0 = \pm \nu_1$.

It should be mentioned that for $\nu_1 \neq \nu_2$ we have, depending on the ratio ν_1/ν_2, two singularities (Fig. 2.18(b)).

It is interesting to note that $f(\nu)$ as given by expression (101) does not depend on the ratio $|\vec{q}_1|/|\vec{q}_2|$ or the angle γ between the two modulation directions $\vec{q}_1$ and $\vec{q}_2$ as long as $\gamma \neq 0$. This is so because the two modulation waves are independent and the corresponding phases and displacements are uncorrelated. If $\gamma = 0$, so that $\vec{q}_1 \parallel \vec{q}_2 \parallel x$-axis and

$$\nu = \nu_0 + \nu_1 \cos(q_1 x) + \nu_2 \cos(q_2 x + \phi_0) \tag{102}$$

the frequency distribution (Figs. 2.18(b) and 2.18(c)) is still given by expression (100), as long as $|\vec{q}_1|$ is incommensurate to $|\vec{q}_2|$

$$\left|\frac{\vec{q}_1}{|\vec{q}_2|}\right| \neq \frac{M}{N}, \quad M, N = 1, 2, 3, \ldots \tag{103}$$

If, however, the ratio $|\vec{q}_1|/|\vec{q}_2|$ is commensurate and the two waves are correlated through the whole crystal, the lineshape depends on $|\vec{q}_1|/|\vec{q}_2|$. For $|\vec{q}_2| = 2|\vec{q}_1|$, for example, where

$$\nu = \nu_0 + \nu_1 \cos\phi_1 + \nu_2 \cos(2\phi_1) \tag{104}$$

the lineshape is the same as for the case of a one-dimensional modulation wave where linear and quadratic terms are present in the expansion of ν in powers of the displacement [2.58].

It should also be noted that in the 'stripe-like' incommensurate phase, where we have domains $\rho_1 = 0, \rho_2 = K$ and $\rho_1 = K, \rho_2 = 0$ associated with freezing of $\pm\vec{q}_1$ and $\pm\vec{q}_2$, respectively, the NMR lineshape is the same as in the 1-q case, i.e. it is given by expressions (94) (Fig. 2.18(a)). This is because the NMR lineshape does not depend on the direction of $\vec{q}$ as long as the associated displacements and frequency changes are the same.

In the stripe phase there is a simple relation between the frequency shift and the value of the relaxation rate T_1^{-1} (Fig. 2.19(a)). In the quilt phase only boundary values of the distribution $F(1/T_1, \nu)$ are simple functions of the frequency shift. These distribution boundaries are illustrated in Fig. 2.19(b) for the case where the spectral densities of the phason and amplitudon fluctuations obey the relations $J_{\phi_1} = J_{\phi_2} = 2J_{A_1} = 2J_{A_2} = J$. At $\nu - \nu_0 = \pm\nu_1$ the distribution shrinks to a delta function with

$$T_1^{-1} = (T_1^{-1})_A \tag{105}$$

At $\nu = \nu_0 = 0$ the distribution is the broadest and

$$T_1^{-1} = (T_1^{-1})_\phi \tag{106}$$

2.2.3 Dispersion relations in incommensurate systems and T_1^{-1}

In this section, we wish to evaluate the temperature and frequency dependence of the nuclear spin-lattice relaxation rate, T_1^{-1}, of nuclei with a non-zero quadrupole moment ($S \geq 1$) in incommensurate systems.

Spin-lattice relaxation is here due to phonon-scattering processes that are accompanied by a spin flip. The probability of such a process increases with increasing phonon occupation number and increasing relative displacement of the lattice nuclei. In translationally periodic crystals undergoing structural phase transitions the main contribution is usually made by low-frequency soft optic modes that are often overdamped [2.72]. The contribution of long-wavelength acoustic modes is negligible as the relative displacements of the nearest neighbours are too small. The situation is quite different in incommensurate systems. Since the energy of an I system is in the continuum limit independent of the phase of the incommensurate distortion wave, these systems have phason branches. The energies of thermally excited phasons is of the order of the acoustic phonons and the relative displacements are not small for the critical wavevector. Spin-lattice relaxation of quadrupolar nuclei should be thus in incommensurate phases mainly determined [2.52] by phasons if competing relaxation channels are not too strong.

In contrast to translationally periodic systems the T_1 will vary over the incommensurate NMR line in a way that allows for the separate determination of the phason and amplitudon contributions [2.1, 2.52] as well as for a discrimination between direct and Raman processes. The average T_1 will be anomalously short and nearly temperature independent in the I phase [2.52].

For direct one-phonon processes that dominate the relaxation rate if the order parameter modes are overdamped, one finds [2.52, 2.75] the phason-induced spin-lattice relaxation rate in the 1D modulated case as:

$$T_1^{-1}{}_\varphi = C\frac{\pi}{2\sqrt{2}}\kappa^{-3/2}\sqrt{\Gamma_\varphi/\omega_\mathrm{L}} \tag{107}$$

for $\Gamma_\varphi \gg \omega > \Delta_\varphi$ and

$$T_1^{-1}{}_\varphi = C\frac{\pi}{4}\kappa^{-3/2}\cdot\Gamma_\varphi/\Delta_\varphi \tag{108}$$

for $\Delta_\varphi \gg \omega_L, \sqrt{\Gamma_\varphi\Delta_\varphi}$. Here, Δ_φ is the phason gap, ω_L the nuclear Larmor frequency and Γ_φ is the phason damping constant.

The amplitudon induced spin-lattice relaxation rate is here

$$T_1^{-1}{}_A = C\frac{\pi}{4}\kappa^{-3/2}\cdot\Gamma_A/\Delta_A \tag{109}$$

where $\Delta_A = \sqrt{2a(T_\mathrm{I}-T)}$ is the amplitudon gap.

If $\Gamma_\varphi = \Gamma_A = \Gamma$ and $\omega_L \ll \Gamma \ll \Delta_A$ one finds

$$\frac{T_{1\varphi}}{T_{1A}} = \frac{\Delta_\varphi}{\Delta_A}, \quad \Delta_\varphi \gg \omega_\mathrm{L} \tag{110}$$

and

$$\frac{T_{1\varphi}}{T_{1A}} = \frac{\sqrt{\omega_L\Gamma}}{\Delta_A}, \quad \Delta_\varphi \leq \omega_\mathrm{L} \tag{111}$$

2.2.4 3-q modulation

Now we turn to the 3-q case, where the structure is simultaneously modulated by three modulation waves and the system is described by a six-component order parameter

$$\nu(\varphi_1,\varphi_2,\varphi_3) = \nu_0 + \sum_{i=1}^{3}\nu_i\cos\varphi_i \tag{112}$$

and where the ν_i are proportional to the amplitude of the ith modulation wave. In the plane-wave limit we have

$$\varphi_i = \vec{q}_i\cdot\vec{r} + \varphi_{0i} \tag{113}$$

Here, the modulation wavevectors $\vec{q}_i$ are incommensurate to the underlying lattice. The phases φ_i take on any value in the interval $[-\infty, +\infty]$ with equal probability. In place of the single sharp NMR line of the high-temperature phase we find a characteristic frequency distribution in the incommensurate phase:

$$f_{3-q}(v) = \int dx_1\,dx_2\,dx_3\,\delta[v - v(\varphi_1(\vec{r}),\varphi_2(\vec{r}),\varphi_3(\vec{r}))] \tag{114}$$

The integration $\int dx_1\,dx_2\,dx_3$ must account for the possible correlation of the phases [2.73, 2.74]. When the modulation waves are independent, the frequency distribution $f(\nu)$ is given by a double convolution

$$f_{3-q}(\nu) = \int\limits_{-\infty}^{+\infty}\int\limits_{-\infty}^{+\infty} d\nu_1'\,d\nu_2'\,f_1(\nu_1')f_2(\nu_2')f_3(\nu-\nu_1'-\nu_2') \tag{115}$$

of frequency distributions $f_i(\nu')$ for single-q modulations.

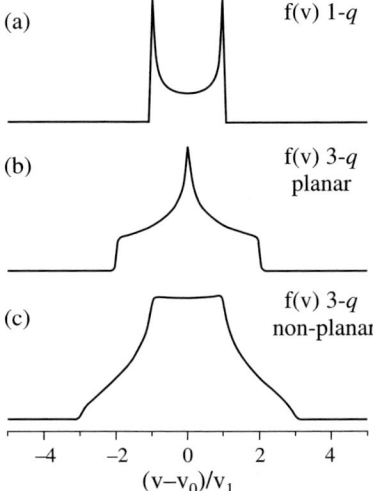

Fig. 2.20 Theoretical NQR lineshapes [2.77] in the plane-wave limit of (a) 1-q, (b) planar 3-q, (c) non-planar 3-q incommensurate systems for a linear coupling between the NQR frequency shift and the displacement. In the 3-q planar case (b), where the modulation waves are confined to a plane, the lineshape is similar to that of a 2-q I system [2.73, 2.74]. In this case the position of the central singularity depends on the relative phases of the three modulation waves. The lineshape is independent of the initial phases of the modulation waves in the non-planar 3-q case (c).

The typical bell-like 3-q lineshape for the linear case with $\nu = \nu_1 = \nu_2 = \nu_3$ is shown in Fig. 2.20(c). In contrast to the 1-q I lineshape, there are no edge singularities in the 3-q I lineshape, but the derivative of the lineshape has four van-Hove-like singular points [2.73]. It is interesting to note that the 3-q I lineshape completely changes if the three modulation waves are correlated (Fig. 2.20(b)). Such is the case in quartz [2.75, 2.76] and berlinite [2.74] where the three modulation waves are coplanar, $\vec{q}_1 + \vec{q}_2 + \vec{q}_3 = 0$ and their relative phase is fixed. As shown in Refs. [2.74] and [2.73], in such a case the lineshape is similar to that of a 2-q I system with a logarithmic singularity at a position that depends on the relative phases.

A comparison of lineshapes in the plane-wave and multisoliton lattice limits for multidimensional modulations is shown in Fig. 2.21.

2.2.5 The determination of the relative phases of the modulation waves

Let us now examine the potential of NMR to discriminate between the different relative phases of the modulation waves. As an example, let us look at the different possible structures of the incommensurate phase of quartz [2.71]. Here the three modulation waves are coplanar $|\vec{q}_1| = |\vec{q}_2| = |\vec{q}_3|$ and $\vec{q}_1 + \vec{q}_2 + \vec{q}_3 = 0$. In addition, their relative phase is fixed by the condition $\psi_1 + \psi_2 + \psi_3 = \psi$, where ψ can be 0, $\pm\pi/2$, or π. Because of the correlation between modulation waves the 3-q case is in fact reduced to the 2-q case where the relation between the frequency shift and the two independent modulation waves is no longer linear.

64 Incommensurate systems

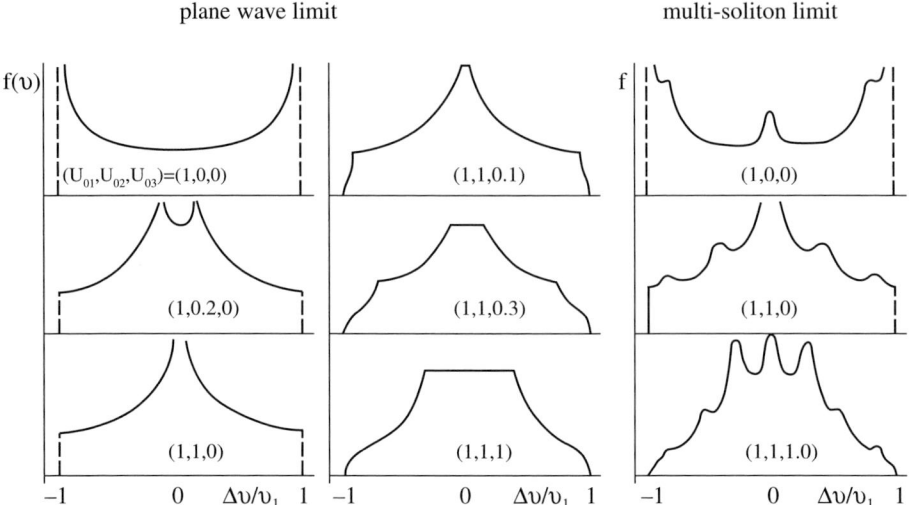

Fig. 2.21 Frequency distribution $f(\nu)$ in the plane-wave limit and the multisoliton limit as calculated for a multidimensional modulation wave with different modulation amplitudes u_{01}, u_{02}, and u_{03}.

The resulting NMR lineshapes are very specific (Fig. 2.22) and can be used to determine the relative phases of the modulation waves and discriminate between the different possible structures of the incommensurate phase of quartz corresponding to the different values of ψ. Thus, NMR can in this case determine the relative phases of the various modulation waves.

The spin-lattice relaxation rate has again several different values at any given frequency. The boundaries of the corresponding distribution $F(1/T_1, \nu)$ are for the case $\nu_1 = \nu_2 = \nu_3$ and $2J_{A_i} = J_{\phi_i}$ shown in Fig. 2.19(c). The upper boundary is the same as in the 2-q case, while the lower one is different and the value $(T_1^{-1})_A$ is reached four times.

2.2.6 Local and non-local case

In the 1-q plane-wave modulation limit the phase of the modulation wave is a linear function of the spatial coordinate x, and the incommensurate distortion is characterized by a single Fourier component q_I:

$$u = A\cos\phi(x) = A\cos(\vec{q}_I x + \phi_0), \phi_0 = \text{constant} \tag{116}$$

In the local approximation the resonance frequency ν_i of a given nucleus depends only on the incommensurate displacement u_i of the resonating nucleus and the displacement of atoms moving in phase with it:

$$\nu_i = \nu[u_i(x_i)] \tag{117}$$

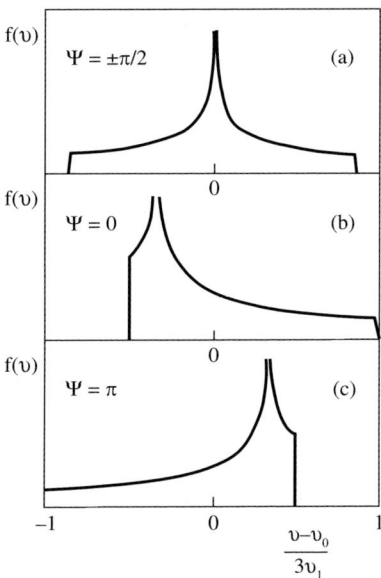

Fig. 2.22 The NMR frequency distributions $f(\nu)$ for the incommensurate phases of quartz [2.73] where correlated modulation waves have $|\vec{q}_1| = |\vec{q}_2| = |\vec{q}_3|$ and $\vec{q}_1 + \vec{q}_2 + \vec{q}_3 = 0$. The spectra for the three possible relative phase shifts $\psi = \psi_1 + \psi_2 + \psi_3$: (a) $\psi = \pm\pi/2$, (b) $\psi = 0$, and (c) $\psi = \pi$ are presented.

This relation can be expanded in powers of the displacements

$$\nu_i = \nu_0 + a_1 u_i(x) + \frac{1}{2} a_2 u_i^2(x) + \ldots \tag{118}$$

yielding

$$\nu_i = \nu_0 + \nu_1 \cos\phi(x) + \frac{1}{2}\nu_2 \cos^2\phi(x) + \ldots \tag{119}$$

where $\cos\phi(x)$ nearly continuously takes on all values between $+1$ and -1, and where $\nu_1 \propto A \propto (T_I - T)^\beta$ and $\nu_2 \propto (T_I - T)^{\tilde{\beta}}$, where $\tilde{\beta} \cong 2\beta$. Here, the phase $\phi(x)$ and the amplitude A of the modulation wave constitute the two-component order parameter of the 1-q normal-incommensurate phase transition.

The intensity of the resonance line in a frequency interval between ν and $\nu + d\nu$ is proportional to the number of nuclei resonating in this interval. Knowing the function $\nu_i = \nu[u_i(x)]$ one can calculate the frequency distribution $f(\nu)$ in I systems from

$$f(\nu)d\nu = C \cdot dx \tag{120}$$

or

$$f(\nu) = \frac{C}{\left|\frac{d\nu}{d\phi}\frac{d\phi}{dx}\right|} \tag{121}$$

In the plane-wave limit $d\phi/dx = \text{const}$ and $f(\nu)$ is determined simply by $d\nu/d\phi$.

66 Incommensurate systems

In the general case the dependence of the resonance frequency on the lattice displacements is non-local [2.58] (i.e. the resonance frequency of the ith nucleus depends not only on its displacement but also on the displacement of other nuclei), as given by

$$\nu_i = \nu[u_i(x_i), u_j(x_j), \ldots], i \neq j \tag{122}$$

This dependence can reflect the non-local nature of such quantities as the electric-field gradient (EFG) tensor or the chemical-shift tensor, which, in addition to the external magnetic field, determine the NMR resonance frequency. Only when the dominant contribution to the NMR resonance frequency comes from a region the dimensions of which are small compared to the wavelength of the incommensurate modulation wave, can the relation between ν_i and u_i be treated as local, so that Eq. (117) becomes strictly valid [2.58].

In the non-local case Eq. (122) yields the spatial dependence of the NMR frequency [2.58] as

$$\nu(x) = \nu_0 + \nu_1 \cos[\phi(x) - \phi_1] + \frac{1}{2}[\nu'_2 + \nu''_2 \cos 2(\phi(x) - \phi_2)] + \ldots \tag{123a}$$

After introducing $\psi = \phi - \phi_2$ and $\tilde{\phi} = \phi_2 - \phi_1$ we find

$$\nu(x) = \nu_0 + \nu_1 \cos[\psi(x) + \tilde{\phi}] + \nu'_2 + \nu_2 \cos^2 \psi(x) + \ldots \tag{123b}$$

where $\nu_2 = \nu''_2$ and $\nu'_2 = \frac{1}{2}(\nu'_2 - \nu_2)$. Expression (123b) is similar to the local description (119) except for the fact that there is now a phase shift $\tilde{\phi} \neq 0$ between the linear and the quadratic terms and there is a uniform frequency shift $\nu'_2 \neq 0$.

In the linear approximation ($\nu_1 \neq 0$, $\nu_2 = \nu'_2 = 0$) there is no difference between the local and the non-local cases. The frequency distribution is obtained as

$$f(\nu) = \frac{\text{const.}}{\sqrt{1 - \left(\frac{\nu - \nu_0}{\nu_1}\right)^2}} \tag{124}$$

and is shown in Fig. 2.18(a). Here, the splitting $\Delta\nu$ between the two edge singularities at

$$\nu_\pm = \nu - \nu_0 = \pm\nu_1 \tag{125}$$

increases with the critical exponent β, since

$$\Delta\nu = 2\nu_1 \propto (T_I - T)^\beta \tag{126}$$

In the quadratic approximation ($\nu_1 = 0$, $\nu_2 \neq 0$, $\nu'_2 \neq 0$) the frequency distribution is different from the one obtained in the local case. It is given by

$$f(\nu) = \frac{\text{const.}}{\sqrt{(\nu - \nu'_2 - \nu_0)(\nu_2 + \nu'_2 + \nu_0 - \nu)}} \tag{127a}$$

The edge singularities occur at
$$\nu - \nu_0 = \nu_2' \tag{127b}$$
and
$$\nu - \nu_0 = \nu_2 + \nu_2' \tag{127c}$$

The positions of both singularities now vary with temperature in contrast to the local case, as ν_1 and ν_2' are T dependent. If both linear and quadratic terms are present, the shape of the spectrum in the non-local case will strongly depend on the ratio between the magnitudes of ν_1 and ν_2' and on the relative phase shift $\tilde{\phi}$ between them. If $\tilde{\phi}$ does not equal $0°$ or $90°$, we will have, at lower temperatures, four and not three edge singularities [2.58] as in the local case.

2.2.7 Multisoliton lattice limit

Let us now consider the case where the phase of the modulation wave ϕ is a non-linear function of x. From
$$f(\nu') = \int \delta[\nu' - \nu(x)]\,dx \tag{128}$$
we get
$$f(\nu) = \frac{\text{const.}}{\left|\frac{d\nu}{du}\right|\left|\frac{du}{d\phi}\right|\left|\frac{d\phi}{dx}\right|} \tag{129}$$

The lineshape will be peaked whenever:

$$\text{i)} \quad d\nu/du \to 0 \tag{130a}$$
$$\text{ii)} \quad du/d\phi \to 0 \tag{130b}$$
$$\text{iii)} \quad d\phi/dx \to 0. \tag{130c}$$

Whereas points i) and ii) give rise to singularities in the spectrum in the plane-wave limit, it is point iii) $d\phi/dx \to 0$, which is specific for the multisoliton lattice limit and that allows for a determination of the soliton density.

The spatial variation of the phase of the modulation wave $\phi(x)$ is here determined from the non-linear sine-Gordon equation
$$\frac{d^2\phi}{dx^2} = -\frac{n}{2}\frac{\Gamma}{\kappa}A_0^{n-2}\sin(n\phi) \tag{131}$$

Here, n represents the lowest power of the 'lock-in' term [2.52, 2.58, 2.59] describing the effect of the discrete crystal lattice, $A_0 = \rho$ is the amplitude of the modulation wave, Γ is a coefficient describing the strength of the lock-in term, and κ is an elastic constant.

It should be noted that $n = 10$ for $[N(CH_3)_4]_2ZnCl_4$, $n = 6$ for Rb_2ZnCl_4, $n = 4$ for $(NH_4)_2BaF_4$, and $n = 2$ for chiral smectic liquid crystals in an external magnetic field [2.58].

68 Incommensurate systems

Close to T_I, where $A_\text{o} \propto (T_\text{I}-T)^\beta \to 0$, $\phi(x)$ is nearly linear in x and the system is in the 'plane-wave' modulation limit. At lower temperatures, where A_o is larger, ϕ becomes non-linear in x and the multisoliton lattice limit is approached. Nearly commensurate regions where $\phi(x) \approx$ constant are separated by soliton-like 'discommensurations' where the phase changes rapidly. The soliton density [2.52]

$$n_s = \frac{d_0}{x_0} \qquad (132)$$

is the ratio of the soliton width d_0 to the intersoliton distance x_0 and corresponds to the volume fraction of the crystal occupied by the phase soliton domain walls (i.e. the 'discommensurations'). The value of n_s is 1 in the plane-wave limit just under T_I and zero at the 'lock-in' transition T_c to the commensurate phase, where $x_0 \to \infty$. By a straightforward integration of Eq. (131) we can obtain $d\phi/dx$ so that $f(\nu)$ becomes:

$$f(\nu) \propto \frac{1}{\left|\frac{d\nu}{du}\frac{du}{d\phi}\right| \sqrt{\Delta^2 + \cos^2\left[\frac{n}{2}(\phi(x) - \phi_0)\right]}} \qquad (133)$$

where ϕ_0 is an initial phase, which depends on the position of the nucleus in the paraelectric unit cell. The integration constant Δ^2 is related to the soliton density n_s via [2.58]

$$n_\text{s} = \frac{\pi/2}{K(k)} \qquad (134)$$

where

$$k = \frac{1}{\sqrt{1+\Delta^2}} \qquad (135)$$

and where $K(k)$ is the complete elliptic integral of the first kind:

$$K(k) = \int_0^{\pi/2} \frac{d\phi}{\sqrt{1 - k^2 \sin^2 \phi}} \qquad (136)$$

As already mentioned Δ^2 is obtained from the experiment by fitting the observed lineshape to Eq. (133).

It should be noted [2.58] that for $T \to T_\text{I}$, $A_0 \to 0$, $k^2 \to 0$, $\Delta^2 \to \infty$, and $K(k) \to \pi/2$ so that $n_\text{s} \to 1$. For $T \to T_\text{c}$, on the other hand, $k^2 \to 1$, $K(k) \to \infty$ and $n_\text{s} \to 0$.

Expression (133) yields the same lineshape as that obtained in the plane-wave limit (where $\Delta^2 \gg 1$), whereas up to n new lines are found [2.58] in the multisoliton limit (where $\Delta^2 \ll 1$). The new lines occur in addition to the singularities formed in the plane-wave limit and appear for $\Delta^2 \ll 1$ when $\cos^2[\frac{n}{2}(\phi(x) - \phi_0)] = 0$, that is when

$$\phi = (2m+1)\pi/n + \phi_0, m = 0, 1, 2, \ldots n-1 \qquad (137)$$

At special values of ϕ_0 the number of new 'commensurate' lines [2.78] will be smaller than n.

The above considerations can be extended to the (2-q) and (3-q) modulation cases [2.73, 2.77]. In the simplest case where the various modulation waves are independent, we find the $\phi_i = \phi_i(x_i)$ relations as solutions of i-uncoupled sine-Gordon equations.

In the linear case ($\nu_2 = 0$) one finds for a 1-q plane-wave-type modulation a frequency distribution limited by two edge singularities at $\nu - \nu_0 = \pm\nu_1$, where $d\nu/d\phi_1 = 0$ (Fig. 2.20). In the 2-q case one finds a lineshape, characterized by a logarithmic singularity in the centre. For the planar 3-q case the lineshape is similar to that of the 2-q case, but the position of the central singularity depends on the relative phases between the modulation waves. For the non-coplanar 3-q phase with three independent modulation waves the NMR spectrum is bell-shaped and does not depend on the initial phases in the plane-wave limit (Figs. 2.20 and 2.21). As long as ϕ_1 is a linear function of $x_1, d\phi_1/dx_1$ is non-zero and constant, and this term does not affect the spectral lineshape. In the non-linear case of a multisoliton lattice, on the other hand, $d\phi_1/dx_1 \to 0$ in the commensurate regions leading to the appearance of the characteristic 'commensurate' lines [2.63] in the spectrum.

The comparison between the plane-wave and multisoliton lattice lineshapes is illustrated in Fig. 2.23 for the 1-q, 2-q and 3-q cases.

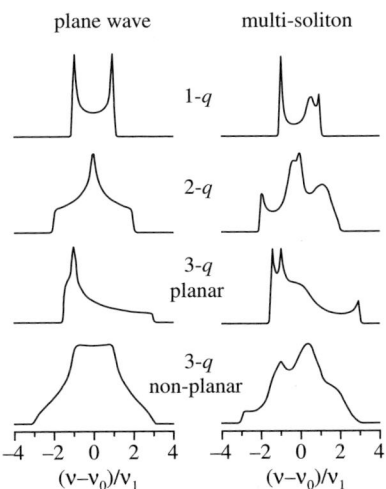

Fig. 2.23 Theoretical NQR lineshapes for the 1-q, 2-q, planar 3-q and nonplanar 3-q incommensurately modulated systems in the plane-wave and multisoliton lattice limits for a linear relation between the resonance frequency and nuclear displacement [2.79]. The lineshapes are obtained from Eq. (92). For the soliton limit the phases are obtained as solutions of the sine-Gordon equation (132). In the planar 3-q case the position of the central logarithmic singularity depends on the relative phases of the modulation waves, even in the plane-wave limit, while in this limit the spectrum is independent of the initial phases for the non-coplanar 3-q case.

2.2.8 Systems with a 6-component order parameter

2.2.8.1 ^{75}As NQR in proustite: The non-planar 3-q case

Proustite structure and phase transition The structure of proustite belongs to the non-centrosymmetric trigonal space group R3c(C_{2V}^6) with two formula units, Ag$_3$AsS$_3$, per primitive rhombohedral unit cell at room temperature [2.80]. The covalently bonded AsS$_3$ pyramids occupy the C$_3$ sites at eight corners of the rhombohedral unit cell and one in the centre, while the Ag atoms reside on the C$_1$ sites in the voids, formed by the S atoms. The unit cell contains 12 positions for Ag$^+$ ions, 6 of them being vacant at room temperature. The AsS$_3$ pyramids are ionically bound by S–Ag–S bonds, which form helices in the direction of the trigonal c-axis. This rather weak bonding leads to a high ionic conductivity and allows Ag$^+$ ions to adopt a quasi-free state. The consecutive AsS$_3$ pyramids along the c-axis undergo a small relative rotation, each pyramid being related to the one above or below by a glide plane σ_V^g.

On cooling below room temperature, proustite exhibits several phase transitions: a transition from the high-temperature normal (N) phase to an incommensurate (IC) phase at $T_{\text{IC}} \approx 60$ K, a lock-in transition to a commensurate phase C$_1$ at $T_{\text{L}} \approx 49$ K [2.81], and a transition to a non-modulated ferroelectric phase C$_2$ at $T_{\text{C}_2} \approx 24$ K. At high pressures the intermediate I and C$_1$ phases are suppressed [2.82]. so that proustite exhibits a direct N–C$_2$ transition. In addition to this, it has been suggested that a photo induced phase transition occurs near 210 K [2.82], connected to a redistribution of Ag$^+$ ions under the action of illumination. In contrast to this, a detailed Raman scattering [2.83] and a combined neutron and X-ray diffraction study [2.84] (both conducted in the dark) failed to reveal any anomalies in the temperature range 80 K to 300 K.

X-ray studies [2.81, 2.85] have shown that the second-order phase transition from the normal to the incommensurate phase is characterized by a star of six wavevectors:

$$\vec{q}_{\pm 1} = \pm \left[\frac{1}{3} - \delta_a, -\left(\frac{1}{3} - \delta_a \right), 0, \frac{1}{3} - \delta_c \right]$$

$$\vec{q}_{\pm 2} = \pm \left[0, \frac{1}{3} - \delta_a, -\left(\frac{1}{3} - \delta_a \right), \frac{1}{3} - \delta_c \right] \qquad (138)$$

$$\vec{q}_{\pm 3} = \pm \left[-\left(\frac{1}{3} - \delta_a \right), 0, \frac{1}{3} - \delta_a, \frac{1}{3} - \delta_c \right]$$

where the hexagonal setting of the rhombohedral cell is adopted. Just below T_{I} the values of δ_a and δ_c are 0.0063 and 0.0125, respectively. Both δ_a and δ_c are temperature dependent and decrease with decreasing temperature. Hence, not just the period, but also the direction of the modulation varies as a function of temperature, while remaining in the glide plane.

The transition into the I phase in proustite has been described within the Landau theory by Pokrovsky and Pryadko [2.86]. The order parameters $\eta_i(\vec{r})$ are the deviations of the displacements $\vec{u}(\vec{r})$ from the commensurate directions:

$$\vec{q}_1^c = \frac{1}{3}\left[1,\bar{1},0,1\right]^*, \vec{q}_2^c = \frac{1}{3}\left[0,1,\bar{1},1\right]^*, \text{ and } \vec{q}_3^c = \frac{1}{3}\left[\bar{1},0,1,1\right]^*,:$$

$$\vec{u}(\vec{r}) = \sum_{i=1}^{3} \eta_i(\vec{r})\xi(\vec{q}_i^C)e i\vec{q}_i^C \vec{r} + c.c. \tag{139}$$

where:
$$\eta_i(\vec{r}) = \rho_i e^{-i\varphi_i(\vec{r})} \tag{140}$$

and $\xi(\vec{q}_i^C)$ stands for normalized polarization vector of the $\vec{q}_i^C$ mode.

The free-energy density has the form:

$$f = A(T - T_I)\sum_i |\eta_i|^2 + \sum_{\alpha,\beta,i} B_{\alpha\beta}^i \left[\frac{\partial \eta_i}{\partial x_\alpha} \cdot \frac{\partial \eta_i^*}{\partial x_\beta} + i\delta_\alpha^i \left(\eta_i \frac{\partial \eta_i^*}{\partial x_\beta} - \eta_i^* \frac{\partial \eta_i}{\partial x_\beta}\right)\right]$$

$$+ C_1 \sum_i |\eta_i|^4 + C_2 \left(\sum_i |\eta_i|^2\right)^2$$

$$+ \left[D_1 \sum_i \eta_i^6 + D_2 a^2 + D_3 ab + D_4 ab^* + D_5 b^2 + D_6 bb^* + c.c.\right] \tag{141}$$

where $a = \eta_1 \eta_2 \eta_3$ and $b = \eta_1^3 + \eta_2^3 + \eta_3^3$.

Keeping only the $D_1 \sum_i \eta_i^6$ 'lock-in' term we get in the constant-amplitude approximation the free-energy density as:

$$f = t\sum_{\alpha,\beta,i} B_{\alpha\beta}^i \left(\frac{\partial \varphi_i}{\partial x_\alpha} - \delta_\alpha^i\right)\left(\frac{\partial \varphi_i}{\partial x_\beta} - \delta_\beta^i\right) + t^3 D_1 \left[\cos(6\varphi_1) + \cos(6\varphi_2) + \cos(6\varphi_3)\right]$$
$$\tag{142}$$

where $t = \frac{T_I - T}{T}$ and $\vec{\delta} = \begin{bmatrix} \delta_x \\ 0 \\ \delta_z \end{bmatrix}$.

Minimizing the free energy $F = \int f d^3 r$ we get three uncoupled sine-Gordon equations for $\varphi(u_i), i = 1,2,3$

$$\phi''(x_i) = -6Dt^2 \sin[6\phi(x_i)]; \quad i = 1,2,3 \tag{143}$$

Here, D is a constant depending on the parameters of the Landau free-energy expansion [2.86]. The solutions for ϕ_i are elliptic amplitudes as in the $(1-q)$ case:

$$\phi(u_i) = \frac{2}{p} am\left(v\frac{n}{2}u_i, k^2\right) \tag{144}$$

The parameter k is here connected to the linear soliton density [2.58], $n_S = \pi/(2K(k))$, $K(k)$ is the complete elliptic integral of the first kind, u_i are the space coordinates in the directions of the modulation wavevectors $\vec{q}_i$, $\nu = \frac{t}{k}\sqrt{\frac{2}{D}}$, while the parameter n is connected with the unit-cell multiplication. In the present case we have $n = 6$ for the transition to the C_1 phase with the triplication of the edge of the unit cell.

Just below T_I one finds that within the constant-amplitude approximation $\eta_i(\vec{r}) = \rho_i e^{-i\varphi_i(\vec{r})}$, the free energy is minimized if the phases of the modulation waves are linear functions of the coordinates:

$$\varphi_i = \vec{k}_i \vec{r} \quad (145)$$

Depending on the signs of the coefficients in the Landau expansion [2.86], either the 3-q or the 1-q 'stripe' phase may be realized just below T_I:

$$3\text{-}q : \rho_1 = \rho_2 = \rho_3 = \frac{\rho}{\sqrt{3}} \propto \sqrt{t} \text{ for } C_1 > 0 \quad (146a)$$

$$1\text{-}q : \rho_2 = \rho_3 = 0, \rho_1 \propto \sqrt{t} \text{ for } C_1 < 0 \quad (146b)$$

Here, t stands for the normalized temperature, $t = (T_{IC} - T)/T$. If the sixth-order invariants [2.79] are included in the Landau expansion, both the 1-q stripe and the 3-q phase may be stable below T_I. Depending on the size of the coefficients, the $N \to (3\text{-}q) \to C_1, N \to (1\text{-}q) \to C_1, N \to (1\text{-}q) \to (3\text{-}q) \to C_1$ or $N \to (3\text{-}q) \to (1\text{-}q) \to C_1$ phase sequences may be realized [2.79].

X-ray measurements [2.81, 2.85] did not reveal any structural change in the temperature region of the incommensurate phase 49–60 K. However, due to the low-temperature resolution of these measurements and the difficulties in discriminating between a true 3-q I phase and a multidomain 1-q stripe phase by X-rays, one cannot exclude the possibility of finding a 1-q stripe phase in a narrow temperature range either between the N and 3-q phase or between the 3-q and C_1 phase.

According to X-ray measurements [2.81, 2.85] the incommensurate parameters δ_a and δ_c vanish on cooling below the lock-in transition temperature $T_L = 49\,\text{K}$ (first order according to [2.85], second order according to [2.81]) to a commensurably modulated phase C_1. The C_1 unit cell has the volume 27 times that of the normal N phase but retains the same space group R3c, with the new crystallographic axes rotated by 180° with respect to the N phase.

On cooling below $T_{C_2} = 24\,\text{K}$ proustite exhibits a strongly first order phase transition to a ferroelectric phase C_2 where the structure is monoclinic with a space group Cc [2.87]. The AsS_3 pyramids are considerably distorted in the C_2 phase with the three As–S bond lengths differing by nearly 10% [2.87]. In addition to a pyroelectric polarization along the c-axis, a spontaneous polarization component appears in the basal plane. In the ferroelectric phase the modulation of the normal phase disappears and the structure can be treated as resulting from a new distortion of the N phase [2.86].

Temperature dependence of NQR frequencies The temperature dependence of the ^{75}As quadrupole resonance frequencies ν_Q and the spin-lattice relaxation time T_1 of proustite are shown in Fig. 2.24. At room temperature a single resonance line

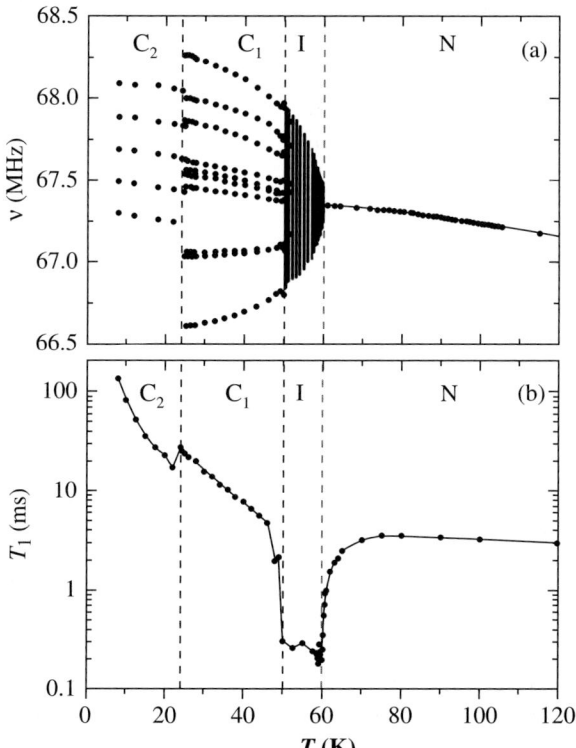

Fig. 2.24 Temperature dependence of the ^{75}As NQR frequencies ν_Q (a) and the average spin-lattice relaxation time T_1 (b) in proustite between 8 and 120 K [2.77]. There is a single resonance line in the high-temperature N phase, an inhomogeneously broadened line (indicated by the vertical lines) in the incommensurate (I) phase, ten resonance lines in the C_1 phase, and five lines in the C_2 phase. Note the critical decrease of the T_1 at the N–I transition and the anomalous short temperature-independent phason-induced T_1 in the I phase.

with a half-height full width of 5 kHz is observed at $\nu_Q = 66.22$ MHz. On cooling, the NQR frequency continuously increases to 67.3 MHz just above the transition to the incommensurate phase at $T_{IC} = 60$ K. There is a slight, but still well-observable change of the slope of the variation of ν_Q with T at about 75 K. The linewidth does not change between room temperature and 60.2 K, which is just above T_{IC}.

Below $T_{IC} = 60$ K an inhomogeneous broadening of the resonance line is observed in the incommensurate phase. The linewidth at half-height increases from 5 kHz just above T_{IC} to about 1.1 MHz in the lower part of the incommensurate phase. On further cooling, a set of sharp lines appears on the broad incommensurate background about 2 K above T_{C1}. The multiplet of sharp lines increases in intensity with decreasing temperature and completely replaces the broad background at the transition to the commensurate phase C_1.

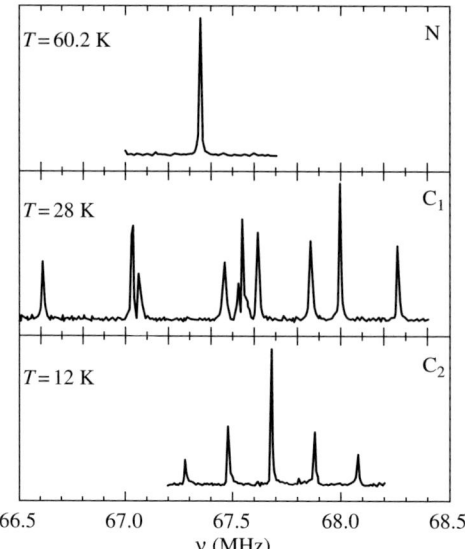

Fig. 2.25 ^{75}As NQR spectra in the three commensurate phases N (just above the N–IC transition), C_1, and C_2 in a proustite single crystal [2.77].

In the C_1 phase one observes 10 narrow lines of different intensity. At 28 K (Fig. 2.25) we have 3 lines of low intensity, 6 lines of medium intensity and a single line of high intensity. The temperature dependence of the positions of these ten lines is shown in Fig. 2.24.

At the transition temperature to the low-temperature commensurate phase $C_2, T_{C_2} = 24$ K, the asymmetric multiplet of ten lines is discontinuously replaced by a symmetric multiplet of five lines (Fig. 2.25), which persists down to the lowest studied temperature (8 K). A similar behaviour of the ^{75}As NQR lines at the C_1 to C_2 transition was observed earlier [2.63, 2.82], but due to the lower sensitivity only the most intense central line was detected.

Temperature dependence of T_1 The ^{75}As spin-lattice relaxation time T_1 (Fig. 2.24(b)), which amounts to 1.5 ms at room temperature, increases monotonically with decreasing temperature and reaches a maximum value of 3.6 ms close to 75 K. There is no observable anomaly in the temperature dependence of either T_1, the resonance frequency ν_Q, or the linewidth $\Delta\nu$, which would indicate the phase transition reported at 210 K [2.88]. The experiments were performed in the dark and no attempt was made to check for the possible effect of the illumination on the parameters $T_1, \nu_Q,$ and $\Delta\nu$. Below 75 K T_1 starts to decrease and shows a cusp-like dip in the vicinity of $T_I = 60$ K, which is characteristic of a displacive structural transition with a condensing soft mode.

The phason-induced T_1 is anomalously short (~ 0.2 ms) and temperature independent in the incommensurate phase. The transition to the C_1 phase is marked by a discontinuous increase of T_1 for an order of magnitude to $T_1 = 2\,\text{ms}$ just below T_{C_1}. In the C_1 phase $\ln(T_1)$ seems to increase linearly with decreasing temperature (Fig. 2.24(b)).

At $T_{C_2} = 25\,\text{K}$ there is a discontinuous decrease from $T_1 = 27\,\text{ms}$ just above T_{C_2} to $T_1 = 17\,\text{ms}$ just below. On cooling, T_1 further increases to reach $135\,\text{ms}$ at $8\,\text{K}$.

The (1-q) to (3-q) transition and the appearance of the multisoliton lattice In the general case. the dependence of the NQR resonance frequency ν_i on the nuclear displacements u_i in the I phase is non-local:

$$\nu_i = \nu\left[u_i(x_i), u_j(x_j), \ldots\right], \quad i \neq j \tag{147}$$

where j runs over all ions that contribute to the resonance frequency at the ith site. Expanding the above relation in powers of the displacements one finds for the ^{75}As NQR frequency in proustite in the 3-q case

$$\nu_i(\phi_1, \phi_2, \phi_3) = \nu_0 + \nu_1 \sum_{i=1}^{3} \cos(\phi_i) + \nu_2 \sum_{i=1}^{3} \cos(2\phi_i + \phi_{20}) + \cdots \tag{148}$$

where ν_1 is proportional to the amplitude of the modulation wave $\nu_1 \propto \rho \propto (T-T)^{\beta}$, i.e. to the order parameter, and $\nu_2 \propto (T-T)^{\tilde{\beta}}$ with $\tilde{\beta} \cong 2\beta$; ϕ_{20} describes the relative phase shift between the linear and quadratic terms.

The incommensurate frequency distribution f_{3q} is now obtained [2.73, 2.79] in the constant-amplitude approximation as

$$f_{3q}(\nu) = K \iiint \delta\left\{\nu - \nu\left[\phi_1(x_1), \phi_2(x_2), \phi_3(x_3)\right]\right\} dx_1 dx_2 dx_3 \tag{149}$$

where x_i are variables along the directions of the modulation wavevectors [2.63]. In the 1-q case expression (58) simplifies to the well known expression:

$$f_{1q}(\nu) = K \int \delta\left\{\nu - \nu\left[\phi_1(x_1)\right]\right\} dx_1 = \frac{\text{const.}}{d\nu/dx_1} = \frac{\text{const.}}{(d\nu/d\phi_1)(d\phi_1/dx_1)} \tag{150}$$

On cooling into the I phase the NQR spectrum becomes inhomogeneously broadened (Fig. 2.24). The lineshape with two edge singularities observed just below T_I at $58.8\,\text{K}$ is typical for a 1-q modulated system in the plane-wave limit rather than for a 3-q modulated system. On further cooling into the I phase an asymmetrical bell-like spectral component appears (Fig. 2.26) which gradually replaces the 1-q spectrum. This component, which dominates the spectrum already $2\,\text{K}$ below T_I, is typical for a 3-q modulated system. We have thus observed a transition from a state, characterized by a superposition of 1-q domains – Eq. (151) – with different directions of modulation vectors to a genuine 3-q state – Eq. (101) – where the crystal is simultaneously modulated along three different directions:

$$\rho_1 \neq 0, \rho_2 = \rho_3 = 0 \text{ etc.} \rightarrow \tag{151}$$

$$\rightarrow \rho_1 \neq 0, \rho_2 \neq 0, \rho_3 \neq 0 \tag{152}$$

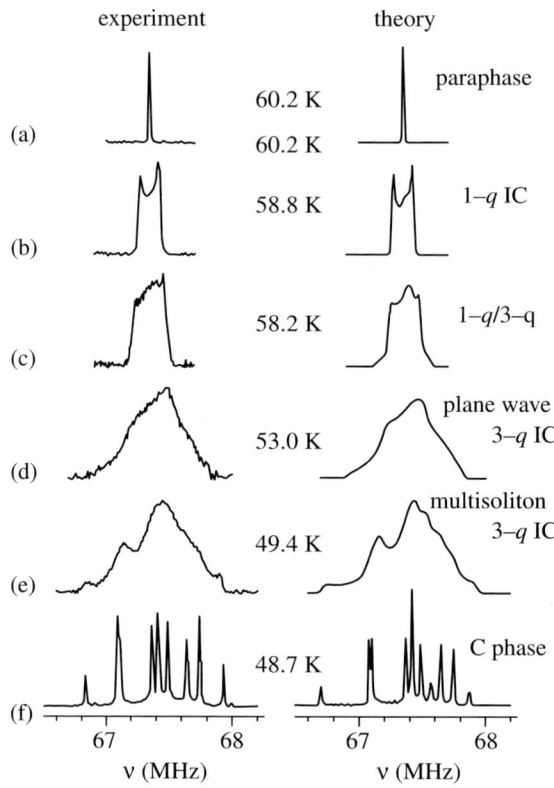

Fig. 2.26 Comparison between the observed and theoretical ^{75}As NQR lineshapes in proustite [2.79]: (a) $T = 60.2$ K, paraphase; (b) $T = 58.8$ K, 1-q incommensurate phase; (c) $T = 58.2$ K, coexistence of 1-q and 3-q phases; (d) $T = 53.0$ K, plane wave regime of 3-q incommensurate phase; (e) $T = 49.4$ K, multisoliton regime of 3-q incommensurate phase; and (f) $T = 48.7$ K, commensurate phase C_1.

The 1-q solution is stable in the high-temperature part of the I phase, whereas the 3-q solution is stable at lower temperatures [2.86]. This has been indeed observed. From the 1-q fit we also obtain the critical exponent for the amplitude of order parameter $\beta_{1q} = 0.3 \pm 0.02, \tilde{\beta}_{1q} \approx 2\beta_{1q} = 0.6 \pm 0.02$, and $\phi_{20} = 80°$. The analogous parameters for the $3-q$ phase are $\beta_{3q} = 0.4 \pm 0.02, \tilde{\beta}_{3q} \cong 2\beta_{31q} = 0.8 \pm 0.02$, and $\phi_{20} = 50°$.

Between 57.5 K and 52 K the NQR lineshape is essentially the one expected for the 3-q case in the plane-wave modulation limit (Fig. 2.26). The non-local 3-q simulation gives an almost perfect fit between the theoretical and the experimentally observed lineshapes. This proves that we are indeed dealing with three independent static modulation waves and not with a superposition of 1-q domains where the direction of the modulation wavevector changes from one domain to another. In this case, the lineshape would be that for a 1-q case since the spectrum does not depend on the direction of the modulation wavevector. The comparison of the ^{75}As NQR lineshape of proustite (non-coplanar 3-q incommensurate phase with $D = 3$) and the ^{27}Al NMR lineshape in berlinite (coplanar 3-q incommensurate phase with $D = 2$) [2.74] further demonstrates the power of NMR and NQR lineshape measurements in determining the number of independent modulation waves and the dimension of the space spanned by the corresponding wavevectors.

Below 52 K the spectrum becomes structured indicating the formation of a 3-q multisoliton lattice (Fig. 2.26) where the phases of the modulation waves are non-linear functions of the spatial coordinates. From a comparison of experimental and theoretical lineshapes the soliton density $n_{si} = d_i/l_{ci}$ has been determined with the help of the first integral of the sine-Gordon [2.13, 2.32, 2.43, 2.58] equations (144) for each of the three modulation waves. The initial phases were chosen to optimize the fit, as well as to give the proper positions of the commensurate lines in the C phase. The fit is therefore unique. In the expression for the soliton density $n_{si} = d_i/l_{ci}$, d_i is the width of the phase soliton and l_{ci} the intersoliton distance, which diverges on approaching the incommensurate–commensurate (I–C) transition at T_c. In view of the point group symmetry we have $\phi_1 = \phi_2 = \phi_3$ and $n_s = n_{s1} = n_{s2} = n_{s3}$. The temperature dependence of the soliton density n_{si} on approaching T_C is shown in Fig. 2.27. It can be relatively well described by the Pokrovsky [2.86] model, which in the simplest approximation predicts

$$n_{si} = \frac{\pi/2}{K(k)} \propto -\frac{1}{\ln\left[(T - T_c)/T_c\right]}, i = 1, 2, 3, \qquad (153)$$

where $K(k)$ is the complete elliptic integral of the first kind. Here k, which depends on the amplitude of the modulation wave, is determined [2.13, 2.32, 2.58] by a minimization of the free energy. The constant k is also directly related to the spatial derivative

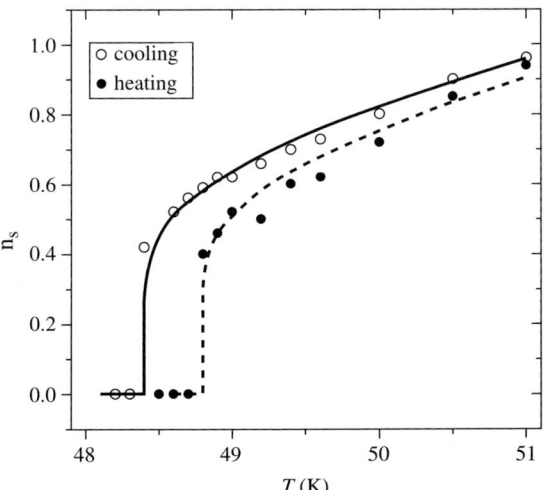

Fig. 2.27 Temperature dependence of the soliton density n_{si} in proustite. The solid line shows the theoretical prediction from expression (153) [2.79].

of the phase of the modulation wave $d\phi_i/dx_i$ and is thus experimentally accessible via NMR and NQR lineshape measurements.

At the I–C transition all three n_{si} vanish and the incommensurate frequency distribution is replaced by a multiplet of narrow lines reflecting the multiplication of the unit cell. Here, the phases of three modulation waves ϕ_i are step functions, taking the commensurate values $\vartheta_1 = \pi/3, \vartheta_2 = \pi$ and $\vartheta_3 = 5\pi/3$. For every combination of commensurate values of phases $\phi_1(\vec{r}) = \vartheta_i, \phi_2(\vec{r}) = \vartheta_j, \phi_3(\vec{r}) = \vartheta_k$ there is a corresponding peak in the NQR spectrum at a frequency $\nu\,(\vartheta_i, \vartheta_j, \vartheta_k)$. The single resonance line of the high-temperature phase should therefore be replaced by $3^3 = 27$ lines. Since, however, ν_{3q} is invariant to the permutation of the phases ϑ_i, ϑ_j and ϑ_k many lines in the C_1 phase overlap. The intensity of the resonance lines is proportional to the number of permutations of phases corresponding to a given frequency. The As NQR spectrum below T_C should thus consist of 3 lines with a relative intensity 1, 6 lines with a relative intensity 3 and a single line with a relative intensity 6. This agrees rather well with the observed spectrum (Fig. 2.26). The hysteresis of 0.6 K, observed at the I–C transition agrees with measurements from Ryan et al. [2.85] and seems to be due to the pinning of the soliton lattice to defects.

2.3 Conclusions

In incommensurate structures two fundamentally new phenomena have been observed:

1) Non-linear soliton structures exist in the ground state.
2) A splitting of the soft mode into amplitudon and phason branches has been observed. The phason represents the gapless Goldstone mode recovering the broken continuous translation symmetry of the normal phase.

3
Ferroelectric liquid crystals

Until recently, ferroelectric liquid crystals were unknown. Ferroelectricity like ferromagnetism were considered to be the property of solids. Ferroelectricity was first reported in chiral *smectic-C** (SmC^*) liquid crystals in 1975 by Meyer and coworkers [3.1]. He stated that SmC^* liquid crystals composed of chiral molecules ought to be ferroelectric (Figs. 3.1–3.3). Chiral molecules differ from others by the absence of a centre of symmetry.

In the tilted chiral smectic phases, molecules are arranged in liquid-like smectic layers. Each layer can possess an inplane spontaneous polarization. The inplane polarization is allowed in a direction perpendicular to the molecular tilt (Fig. 3.1). As one proceeds along the normal to the smectic layers, the molecular director and the spontaneous polarization rotate in a helicoidal fashion. There is a bilinear coupling between the electric polarization and the tilt [3.1, 3.2] that determines the dielectric properties of the homogeneous SmC^* phase. For a general reference see [3.3] and [3.4].

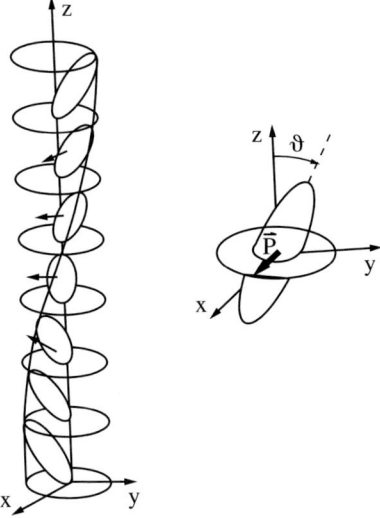

Fig. 3.1 The structure of the chiral ferroelectric SmC^* phase.

The first molecule that showed ferroelectricity in the SmC^* phase was DOBAMBC: p-decyloxybenzylidene p′-amino 2-methyl butyl cinnamate. The polar group is designated by $\uparrow \vec{P}$ and the chiral group by an asterisk*.

$$C_{10}H_{21}O--CH=N--CH=CH-\overset{O\uparrow \vec{P}}{\underset{\|}{C}}-O-CH-\overset{CH_3}{\underset{*}{CH}}-C_2H_5$$

$$Cr \xrightarrow{76\,°C} SmC^* \xrightarrow{95\,°C} SmA \xrightarrow{117\,°C} I$$

The existence or non-existence of ferroelectricity in a given phase can be deduced from Neumann's principle that states: 'The symmetry of any physical property of the medium must be higher but not lower than the point group symmetry of the medium.' Of the 32 crystal point groups, 22 are non-polar, ten are polar.

Ferroelectricity is also allowed in the chiral tilted SmI^* and SmF^* phases. Here, the molecular arrangement is similar to that of the SmC^* phase (Fig. 3.2). The isotropic, nematic and SmA phases do not allow for ferroelectricity though they are composed of chiral molecules.

The tilted chiral SmC^* phase [point group C_2] is shown in Figs. 3.2 and 3.3 and compared with the point group D_∞ of the chiral SmA phase. To be as simple as possible we first consider the homogeneous case of a thin film a few smectic layers thick. So $p \gg d$. Here, d designates the thickness of the smectic layers and p stands for the period of helix. Typically, $d/p \sim 10^{-3}$ thus supporting the above assumption.

Chiral molecules are not mirror images of each other. If mirror symmetry exists such as in C_{2h} smectic C liquid crystals, a 180° rotation around the y-axis yields

$$\vec{P} = \begin{Bmatrix} P_x \\ P_y \\ P_z \end{Bmatrix} \rightarrow \begin{Bmatrix} -P_x \\ P_y \\ -P_z \end{Bmatrix} \Rightarrow \vec{P} = \begin{Bmatrix} 0 \\ P_y \\ 0 \end{Bmatrix} \quad (1)$$

This is an identity operation and in view of the mirror symmetry the left-hand side must be equal to the right-hand side. This is possible only if P_x and P_z are zero.

A rotation of 90° around the z-axis is also an identity operation in the C_{2h} point group. It transforms P_y into P_x so that in view of Eq. (1)

$$\begin{Bmatrix} 0 \\ P_y \\ 0 \end{Bmatrix} \rightarrow \begin{Bmatrix} P_x \\ 0 \\ 0 \end{Bmatrix} \Rightarrow \vec{P} = 0 \quad (2)$$

As P_x is zero (1), the total polarization in C_{2h} smectics is zero (Fig. 3.2).

An isotropic liquid can be optically active, but not ferroelectric. This is so as $\vec{P}$ is a polar vector, whereas the optical activity coefficient is a pseudoscalar.

The point groups of a chiral nematic and the smectic A phase consisting of chiral molecules is D_∞ and similarly no ferroelectricity is allowed. This is so as the

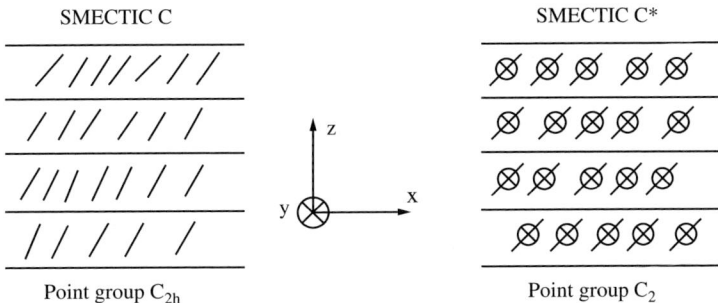

Fig. 3.2 The structure and the point group of chiral and achiral polar tilted smectics. The cross on the right picture represents the direction of polarization, which is in smectic layers pointing into the plane of the paper.

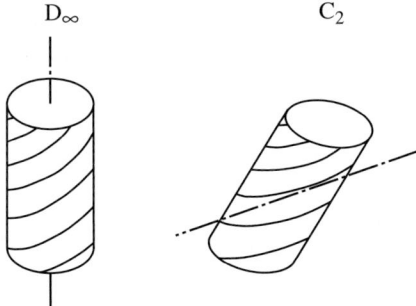

Fig. 3.3 The chiral *smectic A–smectic C* phase transition, leading to inplane ferroelectricity.

same symmetry arguments can be applied as in the isotropic phase: the 180° rotation around the y-axis and the 90° rotation around the z-axis are allowed symmetry operations.

3.1 Modulated ferroelectric liquid crystals

The Landau theory of the SmA–SmC^* phase transition was first derived by Blinc [3.2, 3.5] for the homogeneous case and by Indenbom [3.6] for the inhomogeneous case.

The order parameter of the transition is the two-component combination of the molecular tilt $(n_x n_y, n_y n_z)$ or the electric polarization (P_x, P_y). The primary order parameter of the transition is the tilt vector and the secondary is the spontaneous polarization. This is so as the dipole–dipole interactions are too small to drive the transition. The lack of mirror symmetry in these systems allows for the existence of the Lifshitz term in the free energy

$$\xi_x \frac{\partial \xi_y}{\partial z} - \xi_y \frac{\partial \xi_x}{\partial z} \tag{3}$$

A similar term exists for the spontaneous polarization

$$P_x \frac{\partial P_y}{\partial z} - P_y \frac{\partial P_x}{\partial z} \tag{4}$$

The order parameter of the SmA–SmC^* transition is thus the inplane projection of the molecular tilt $\vec{\xi} = (\xi_x, \xi_y)$ and the electric polarization $\vec{P} = (P_x, P_y)$. The Lifshitz term induces a helicoidal precession of the molecular tilt as we move along the layer normal. The period of the precession is of the order of visible light. The period of the helix is in general incommensurate with the one-dimensional translational period of the *smectic A* phase (Fig. 3.4).

There are two types of couplings between the molecular tilt and the polarization. A bilinear coupling that is 'piezoelectric'-like

$$P_x \xi_y - P_y \xi_x \tag{5a}$$

and that results in a proportionality between the tilt and the polarization, and a flexoelectric term

$$P_x \frac{\partial \xi_x}{\partial z} + P_y \frac{\partial \xi_y}{\partial z} \tag{5b}$$

which results in a proportionality between the polarization and bending and twisting of the director field in space.

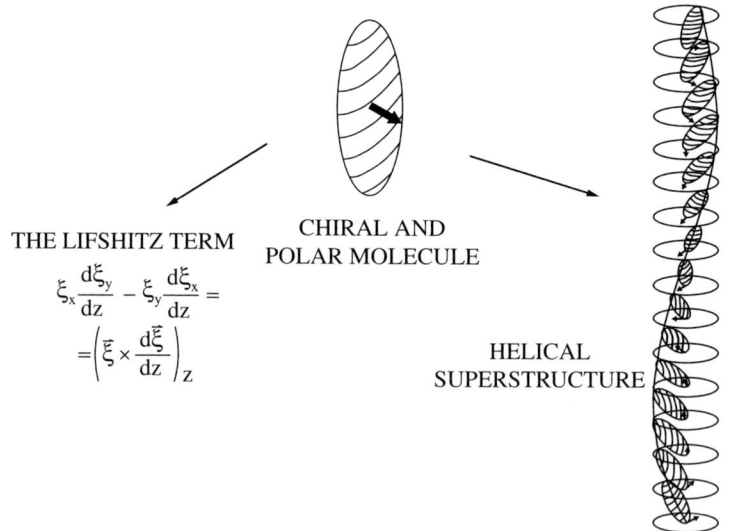

Fig. 3.4 Molecular chirality, helical modulation and the Lifshitz term. Note that the Lifshitz term is of chiral nature.

The expansion of the non-equilibrium free-energy density g in terms of the scalar invariants of the chiral SmA phase is [3.7]:

$$g = g_A + \frac{1}{2}a\left(\xi_x^2 + \xi_y^2\right) + \frac{1}{4}b\left(\xi_x^2 + \xi_y^2\right)^2 - \Lambda\left(\xi_x \frac{d\xi_y}{dz} - \xi_y \frac{d\xi_x}{dz}\right)$$
$$+ \frac{1}{2}K_{33}\left[\left(\frac{d\xi_x}{dz}\right)^2 + \left(\frac{d\xi_y}{dz}\right)^2\right]$$
$$+ \eta\left(\xi_x^2 + \xi_y^2\right)\left(\xi_x \frac{d\xi_y}{dz} - \xi_y \frac{d\xi_x}{dz}\right) + \frac{1}{2\varepsilon}(P_x^2 + P_y^2) - \mu\left(P_x \frac{d\xi_x}{dz} + P_y \frac{d\xi_y}{dz}\right) \quad (6)$$
$$+ C\left(P_x \xi_y - P_y \xi_x\right)$$

Here, g_A is the equilibrium free energy density of the SmA phase, $\xi_x = n_x n_z$, $\xi_y = n_y n_z$ and $\vec{n} = (n_x, n_y, n_z)$ is the director. The coefficient $a = \alpha(T - T_0)$ is the only T-dependent coefficient. K_3 is the twist elastic constant, ε is the dielectric constant, μ and C are coefficients of the flexoelectric and piezoelectric bilinear coupling between the tilt and the polarization.

The presence of the Lifshitz term allows for a modulation of the structure of the SmC^* phase

$$T < T_C : \xi_x = \theta_0 \cos q_c z, \quad \xi_y = \theta_0 \sin q_c z \quad (7a)$$
$$P_x = -P_0 \sin q_c z, \quad P_y = P_0 \cos q_c z \quad (7b)$$

This solution has a finite value of the tilt $\vartheta(T) = \theta_0$ and the polarization P_0 below T_C. Here, q_c is the critical wavevector, $q_c = \frac{2\pi}{p}$, where p is the period of the helix. The temperature dependence of ϑ, P and the period of the helix p is shown in Fig. 3.5.

A more general form is to expand the fluctuations of the tilt $\xi(z,t)$ and polarization $P(z,t)$ in terms of helical, elliptically polarized waves of the SmA phase with a wavevector $\vec{q} = (0, 0, q)$ along the layer normal.

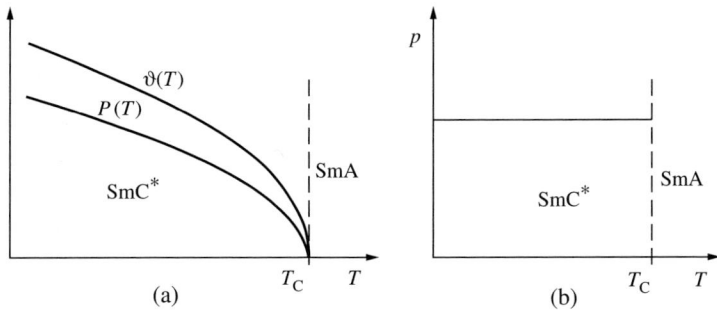

Fig. 3.5 The temperature dependence of the tilt angle ϑ, the spontaneous polarization P and the period of the helix p.

84 Ferroelectric liquid crystals

The inhomogeneous fluctuations of the SmA phase are:

$$\xi_x = \sum_q (x_q \cos qz - y_q \sin qz) \qquad (8a)$$

$$\xi_y = \sum_q (x_q \sin qz + y_q \cos qz) \qquad (8b)$$

$$P_x = \sum_q (-u_q \sin qz - v_q \cos qz) \qquad (8c)$$

$$P_y = \sum_q (u_q \cos qz - v_q \sin qz) \qquad (8d)$$

3.2 Excitations in ferroelectric liquid crystals: The Goldstone mode, the amplitudon mode and the soft mode

The dynamics of helicoidal ferroelectric *smectic C^** liquid crystals was first treated by Blinc and Žekš [3.7].

In view of the existence of four order parameters, there are four overdamped orientational elementary excitations of the system. Two degrees of freedom correspond to the director-like (low-frequency) and two to the polarization-like (high-frequency) modes (Figs. 3.6 and 3.7).

In view of the cylindrical symmetry of the problem we have above T_c for each q two degenerate helicoidal 'tilt-polarization' waves. One is 'left-handed' and the other 'right-handed'. The breaking of the cylindrical symmetry below T_c will remove the degeneracy of the above two modes (Fig. 3.6).

The time-fluctuating parts of $\xi(z,t)$ and $P(z,t)$ can be decomposed into components parallel to the equilibrium value of the order parameters ($\delta\xi_\parallel$ and $\delta P_\parallel$) and components perpendicular to the order parameter (Fig. 3.7). The parallel components represent 'amplitude excitations' as they change the magnitude of the order parameters, i.e. the 'tilt' and the 'polarization'. The perpendicular components represent 'phase excitations' as they change the phase profile and consequently the director field and the polarization field.

In the SmA phase there are thus two branches of doubly degenerate excitations. The 'director' branch and the 'polarization' branch. The polarization branch represents the out-of-phase fluctuations of the polarization and the tilt. This branch is nearly temperature independent and has a large relaxation rate ($\Delta\nu/\nu \approx 10^{-2}$). It splits into two branches that are nearly temperature independent with relaxation rates of the order of 100 MHz (Figs. 3.6 and 3.7).

The phason branch, on the other hand, represents the inphase fluctuations of the tilt and the polarization.

The polarization in ferroelectric liquid crystals is a secondary order parameter. In view of that it is non-critical with a relaxation rate that is much higher than the relaxation rates of the director modes. Ferroelectric liquid crystals are thus 'improper' ferroelectrics driven by intermolecular rather than dipole–dipole forces.

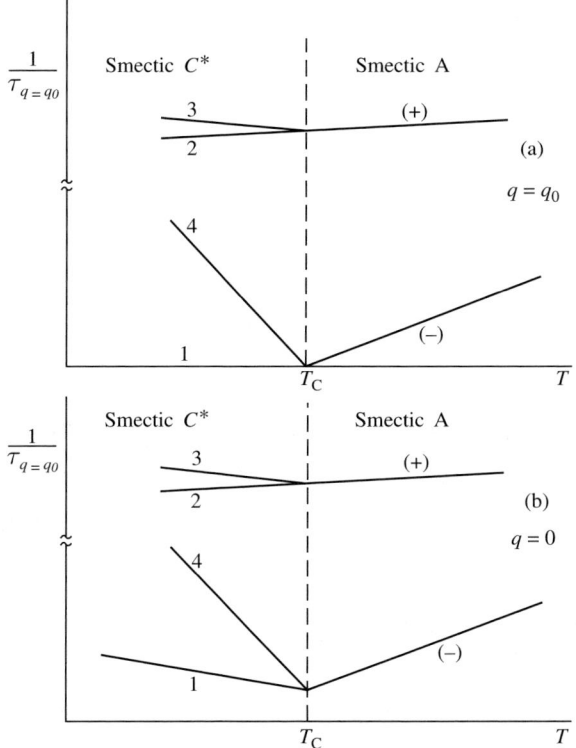

Fig. 3.6 Temperature dependence of the order-fluctuation modes. (a) At the critical wavevector $q_0 = q_c = \frac{2\pi}{p}$ (light scattering) and (b) at the wavevector $q = 0$ (dielectric measurements) at the helicoidal ferroelectric *smecticA* → *smectic C** transition. $1/\tau_1$ represents the Goldstone mode of this transition that recovers the continuous symmetry group broken at $T_C(D_\infty \to C_2)$.

The structure and dynamics of liquid and solid crystals are compared in Fig. 3.8. Whereas we deal in solid crystals with positional order we deal in liquid crystals with orientational order. Instead of phonons we have here orientational fluctuation modes.

Within the Landau–Khalatnikov dynamics, the non-equilibrium part of the order parameter is driven back to equilibrium by the 'thermodynamical restoring force':

$$\frac{d\delta\xi}{dt} = -\Gamma \frac{\partial G(\xi,t)}{\partial \xi} \tag{9}$$

Here, Γ is a non-critical Khalatnikov parameter and $G = \int g dz$ is the free energy (6).

In agreement with the above statement we have for a given wavevector q four different eigenmodes with four different relaxation rates (Fig. 3.6).

The critical elementary excitations are overdamped waves of the form

$$\delta\xi(z,t) = e^{-t/\tau(q=q_{\text{crit}})}(\cos qz, \sin qz) \tag{10}$$

86 Ferroelectric liquid crystals

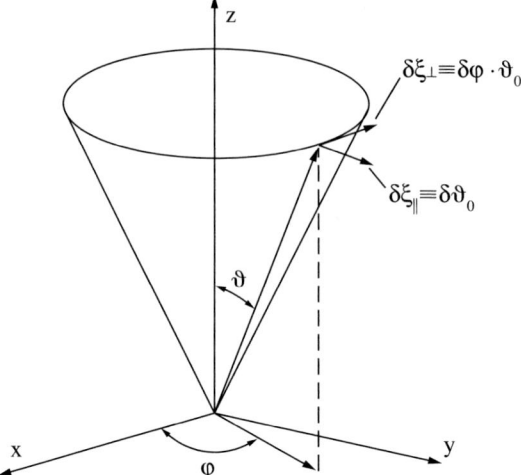

Fig. 3.7 The fluctuations of the tilt can be decomposed into two components.

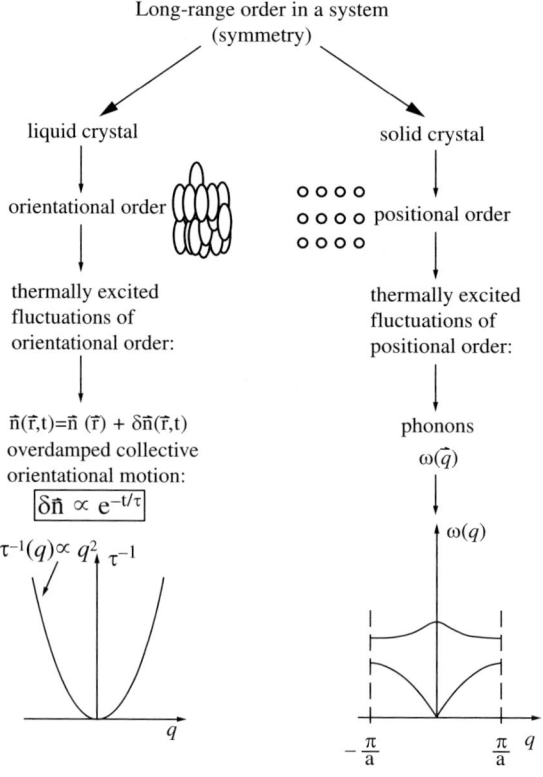

Fig. 3.8 Positional and orientational elementary excitations in solids and liquid crystals.

The high-temperature *smectic A* phase is thus characterized by the existence of a soft, strongly temperature-dependent director mode, the frequency of which goes to zero for the critical wavevector as the $SmA \to SmC^*$ transition T_C is approached (Fig. 3.8):

$$\tau_S^{-1} = A(T - T_C) + B(\vec{q} - \vec{q}_C)^2 \qquad (11)$$

It should be stressed that the condensation of the soft mode occurs at a general point of the wavevector space.

The spontaneous breaking of the point symmetry ($D_\infty \to C_2$) of the *smectic A* phase at T_C is connected with the breaking of a continuous point symmetry group. This leads to the appearance of a zero-frequency symmetry restoring Goldstone mode designated τ_1^{-1} in Fig. 3.9. This phason mode exists in the *smectic C^** phase and tries to restore the broken continuous symmetry. The excitation spectrum of one branch of the order-parameter spectrum of the SmC^* phase is thus gapless (Fig. 3.9(a)). The other branch of the director spectrum below T_C is the temperature-dependent amplitudon excitation (τ_2^{-1} in Fig. 3.9).

There are thus two characteristic modes: the soft mode that critically slows down when the $SmA \to SmC^*$ phase transition temperature T_C is approached from above and the zero-frequency Goldstone mode of the low-T phase that tries to restore the lost symmetry (Figs. 3.6 and 3.10). The concept of the soft mode in ferroelectric liquid crystals below T_C was first introduced by Blinc [3.2].

The zero-frequency Goldstone mode appears when the continuous symmetry of the system is spontaneously broken. The zero-frequency symmetry restoring Goldstone

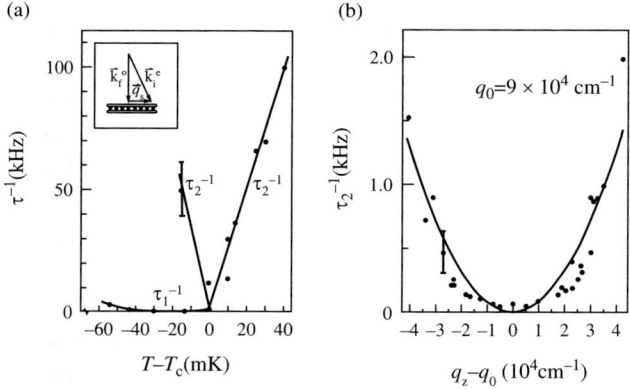

Fig. 3.9 (a) The splitting of the soft director mode of the SmA phase into the phase and amplitude modes near the $SmA \to SmC^*$ transition in the ferroelectric liquid crystal DOBAMBC [3.8, 3.9]. (b) The dispersion relation for the hydrodynamic director (phason) modes in the ferroelectric SmC^* phase. Note that the nature of the soft mode here is different as in the nematic liquid crystals. The soft mode in ferroelectric liquid crystals represents collective fluctuations of the tilt angle, whereas the degree of the orientational order (i.e. nematic order) remains nearly the same in SmA and SmC^* phases.

88 Ferroelectric liquid crystals

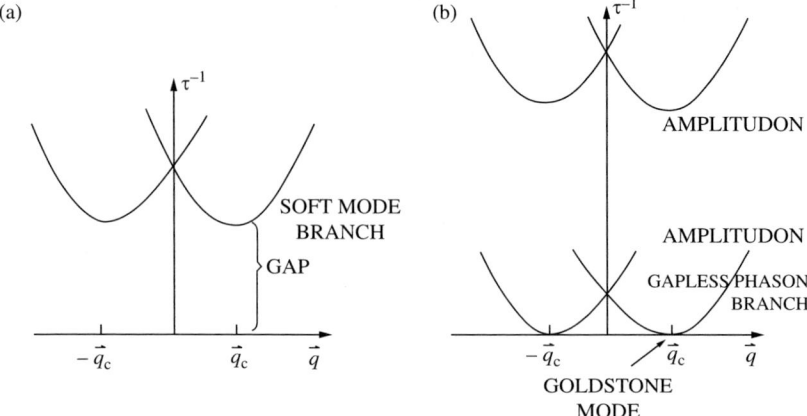

Fig. 3.10 The dispersion of the (a) soft-mode excitations for $T \geq T_c$. Whereas the dielectric experiment measures the linear response at $\vec{q} = 0$, the quasi-elastic light-scattering spectroscopy measures the dynamics at finite $\vec{q}$. (b) The dispersion of the amplitudon and Goldstone modes below T_C.

mode exists in the unperturbed SmC^* phase as well as in the soliton-like distorted SmC^* phase. It does not exist in the unwound SmC phase.

Both the continuous rotational and the continuous translational symmetries are spontaneously broken by the appearance of the helical C^* structure. The symmetry-breaking soft mode, on the other hand, appears whenever a phase transition point or line is approached. It softens when $T \to T_c$, $\vec{q} \to \vec{q}_C$ (Fig. 3.6(a)).

The point group of the chiral nematic and chiral *smectic A*(Fig. 3.3) phase is D_∞. In view of the above symmetry arguments no spontaneous polarization exists.

In the presence of an external magnetic field applied perpendicularly to the helix, the D_∞ symmetry of the SmA phase is reduced to D_2 (Fig. 3.11).

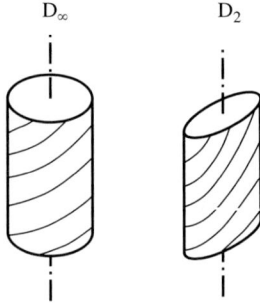

Fig. 3.11 Magnetic-field-induced $D_\infty \to D_2$ phase transition from the chiral *smectic A* phase, leading to an inplane quadrupolarly ordered structure.

3.3 The magnetic-field-induced Lifshitz point

This was first discussed by Hornreich *et al.* [3.10]. It was introduced into the liquid-crystal field by Michelson [3.11, 3.12]. The Lifshitz point is a triple point where the SmA, the homogeneously ordered SmC^* and the inhomogeneously ordered SmC^* phases meet (Fig. 3.12).

Michelson [3.11, 3.12] proposed the physical realization of the Lifshitz point in the H–T phase diagram of a ferroelectric liquid crystal. If here, an external magnetic field is applied perpendicularly to the helical axis, a π-soliton-like structure appears (Fig. 3.13). We have to add to the free energy density $g(z)$ a magnetic field term

$$g_H = -\frac{1}{2}\Delta\chi H^2 \xi_y^2 \tag{12a}$$

Here $\Delta\chi = \chi_\parallel - \chi_\perp$ is the diamagnetic anisotropy and the magnetic field is applied in the y direction $\vec{H} = (0, H, 0)$.

If the diamagnetic anisotropy is positive, the molecules align along the direction of the magnetic field. If it is negative they align in a plane perpendicular to the field. When the external field approaches a critical value, $H \to H_c$, the distorted helix unwinds into a spatially homogeneous SmC^* structure.

It can be shown that at the Lifshitz point two phase boundaries merge tangentially into each other (Figs. 3.12 and 3.14) and into the line of the critical magnetic field that separates the modulated and unwound phases. Near the Lifshitz point the line of the critical magnetic field is of first order $H_C = H_C(T)$. Below the Lifshitz field H_L the SmA and the modulated SmC^* phases are separated by the λ-line. Along the λ-line the transition is of second order.

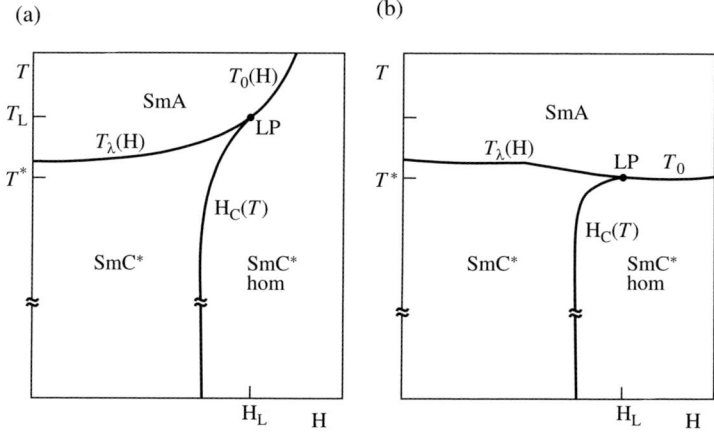

Fig. 3.12 Lifshitz point (H_L, T_L) in the (H, T) phase diagram of a ferroelectric liquid crystal in an external magnetic field. (a) Positive diamagnetic anisotropy. (b) Negative diamagnetic anisotropy.

90 Ferroelectric liquid crystals

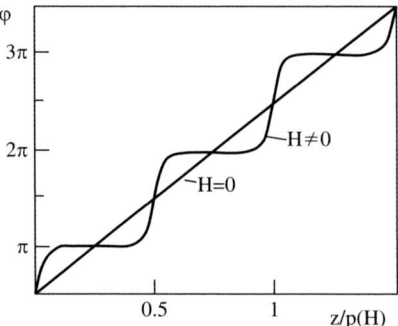

Fig. 3.13 The soliton phase profiles $\phi_0(z)$ that are the solutions of the sine-Gordon equation for a magnetic field applied perpendicularly to the SmC* helix: $H \neq 0$. π-soliton walls are shown. The plane-wave profiles for $H = 0$ are presented for comparison.

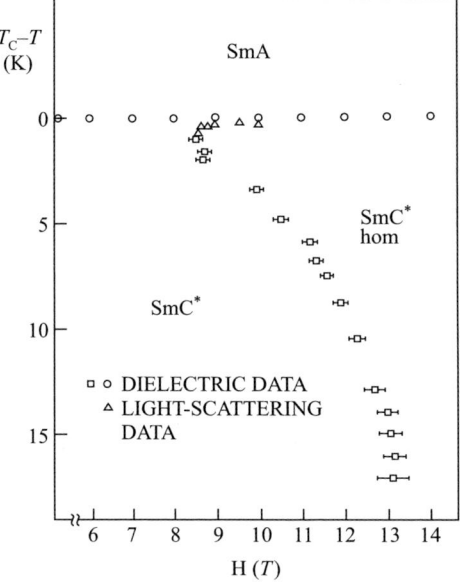

Fig. 3.14 Experimental H–T phase diagram of DOBAMBC in an external magnetic field [3.13], showing the presence of the Lifshitz point.

The phase profiles $\phi(z)$ that minimize the free energy and form the π-soliton lattice are described by Jacobi's elliptic functions

$$sn(uk) = \sin \phi_0 \tag{13}$$

Here $sn(uk)$ is the Jacobi elliptic sine of the reduced coordinate,

$$u = z/(\xi k) \tag{14}$$

and
$$\xi = \sqrt{K_3/(\Delta\chi \cdot H^2)} \qquad (15)$$
is the magnetic coherence length.

The modulus of k is defined by
$$k = \frac{H}{H_C} E(k) \qquad (16)$$
where
$$E(k) = \int_0^{\pi/2} \sqrt{1 - k^2 \sin^2 x}\, dx \qquad (17)$$
is the complete elliptic integral of the second kind and H_C is the critical magnetic field for the unwinding of the helical structure:
$$H_C = \frac{\pi^2}{p_0} \sqrt{\frac{K_3}{|\Delta\chi|}}, \qquad (18)$$

The Lifshitz field is $H_L = \frac{\pi}{4} H_C$.

Minimization of the free energy G yields the sine-Gordon equation for $\phi(z)$:
$$\frac{\partial^2 \phi(z)}{\partial z^2} + \left(\frac{\Delta\chi \cdot H^2}{2K_3}\right) \sin 2\phi(z) = 0 \qquad (19)$$

The resulting phase profiles are shown in Fig. 3.13. The $H-T$ phase diagram is shown in Figs. 3.12 and 3.14.

3.4 The electric-field-induced Lifshitz point

The electric-field-induced Lifshitz point was treated by Michelson and Cabib [3.14].

The phase-dependent part of the free energy now contains a $g_E(z) = -EP\cos\phi(z)$ term. The sine-Gordon equation is in the electric case
$$\frac{d^2\phi(z)}{dz^2} - \frac{EP}{K_3\theta^2} \cdot \sin\phi(z) = 0 \qquad (20)$$

The (E,T) phase diagram in an applied external field is here fundamentally different from its magnetic counterpart. We have here in the sine-Gordon equation a $\sin\Phi(z)$ term instead of the $\sin 2\Phi(z)$ term as in the magnetic case. This results in a 2π-soliton lattice instead of a π-lattice found in the magnetic case (Figs. 3.15 and 3.16).

Another difference is that even a very small electric field applied along the layers of the SmA phase breaks the symmetry of this phase and induces a spatially uniform polarization and tilt of the molecules. This means that the SmA phase is stable only for $E = 0$, whereas for any finite field the SmC^* is stable (Fig. 3.16). In the magnetic

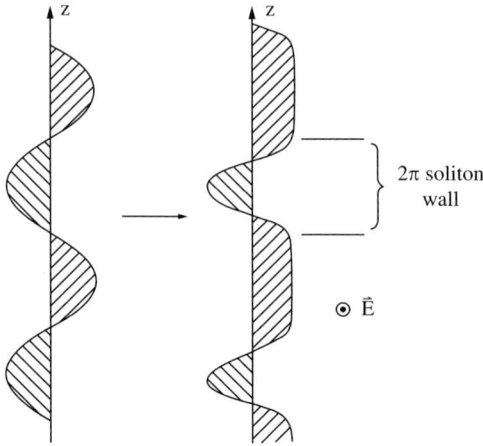

Fig. 3.15 The application of a DC electric field distorts the originally 'smoothly' modulated helical *smectic-C** phase into a so-called 2π-soliton lattice.

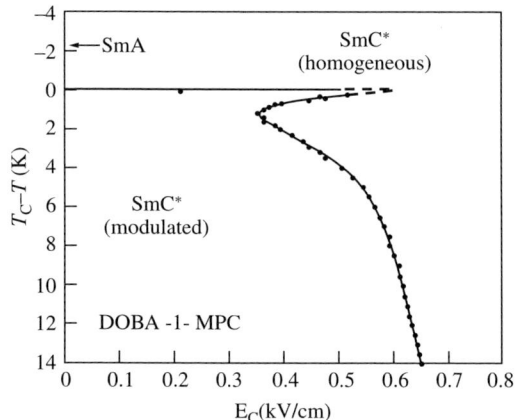

Fig. 3.16 (E, T) phase diagram of DOBA-1-MPC in a DC electric field applied parallel to the smectic layers.

case on the other hand, a finite critical magnetic field is required to induce these changes.

The electric-field-induced Lifshitz point is also significantly different from the magnetic-field-induced one. As already mentioned, a 2π soliton lattice is formed in the electric case, whereas in the magnetic field case a π-soliton lattice is formed.

The response of a helix of the ferroelectric SmC^* phase to an external electric field applied in a direction parallel to the smectic layers is shown in Figs. 3.17 and 3.18.

As a result of the electric field there is only a phase boundary $T_c(E)$ that separates a spatially uniform SmC^* phase from the spatially modulated SmC^* phase. A triple

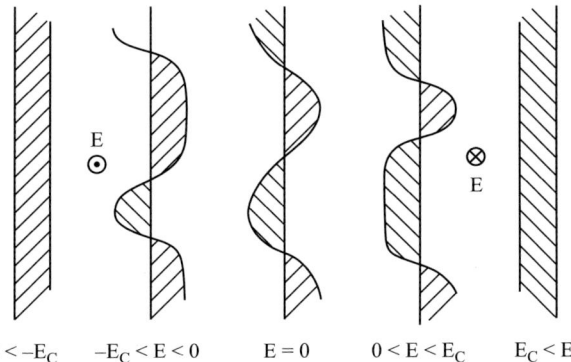

Fig. 3.17 Response of the ferroelectric SmC^* helix to an external electric field.

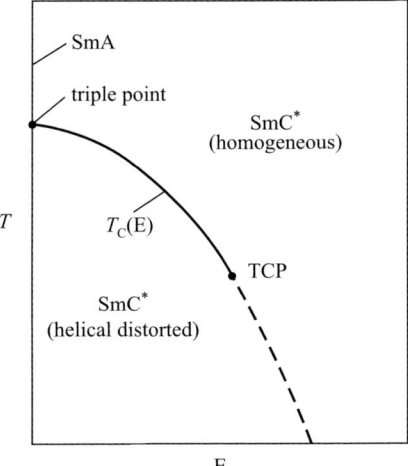

Fig. 3.18 Tricritical point (TCP) and triple point in the (E, T) phase diagram of a ferroelectric liquid crystal in a dc electric field.

point where the SmA phase, the homogeneous *smectic C^** and the helically distorted SmC^* phases meet is located at $T_C(E=0)$ (Fig. 3.18) and not at a finite field.

Finally, one should mention that ferroelectric liquid crystals in an electric field can be used as submicrosecond bistable electro-optic displays in the surface-stabilized geometry, as shown by Clark and Lagerwall [3.15].

3.5 Freely suspended ferroelectric smectic thin films

Freely suspended smectic films represent a system where the cross-over from the 3D to 2D universality class can be studied as the number of smectic layers N decreases.

94 Ferroelectric liquid crystals

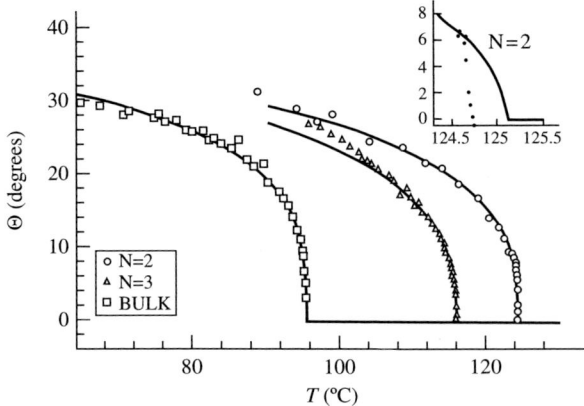

Fig. 3.19 The temperature dependence of the tilt angle (which is proportional to the spontaneous polarization) for different numbers of layers of freely suspended ferroelectric liquid-crystal films of DOBAMBC. The solid curves are fits to a discrete mean-field theory [3.16, 3.17].

The interior layers are still SmA-like, whereas the two surface layers already show a finite tilt and finite polarization characteristic for the ferroelectric SmC^* phase. Ferroelectricity here is surface induced.

The temperature dependence of the tilt angle for a different number of layers from $N = 2$ to $N = 10$ is shown in Fig. 3.19 together with bulk results [3.16, 3.17].

The exterior layers undergo a phase transition into the ferroelectric phase at much higher temperatures than the interior layers [3.16, 3.17].

It is also interesting to note that in the bulk the SmA-SmC^* phase transition is of first order, whereas it is of second order in the $N = 2$ layer system.

4
Dipolar glasses

4.1 Introduction

The nature of the glass transition [4.1] and the origin of certain universal features in spin glasses, proton and deuteron glasses, quadrupolar glasses and other structural glasses [4.1] is still one of the great open problems of condensed-matter physics. Here, we look at dipolar glasses that are close relatives of ferroelectrics. Examples of such systems are mixed solid solutions of ferroelectric KH_2PO_4 (KDP) and anti-ferroelectric $NH_4H_2PO_4$ (ADP) and other systems where competing ferroelectric and anti-ferroelectric interactions [4.2] and random electric fields lead to frustration and to a random freeze-out of the protons or deuterons into one of the two potential minima in the O—H$\cdots$O bonds. In KDP-type systems [4.2] there are six possible H_2PO_4 Slater orientations with differently oriented dipole moments. The two ferroelectric (FE) Slater configurations $FE_\uparrow$ and $FE_\downarrow$ with dipole moments along the $+c$ and $-c$ axes, respectively, of the tetragonal a, b, c coordinate system are realized in the two 180° ferroelectric domains of the FE phase of KDP. The four anti-ferroelectric (AFE) configurations $AFE_{a\uparrow}$, $AFE_{a\downarrow}$, $AFE_{b\uparrow}$ and $AFE_{b\downarrow}$ with dipole moments along the $+a, -a, +b$ and $-b$ directions, are found in the four 90° domains of the AFE phase of ADP. In mixed KDP–ADP systems frustration prevents the system becoming long-range ordered and leads to a disordered state with glass-like static and dynamic properties [4.3]. It can be compared to the situation in hexagonal ice [4.1] where the H_2O molecules are also frozen more or less randomly into one of the six possible orientations with equal probability.

In contrast to magnetic spin glasses where we deal with random bond-type interactions, proton and deuteron glasses are characterized by the presence of both random bonds and random fields [4.3, 4.4]. This difference is due to the fact that magnetic spins are essentially uncoupled from the lattice, whereas the O—D$\cdots$O or O—H$\cdots$O bonds are parts of the lattice and thus more strongly affected by the substitutional disorder. Since the variance of the random field Δ acts as an effective ordering field for the deuteron glass, the corresponding Edwards–Anderson order parameter q_{EA} is non-zero at all temperatures [4.3, 4.4]. A conventional phase transition where q_{EA} would change from zero to a non-zero value is thus here – in contrast to magnetic spin glasses – conspicuously absent.

A well-defined freezing transition may, however, exist in deuteron glasses according to the random-bond–random-field Ising model [4.3]. In analogy to the Almeida–Thouless (AT) line [4.3] for spin glasses in an external magnetic field, an instability

96 Dipolar glasses

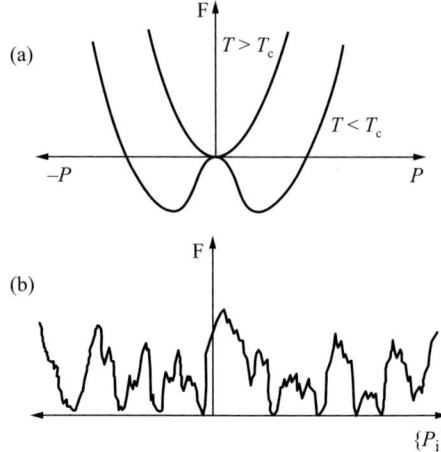

Fig. 4.1 Free-energy surface in (a) a homogeneous ferroelectric and (b) a deuteron pseudospin glass.

line $T_f(\Delta)$ can be defined in the temperature–random field variance plane [4.3] which separates the replica-symmetric pseudospin glass phase from the phase where the replica symmetry is broken, i.e. the ergodic from the non-ergodic deuteron glass phase [4.3].

4.2 Local structure determination and local polarization distribution function

The free energy of a ferroelectric with a one-component order parameter has a single minimum at $P = 0$ at $T > T_c$ and two minima at $+P$ and $-P$ below T_c corresponding to oppositely polarized domains (Fig. 4.1), whereas a rough fractal-like free-energy surface occurs in glasses.

Solid solutions of ferroelectric (FE) and anti-ferroelectric (AFE) crystals form in the intermediate concentration range $x_{FE} \leq x \leq x_{AFE}$ (Fig. 4.2) a pseudospin-like dipolar glassy phase where no macroscopic symmetry breaking takes place at any temperature and no long-range order parameter appears. Instead, we have a random local freeze-out of the dipoles and possibly an ergodic–non-ergodic transition. The existence of the glassy phase is due to randomly competing ferroelectric and anti-ferroelectric interactions and the resulting random frustration of the system. The free-energy surface of such a system is rough and fractal-like and exhibits a large number of nearly degenerate minima separated by high potential barriers [4.1] (Fig. 4.1(b)).

In the concentration range $0 \leq x \leq x_{FE}$ and $x_{AFE} \leq x \leq 1$ (Fig. 4.2) we have an inhomogeneous ferroelectric and anti-ferroelectric phase, respectively, where long-range order and glassy order coexist at $T \neq 0$.

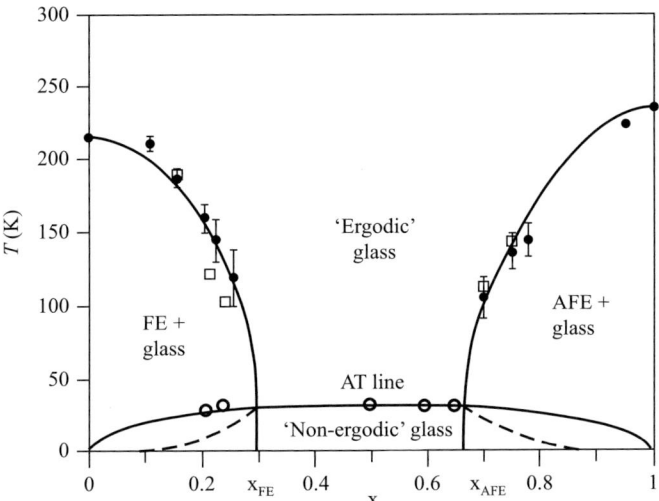

Fig. 4.2 Phase diagram [4.5] of the ferroelectric solid solution $Rb_{1-x}(ND_4)_xD_2PO_4$ that forms a deuteron glass in the intermediate concentration range $x_{FE} \leq x \leq x_{AFE}$. Also shown are the inhomogeneous ferroelectric and inhomogeneous anti-ferroelectric regions for $0 \leq x \leq x_{FE}$ and $x_{AFE} \leq x \leq 1$. The solid lines are calculated from the random-bond–random-field pseudospin Ising model. Also shown is the Almeida–Thouless (AT) line separating the ergodic from the non-ergodic glassy phase.

The local dipolar structure of deuteron glasses and inhomogeneous ferroelectrics is characterized by a local polarization distribution function [4.3],

$$W(p) = \frac{1}{N} \sum_i \delta\left(p - \langle S_i^z \rangle\right) \tag{1}$$

where the pseudospin $S_i^z = \pm 1$ describes the two orientational states of the dipole at site i, and the brackets $\langle \ldots \rangle$ stand for the thermal average. In a homogeneous ferroelectric the first moment of $W(p)$ is different from zero below T_c and is zero above T_c:

$$P = \int W(p) p \, dp \neq 0, \quad T < T_c \tag{2a}$$

$$P = 0, \quad T > T_c \tag{2b}$$

In a homogeneous ferroelectric the polarization distribution function is a delta function at $p = P$

$$W(p) = \delta(p - P) \tag{2c}$$

and the second moment of $W(p)$ is zero,

$$\tilde{q} = q - P^2 = \int W(p) p^2 \, dp - P^2 = 0. \tag{2d}$$

In a proton or deuteron dipolar glass the first moment of $W(p)$ is zero as there is no long-range order,

$$P = \int W(p) p \, dp = 0 \tag{3a}$$

but the second moment of $W(p)$ is different from zero,

$$q = \int W(p) p^2 \, dp \neq 0 \tag{3b}$$

Here, q is just the well known Edwards–Anderson glass order parameter that is also defined [4.1] as

$$q = \lim_{t \to \infty} \lim_{N \to \infty} \frac{1}{N} \sum_i \langle S_i^z(0) S_i^z(t) \rangle \tag{3c}$$

If there are no random fields present, we have a paraelectric–glass transition [4.3] at T_g:

$$q = 0, T > T_g \tag{4a}$$

$$q \neq 0, T < T_g \tag{4b}$$

In the presence of a random field, which is conjugate to this glass order parameter, $q \neq 0$ at all temperatures [4.3, 4.4]. We have, however, as already mentioned, at lower temperatures an ergodic-non-ergodic transition below the Almeida–Thouless line in the random field variance-temperature plane where the pseudospin glass susceptibility, as well as the longest glass relaxation time, diverge [4.1, 4.5]. Here, the replica symmetry of the 'ergodic' glassy phase is broken [4.1] and instead of a single order parameter q we have an infinite number of order parameters represented by the order-parameter function [4.1] $q(x)$.

In an inhomogeneous ferroelectric, on the other hand, both the first and the second moment of $W(p)$ are different from zero:

$$P \neq 0 \tag{5a}$$

and

$$\tilde{q} = q - P^2 \neq 0 \tag{5b}$$

The important point is that $W(p)$ can be directly measured by local techniques, e.g., NMR or EPR. The local polarization distribution function $W(p)$ is reflected in the inhomogeneous NMR frequency spectrum $f(\omega)$ via

$$f(\omega) d\omega = W(p) dp \tag{6}$$

A knowledge of the inhomogeneous NMR frequency spectrum $f(\omega)$, and the relationship $\omega = \omega(p)$, thus enable one to determine $W(p)$ and all its moments. This is of crucial importance in glasses and inhomogeneous ferroelectrics and anti-ferroelectrics.

4.3 Dielectric properties

The deuteron glass transition would thus correspond to a change from a single replica symmetric phase with a spin-glass order parameter q at $T > T_f$ to a replica symmetry broken phase at $T < T_f$ represented by the Parisi order-parameter function $q(x), 0 \leq x \leq 1$. Here, T_f is the freezing temperature. The response of the system to an external field will now depend on the history of the system. Specifically, the field-cooled (FC) dielectric susceptibility χ_{FC} will differ from the zero field cooled (ZPC) one. Above the instability line, $T_f(\Delta)$, the FC and ZFC susceptibilities should coincide, $\chi_{FC} = \chi_{ZFC}$, whereas below T_f, $\chi_{FC} > \chi_{ZFC}$. Another striking phenomenon is that in analogy to spin glasses a non-zero remanent polarization – obtained after switching off the electric field in a FC experiment – should be observed over macroscopic time scales.

Both of these effects have been observed in the deuteron glass $Rb_{0.4}(ND_4)_{0.6}D_2PO_4$ (DRADP) [4.5], as shown in Fig. 4.3. Here, $\chi_{FC} = \chi_{ZFC}$ above $T_f = 61\,K$, whereas χ_{FC} becomes larger than χ_{ZFC} below this temperature. It should be stressed that this effect is found in truly static measurements where the experimental time scales are longer than the dielectric relaxation times. The field-cooled susceptibility was determined by slowly ($\approx 1\,K/min$) cooling the crystal from room temperature to $5\,K$ in a small external electric field and measuring the static dielectric polarization $P_{FC}(E,T)$ with an electrometer. The zero-field-cooled susceptibility $\chi_{ZFC} = \lim_{E \to 0} \frac{P_{ZFC}(E,T)}{\varepsilon_0 E}$ was, on the other hand, obtained by cooling in $E = 0$, applying an electric field step pulse E at T and integrating the resulting polarization current to obtain $P_{ZFC}(E,T)$.

Below T_f a remanent polarization P_R was also observed when the sample was cooled down in an electric field and then the field was switched off, and the sample heated. P_R vanishes when $T \to T_f$.

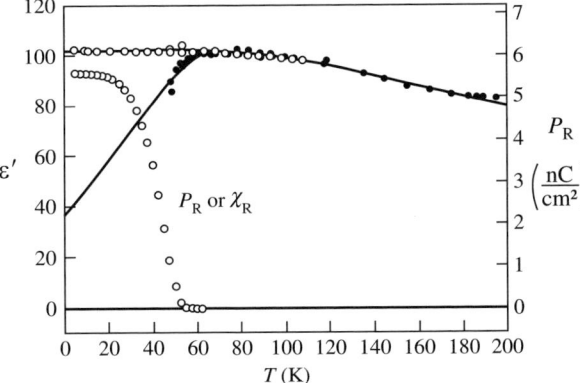

Fig. 4.3 Temperature dependence of the field-cooled (○) and zero-field-cooled (●) static dielectric constant ε' of DRADP with $x = 0.6$ measured along the a-axis. Solid lines are calculated from the random-bond–random-field model [4.3, 4.5]. Also shown is the remanent polarization $P_R(\Diamond)$ obtained in a heating run after field cooling in $E = 500\,V/cm$.

The above results are highly reminiscent of spin glasses. Of particular importance is the fact that the splitting between χ_{FC} and χ_{ZFC} is observed below T_f though the Edwards–Anderson order parameter is non-zero both above and below T_f [4.4].

It should be noticed that the splitting between χ_{FC} and χ_{ZFC} is a dynamic effect characteristic of the glassy state.

It should be also mentioned that, as shown before, a classical to quantum crossover (Fig. 1.16) has been found [1.58] in proton glasses RADP-72 and RADP-35 as $T \to 0$. Here, tunneling of Takagi groups in a symmetric double-well potential – as determined by Compton neutron scattering – becomes rate determining at low T.

4.4 NMR in homogeneous ferroelectrics and anti-ferroelectrics in the fast-motion regime

Conventional methods of crystal structure determination such as X-ray and neutron scattering yield the unit-cell structure averaged over space and time. While the knowledge of such a space- and-time averaged structure is sufficient in most systems this is not the case in inhomogeneous or order–disorder-type ferroelectrics and proton glasses where local deviations from the averaged structure determine the specific properties of these systems. One-dimensional (1D) and two-dimensional (2D) NMR of nuclei with a non-zero quadrupole coupling or chemical-shift tensor interaction as well as NQR provide a unique tool for the determination of the local atomic structure and dynamics, which is unobtainable by other techniques.

In a classical system the use of NMR or EPR is based on the following facts:

In the fast-motion regime, $\omega\tau \ll 1$, the nucleus at site i 'sees' the time averaged value of the quadrupole coupling or chemical-shift tensor interaction so that [4.2] in NMR experiments the effective nuclear frequency ω_i is

$$\omega_i = \omega_0 + \omega_1 \langle \eta_i \rangle; \langle \eta_i \rangle \neq 0, \ T < T_c \tag{7a}$$

whereas

$$\omega_i = \omega_0; \langle \eta_i \rangle = 0, \ T > T_c \tag{7b}$$

Here, ω_0 is the nuclear Larmor frequency, $\langle \eta_i \rangle$ is the time-averaged value of the order parameter at site i, and τ is the characteristic nuclear spin correlation time. It is $\langle \eta_i \rangle$, which is zero above T_c and non-zero below T_c in a classical phase transition characterized by long-range order.

Close to the phase-transition temperature T_c, the nuclear spin-lattice relaxation rate will be determined by the order-parameter fluctuations [4.2]

$$T_1^{-1} \propto \sum_q \int_{-\infty}^{\infty} \langle \delta\eta(\vec{q},0)\delta\eta(\vec{q},t)\rangle e^{i\omega t} dt \tag{8a}$$

where $\delta\eta(\vec{q},t) = \eta(\vec{q},t) - \langle \eta(\vec{q},t)\rangle$ and $\langle \ldots \rangle$ denotes the ensemble average.

Applying the fluctuation dissipation theorem, Eq. (8a) can be rewritten as [4.2]

$$T_1^{-1} \propto \sum_q \frac{\chi(\vec{q},0)\tau(\vec{q})}{1+\omega^2\tau(\vec{q})^2} \qquad (8b)$$

where $\chi(\vec{q},0)$ is the wavevector-dependent static susceptibility of the order parameter and $\tau(\vec{q})$ is the corresponding wavevector-dependent correlation time [4.2]. For the Ising model one finds

$$\chi(\vec{q}) = \frac{\chi(\vec{q}=0)}{1+\vec{q}^2\xi^2} \qquad (9a)$$

and

$$\tau(\vec{q}) = \frac{\tau(\vec{q}=0)}{1+\vec{q}^2\xi^2} \qquad (9b)$$

where

$$\tau(\vec{q}=0) \propto \chi(\vec{q}=0) \propto \frac{C}{T-T_c}; T > T_c \qquad (9c)$$

and ξ is the correlation length, which diverges at T_c. In the limit $\omega\tau \ll 1$, T_1^{-1} thus exhibits a critical behavior near T_c, while for $\omega\tau \gg 1$, T_1^{-1} should stay nearly constant. Another point to be stressed is that in the slow-motion regime the nucleus 'sees' the instantaneous value of the order parameter and not the time-averaged value.

Ferroelectric and anti-ferroelectric phase transitions [4.2] in crystals are thus characterized by:

1. The breaking of a discrete symmetry group at the paraelectric to ferroelectric or anti-ferroelectric transition.
2. The appearance of a long-range order parameter below T_c: the spontaneous polarization P in ferroelectrics or the sublattice polarization $P_a = -P_b$ in anti-ferroelectrics.
3. The critical slowing down of the order-parameter dynamics $\tau(\vec{q}_{\text{crit}}) \to \infty$ as $T \to T_c$ from above or below.

The breaking of the symmetry of the high-temperature paraelectric phase is the result of the condensation of a polar soft phonon or a – usually overdamped – soft pseudospin wave mode, the frequency of which decreases with decreasing temperature in the high-temperature phase and vanishes at T_c. The frozen-out eigenvector of the soft mode represents the order parameter of the transition. In ferroelectrics the condensation of the soft mode occurs at the Brillouin-zone centre, whereas it occurs in anti-ferroelectrics at the Brillouin-zone boundary. Alternatively, it is the result of ion ordering in a multisite potential.

4.4.1 Deuteron NMR and relaxation

In hydrogen-bonded ferroelectrics and anti-ferroelectrics the O—H···O bonds are of the double-well type (Fig. 4.4) and are the main reversible dipoles in the structure.

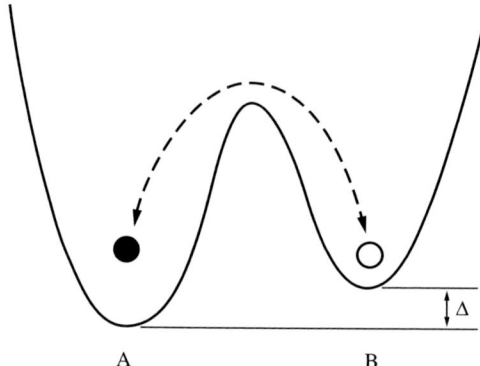

Fig. 4.4 Double-well O—H···O hydrogen-bond potential in hydrogen-bonded ferroelectrics. The two positions of the proton or deuteron in this potential can be described by a pseudospin $S^z = \pm 1$.

The two positions of the proton ($I = \frac{1}{2}$) or, respectively, deuteron ($I = 1$) can be represented by a pseudospin $S^z = \pm 1$. The quadrupole perturbed NMR frequency of the deuteron ω_i changes when the deuteron is transferred from the 'left' to the 'right' position, O—H···O ↔ O···H—O, in the double-well potential:

$$\omega_i = \omega_0 + \omega_1 S_i^z(t) \tag{10a}$$

In the 'fast-motion' limit, where $\tau \ll \tau_{\text{NMR}} = 1/2\omega_1$ (Fig. 4.5) the nucleus 'sees' just the time averaged value of $S_i^z(t)$ so that equation (10a) becomes

$$\omega_i = \omega_0 + \omega_1 p_i \tag{10b}$$

where $p_i = \langle S_i^z(t) \rangle$ is the local polarization. In a homogeneous ferroelectric, e.g. KD$_2$PO$_4$, where Eq. (2c) holds, $p_i = p_j = p = P$, so that ^{2}H NMR provides a method for a microscopic measurement of the long-range order parameter P as well as yielding direct information about the deuteron order–disorder transition at T_c [4.2]. As can be

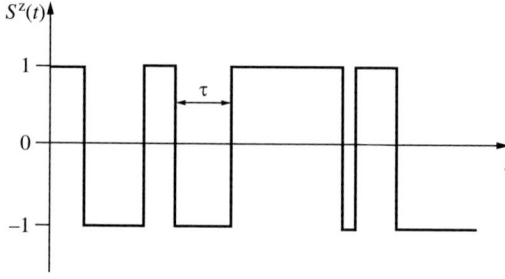

Fig. 4.5 Time dependence of the pseudospin variable $S^z(t)$ reflecting the deuteron motion between the two equilibrium sites in the O—H···O bond.

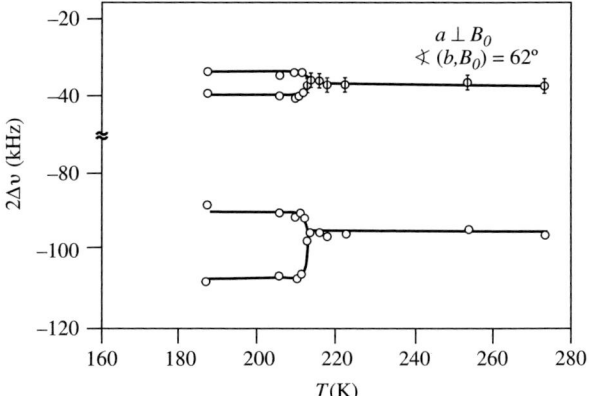

Fig. 4.6 Temperature dependence of the quadrupole splitting of the deuteron NMR lines in CsD_2AsO_4.

seen from Fig. 4.6, $\langle S^z(t) \rangle = 0$ above T_c and $\langle S^z(t) \rangle \neq 0$ below T_c. The deuterons are dynamically disordered between the two equilibrium sites above T_c and order into one of the two positions below T_c [4.6]. The T dependence of $p = P$ is equal to the one determined by macroscopic polarization measurements.

The fluctuating part of the local polarization $p_i = p + \delta p_i(t)$ and the related part of the quadrupole coupling Hamiltonian $\hat{\mathcal{H}}_Q^{(1)}(t)$,

$$\hat{\mathcal{H}}_Q(t) = \langle \hat{\mathcal{H}}_Q \rangle + \hat{\mathcal{H}}_Q^{(1)}(t) \tag{11a}$$

produce, in agreement with Eqs. (8) and (9), a characteristic dip in T_1 at T_c (Fig. 4.7). The form of the T_1^{-1} anomaly is the one expected [4.2] for the condensation of an overdamped soft pseudospin wave mode:

$$T_1^{-1} \propto \ln \frac{T - T_c}{T_c} + \text{constant}, \quad T > T_c \tag{11b}$$

$$T_1^{-1} \propto 1 - p^2, \quad T < T_c \tag{11c}$$

From the above data the non-interacting deuteron correlation time, i.e. the time the deuteron stays at a given site before it jumps to the other site in the O—D···O potential, is determined as $\tau_0 = 2.7 \times 10^{-13}$ s at room temperature. This justifies the fast-motion approximation. Similar results were obtained in other hydrogen-bonded ferroelectrics [4.2, 4.7].

4.4.2 Oxygen-17–proton nuclear quadrupole double resonance

In view of the 'geometrical' isotope effect [4.8] and the resulting expansion and changes in the shape of the hydrogen-bond potential on deuteration one cannot simply transfer the conclusions obtained for deuterated crystals to the undeuterated ones. Since

104 Dipolar glasses

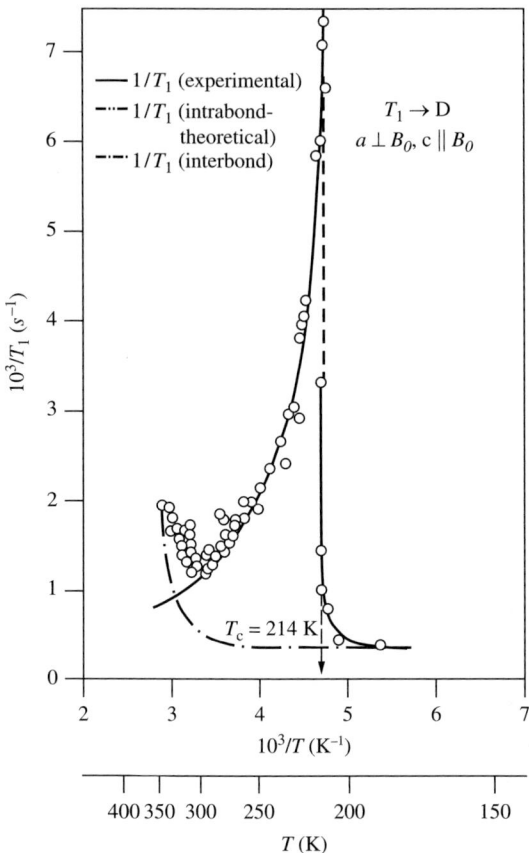

Fig. 4.7 Temperature dependence of the deuteron spin-lattice relaxation rate in KD_2PO_4 at $B_0 = 1.65$ T.

the distance between the two equilibrium sites of the hydrogen in short O—H···O bonds ($R_{O-O} \leq 2.49$ Å) may be 0.35 Å or less, it is difficult to determine by neutron diffraction whether the proton is located in the centre of the hydrogen bond or dynamically disordered between two separate, off-centre equilibrium sites. For $T > T_c$ a single-peaked proton density distribution in the centre of the hydrogen bond is usually observed by neutron scattering. Below T_c the diffraction data show an off-centre position of the proton.

The answer is provided by ^{17}O—^{1}H quadrupole double resonance spectroscopy and the direct determination of the ^{17}O—^{1}H distance from the magnetic dipole–dipole fine structure of the ^{17}O NQR lines [4.9].

As an example, let us consider [4.7] RbH_2PO_4, which is isomorphous with KH_2PO_4 and has all O—H···O bonds equivalent. In this system, the PO_4 groups are linked by O—H···O bonds. In view of the low natural abundance of ^{17}O(0.037%) there is at most one ^{17}O nucleus ($I = \frac{5}{2}$) per O—H···O bond.

The energies of the three doubly degenerate ($m = \pm\frac{1}{2}, m = \pm\frac{3}{2}, m = \pm\frac{5}{2}$) ^{17}O quadrupole energy levels $E = xA, A = e^2qQ/4I(2I-1), eq = V_{zz}$ can be obtained from

$$x^3 - 7(3+\eta^2)x - 20(1-\eta^2) = 0 \tag{12}$$

where $\eta = (V_{xx} - V_{zz})/V_{zz}$ is the asymmetry parameter of the EFG tensor and where $V_{zz} \geq V_{yy} \geq V_{xx}$. For small η we have

$$\nu_{1/2 \to 3/2} = \frac{3e^2qQ}{20h}\left(1 + \frac{59}{54}\eta^2\right) \tag{13a}$$

$$\nu_{1/2 \to 5/2} = \frac{3e^2qQ}{10h}\left(1 - \frac{11}{54}\eta^2\right) \tag{13b}$$

and $\nu_{1/2 \to 5/2} = \nu_{1/2 \to 3/2} + \nu_{3/2 \to 5/2}$.

If the proton is centrally located we expect just one set of ^{17}O NQR lines ($\nu_{1/2 \to 3/2}, \nu_{3/2 \to 5/2}, \nu_{1/2 \to 5/2}$) per ^{17}O—H $\cdots$ O bond. If the proton is located in an 'off-centre' site we expect, in view of the random occurrence of the ^{17}O isotope, two sets of ^{17}O NQR lines per O—H $\cdots$ O bond. One of them would correspond to the proton in the 'close' position, i.e. ^{17}O—H $\cdots$ O, and the other to the proton in the 'far' position, i.e. ^{17}O $\cdots$ H—O. If the protons are dynamically disordered between the two 'off'-centre sites, the observed ^{17}O EFG tensor should be an average of the tensors for the two separate 'off'-centre sites, the 'far' and the 'close' one. It should be noted that the ^{17}O quadrupole coupling constant increases with decreasing O–H bond distance [4.8], thus allowing for a discrimination between 'close' and 'far' sites.

The ^{17}O nuclear quadrupole double resonance results for RbH$_2$PO$_4$ are shown in Fig. 4.8. Above T_c only three ^{17}O lines ($\nu_{1/2 \to 3/2}, \nu_{3/2 \to 5/2}, \nu_{1/2 \to 5/2}$) are observed, demonstrating that all PO$_4$ oxygen sites are equivalent on the NQR time scale due to fast proton motion. Below T_c each line splits symmetrically into two components, demonstrating that we are dealing with dynamic disorder between two equivalent sites above T_c and static long-range order below T_c. The average value of the two components is constant while the difference is proportional to the local order parameter p and measures the time-averaged asymmetry in the hydrogen-bond potential below T_c. We thus have above T_c

$$\langle\nu(^{17}\text{O} - \text{H})\rangle = \langle\nu(^{17}\text{O} \cdots \text{H})\rangle = \frac{1}{2}[\nu(^{17}\text{O} - \text{H}) + \nu(^{17}\text{O} \cdots \text{H})], \quad T > T_c \tag{14a}$$

and below T_c

$$\langle\nu(^{17}\text{O} - \text{H})\rangle = \frac{1+p}{2}\nu(^{17}\text{O} - \text{H}) + \frac{1-p}{2}\nu(^{17}\text{O} \cdots \text{H}), \quad T < T_c \tag{14b}$$

$$\langle\nu(^{17}\text{O} \cdots \text{H})\rangle = \frac{1-p}{2}\nu(^{17}\text{O} - \text{H}) + \frac{1+p}{2}\nu(^{17}\text{O} \cdots \text{H}), \quad T < T_c \tag{14c}$$

The difference

$$\langle\nu(^{17}\text{O} - \text{H})\rangle - \langle\nu(^{17}\text{O} \cdots \text{H})\rangle = p[\nu(^{17}\text{O} - \text{H}) - \nu(^{17}\text{O} \cdots \text{H})] \tag{14d}$$

106 Dipolar glasses

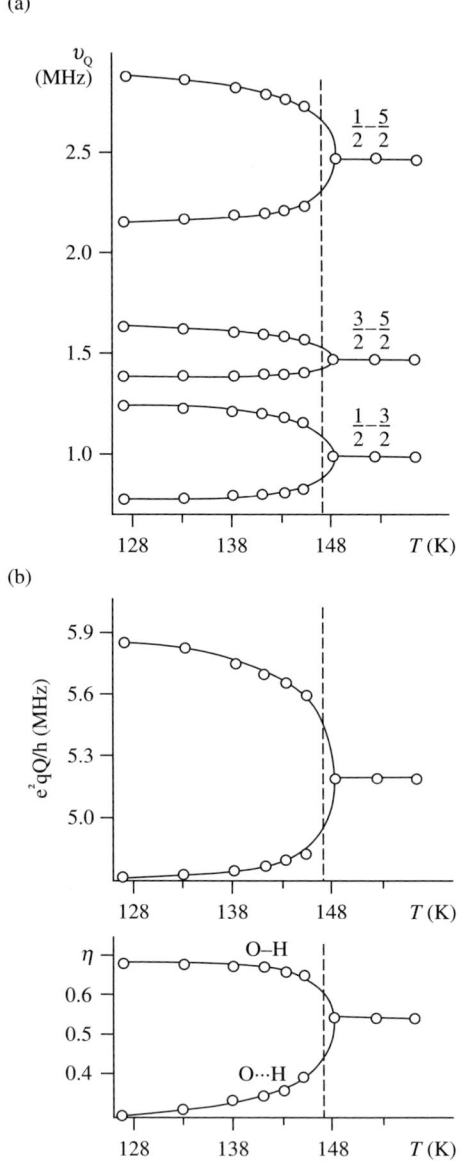

Fig. 4.8 Temperature dependence of (a) the ^{17}O NQR transition frequencies in RbH$_2$PO$_4$ and (b) the quadrupole coupling constant and asymmetry parameter.

is here indeed proportional to p and allows for a measurement of the local asymmetry of the hydrogen bond potential in frequency units.

The corresponding measuring sequence for the ^{17}O—H double resonance experiment in the laboratory frame is schematically shown in Fig. 4.9.

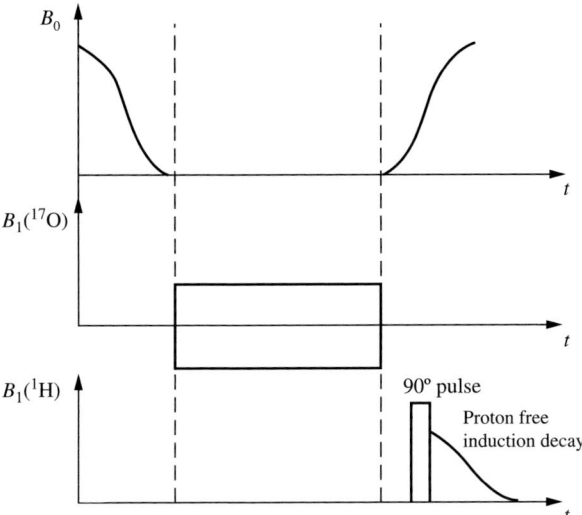

Fig. 4.9 Field sequence for the ^{17}O–^{1}H double resonance in the laboratory frame experiment.

An independent confirmation of the above results is provided by the proton magnetic dipolar fine structure of the ^{17}O lines obtained with a double-frequency irradiation technique [4.9].

Above T_c the proton spends an equal amount of time at the 'close' ($R_{O-H} = 1.06$ Å) and 'far' ($R_{O\cdots H} = 1.45$ Å) equilibrium sites so that the dipolar splitting is characterized by an 'apparent' ^{17}O–H distance

$$\langle R^{-3}\rangle = \frac{1}{2}\left[R_{O-H}^{-3} + R_{O\cdots H}^{-3}\right], \quad T > T_c \tag{15a}$$

yielding $R = 1.18$ Å. If, on the other hand, the proton is above T_c, situated in the centre of the O–H–O bond, the effective O–H distance obtained from the dipolar fine structure would be

$$R_{\text{symm}}(O-H) = 1.225 \text{Å}, \quad T > T_c \tag{15b}$$

which would contradict the experimental data. The ^{17}O quadrupole coupling as well as the ^{17}O–H magnetic dipolar fine structure thus provide direct evidence for the proton intra-^{17}O–H$\cdots$O $\leftrightarrow$ ^{17}O$\cdots$H–O motion between the two sites and the dynamic order–disorder nature of the phase transition in RbH$_2$PO$_4$.

Similar results were obtained for KH$_2$PO$_4$, CsH$_2$PO$_4$, TlH$_2$PO$_4$, PbHPO$_4$, as well as squaric acid (C$_4$H$_2$O$_2$) and KH$_3$(SeO$_3$)$_2$[4.7, 4.10]. In partially deuterated RbH$_{2x}$D$_{2(1-x)}$PO$_4$ the results show that the O–H$\cdots$O bonds are practically identical to the ones in pure RbH$_2$PO$_4$, whereas the O–D$\cdots$O bonds expand and show a nearly 25% higher spontaneous asymmetry below T_c than the O–H$\cdots$O bonds [4.7].

4.4.3 Proton chemical-shift tensors

For protons ($I = \frac{1}{2}$) in solids, the chemical-shift term $\hat{\mathcal{H}}_\sigma$ in the spin Hamiltonian

$$\hat{\mathcal{H}} = \hat{\mathcal{H}}_{\rm Z} + \hat{\mathcal{H}}_{\rm D} + \hat{\mathcal{H}}_\sigma \tag{16a}$$

$$\hat{\mathcal{H}}_\sigma = \gamma \sum_i \vec{I}_i \cdot \underline{\sigma}_i \cdot \vec{B}_0 \tag{16b}$$

is usually much smaller than the dipolar term so that proton chemical shifts are lost in the dipolar width of the NMR line. Using high-resolution NMR-in-solids techniques like the WAHUHA sequence [4.11],

$$(\tau, P_x, 2\tau, P_{-x}, \tau, P_y, 2\tau, P_{-y})$$

where P_x stands for a 90° radiofrequency pulse in the x-direction, the homonuclear dipolar broadening can be eliminated, while the chemical shifts are reduced by $\sqrt{3}$. This allows for a determination of the proton chemical-shift tensors $\underline{\sigma}$ from the angular dependence of the spectra. Such measurements were performed in KH_2PO_4 and other hydrogen-bonded systems [4.12]. The information obtained is in principle similar to that obtained from the deuteron quadrupole coupling data but the experiments are more tedious and the resolution is lower.

4.4.4 Phosphorus-31 chemical-shift tensors

Whereas ^{17}O NQR and deuteron quadrupole perturbed NMR measure the single-particle intra-hydrogen-bond dynamics, the ^{31}P chemical-shift tensor $\underline{\sigma}$ interaction in KH_2PO_4-type crystals measures the local arrangement of the four O—H$\cdots$O bonded protons in the hydrogen bonds linking a given PO_4 group to its neighbours. The chemical shift term is

$$\hat{\mathcal{H}}_\sigma = \gamma \sum_i \vec{I}_i \cdot \underline{\sigma}_i \cdot \vec{B}_0 \tag{17a}$$

where

$$\underline{\sigma} = \underline{\sigma}(p_1, p_2, p_3, p_4) \tag{17b}$$

The temperature dependence of the ^{31}P chemical shift on going through T_c in KD_2PO_4 is shown [4.7, 4.10] in Fig. 4.10(a). There are four ^{31}P nuclei per unit cell of KH_2PO_4 both above and below T_c. The edges of two PO_4 tetrahedra are rotated by $+13°$ with respect to the orthorhombic a-, b-axes (A sites), whereas the other two are rotated by $-13°$ (B sites). Above T_c all ^{31}P sites in the unit cell are chemically and physically equivalent due to dynamic proton disorder in the O—H$\cdots$O bonds. The ^{31}P chemical-shift tensor is axially symmetric around the fourfold c-axis, which passes through the phosphorus sites. Below T_c there are four physically nonequivalent ^{31}P sites, which are all chemically equivalent and that reflect the ordering of the four protons into two 'close' and two 'far' sites in the O—H$\cdots$O bonds: $\sigma_1 = \sigma_A(+\vec{P}), \sigma_2 = \sigma_A(-\vec{P})$, $\sigma_3 = \sigma_B(+\vec{P}), \sigma_4 = \sigma_B(-\vec{P})$. The paraelectric $\underline{\sigma}$ tensor is the average of the four

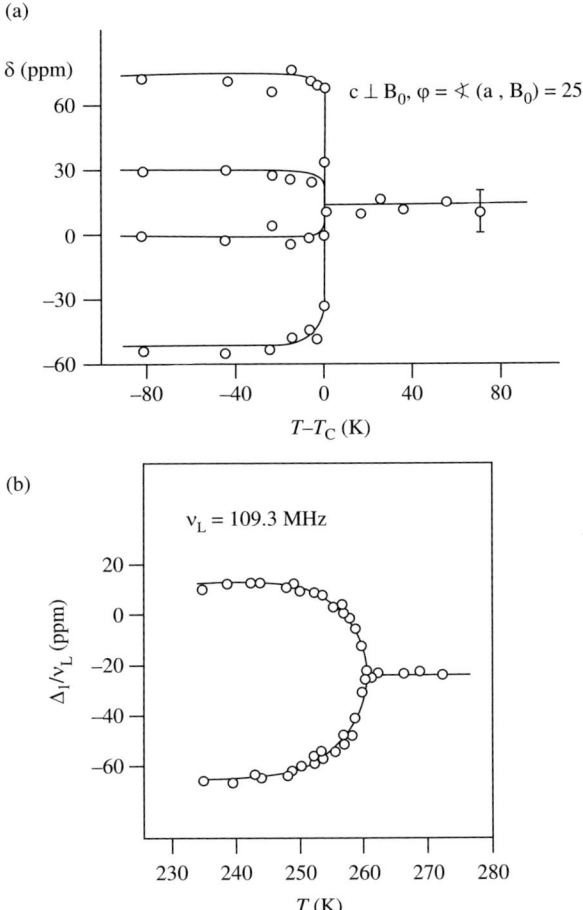

Fig. 4.10 (a) Temperature dependence of the ^{31}P chemical shift on cooling through T_c in KD$_2$PO$_4$. (b) Splitting of the ^{31}P NMR line [4.11] in pseudo-1D CsD$_2$PO$_4$ where the O—H(1)$\cdots$O bonds are ordered above and below T_c whereas the O—H(2)$\cdots$O bonds are disordered above T_c and ordered below T_c and where $\underline{\sigma}_{T>T_c} = \frac{1}{2}(\underline{\sigma}_1 + \underline{\sigma}_2)_{T<T_c}$.

ferroelectric ones [4.7, 4.10]. This confirms that the phase transition in KH$_2$PO$_4$ is a protonic order–disorder one and is not driven by electronic instabilities.

The largest principal axis of the ^{31}P chemical-shift tensor is nearly exactly parallel to the line connecting the two PO$_4$ oxygens to which the protons are directly attached. A change in the H$_2$PO$_4$ proton 'arrangement' from the 'top' to the 'bottom' oxygens, as viewed along the c-axis, results in the polarization reversal $(\vec{P} \to -\vec{P})$ and in a rotation of the two largest principal axes of the ^{31}P $\underline{\sigma}$ tensor by 90°. The ^{31}P $\underline{\sigma}$ tensor is thus a very sensitive indicator of the arrangement of the four protons around a given PO$_4$ group in KH$_2$PO$_4$ crystals.

110 Dipolar glasses

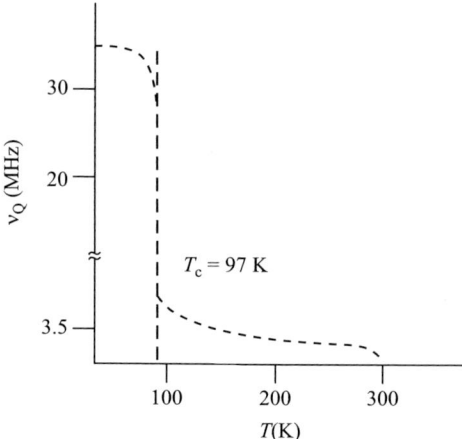

Fig. 4.11 Temperature dependence of the ^{75}As NQR frequency ν_Q in KH$_2$AsO$_4$ on cooling through T_c.

Similar results were also obtained in pseudo-1D CsD$_2$PO$_4$ [4.13] (Fig. 4.10(b)), PbHPO$_4$ [4.14], squaric acid and other systems [4.14] using high-resolution NMR-in-solids techniques.

4.4.5 Arsenic-75 quadrupolar coupling

Information similar to that from the ^{31}P chemical-shift tensor can be obtained from the ^{75}As quadrupole coupling tensor in KH$_2$AsO$_4$-type arsenates.

Here, however, the resolution of the experiment is much larger as the ^{75}As NQR frequency changes [4.10] from ≈ 3.5 MHz at $T > T_c$ to ≈ 35 MHz at $T < T_c$ (Fig. 4.11). The ^{75}As EFG tensor also depends on the arrangement of the four protons around a given AsO$_4$ group:

$$\underline{V} = \underline{V}(p_1, p_2, p_3, p_4) \qquad (18)$$

The direction of the largest principal axis (V_{zz}) of this tensor is parallel to the line connecting the two oxygens in the AsO$_4$ group to which the two protons are directly attached, whereas the smallest principal axis (V_{xx}) is parallel to the direction of the electric dipole moment of the H$_2$AsO$_4$ group.

The ^{75}As EFG tensors determined [4.2, 4.7, 4.10, 4.15] in ferroelectric KH$_2$AsO$_4$ and anti-ferroelectric NH$_4$H$_2$AsO$_4$ could be thus assigned to the six Slater [4.16] H$_2$AsO$_4$ configurations predicted by the Slater–Nagamiya model [4.16] of the phase transition in KH$_2$PO$_4$. In ferroelectric KH$_2$AsO$_4$ both 'close' hydrogens are either attached to the 'upper' ($\underline{V}^{(1)}$) or to the 'lower' two H$_2$AsO$_4$ oxygens ($\underline{V}^{(2)}$) as viewed along the c-axis, so that the dipole moment points along $\pm c$. In anti-ferroelectric $NH_4H_2AsO_4$, on the other hand, one hydrogen is attached to the 'upper' and one to the 'lower' oxygen,

$$\begin{matrix} O & & O & & & & O & O \\ \Box & , & \Box & , & \Box & , & \Box \\ O & & & O & O & & & O \end{matrix}$$

resulting in four 'anti-ferroelectric' H_2AsO_4 arrangements - $(\underline{V}^{(3)}, \underline{V}^{(4)}, \underline{V}^{(5)}, \underline{V}^{(6)})$ – with dipole moments along the $\pm b$ and $\pm a$ crystal axes. In the anti-ferroelectric domain, where the sublattice polarization $\overleftrightarrow{P} \parallel [100]$ is along the $\pm a$-axis, one sublattice corresponds to $\underline{V}^{(3)}$ and the other to $\underline{V}^{(4)}$, whereas for $\overleftrightarrow{P} \parallel [010]$ the corresponding EFG tensors are $\underline{V}^{(5)}$ and $\underline{V}^{(6)}$. These results represented the first direct proof that the anti-ferroelectric proton ordering suggested by Nagamiya [4.16] is correct.

In the paraelectric phase each H_2AsO_4 group fluctuates between the six Slater configurations. The fluctuations result in a time averaging of the EFG tensors and in an anomalous increase in the ^{75}As spin-lattice relaxation rate T_1^{-1} as $T \to T_c$ [4.17] (Fig. 4.12). The ^{75}As T_1^{-1} has been measured by the proton–arsenic cross-relaxation technique in the dipolar frame [4.17].

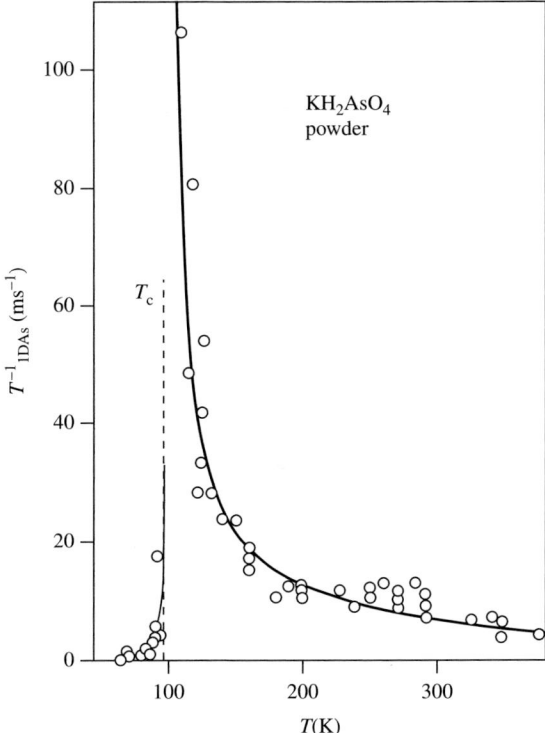

Fig. 4.12 Temperature dependence of the arsenic spin-lattice relaxation rate in the dipolar (rotating) frame as determined by proton-arsenic cross-relaxation.

4.5 NMR in proton and deuteron glasses

4.5.1 Determination of the Edwards–Anderson order parameter q_{EA} in the fast-motion limit

As already mentioned solid solutions $Rb_{1-x}(ND_4)_xD_2PO_4$ (RADP) of ferroelectric RbH_2PO_4 and anti-ferroelectric $ND_4D_2PO_4$ (Fig. 4.3) represent a randomly frustrated hydrogen-bonded system with competing ferroelectric and anti-ferroelectric interactions. For $0.22 \leq x \leq 0.78$ the system forms a deuteron glass [4.5, 4.18, 4.19] with no macroscopic symmetry breaking and no spontaneous polarization at any temperature. Here too, the O—D···O bond represents a two-position reorientable dipole (see Fig. 4.4) that can be described by an Ising pseudospin, $S^z = \pm 1$. In contrast to magnetic spin glasses, where we deal with random bond interactions, proton and deuteron glasses are – as already mentioned – characterized by the presence of both random bonds and random fields [4.3]

$$\hat{\mathcal{H}} = -\sum_{i<j} J_{ij} S_i^z S_j^z - \sum_i f_i S_i^z \qquad (19)$$

Here, the J_{ij} are quenched random interactions and f_i are quenched random fields characterized by their second cumulants $[J_{ij}^2]_{AV} = J^2/N$ and $[f_i f_j]_{AV} = \delta_{ij}\Delta$. The random-bond distribution is assumed to be Gaussian and centred around a mean value $J_0 = J_{AFE}x + J_{FE}(1-x)$ where x represents the NH_4 concentration. J^2 and Δ, on the other hand, are proportional to $x(1-x)$ and vanish for $x \to 0$ and $x \to 1$. For $|J_0| < J$ the system forms a glass phase without long-range order, whereas for $|J_0| > J$ a long-range-ordered inhomogeneous ferroelectric or anti-ferroelectric phase is predicted to occur. Equation (19) describes the random-bond–random-field Ising model Hamiltonian.

The basic question in deuteron glasses is how to measure the Edwards–Anderson order parameter [4.1]

$$q_{EA} = \frac{1}{N} \sum_i \langle S_i^z \rangle^2 \qquad (20)$$

as there is no macroscopic conjugate field attached to it. The answer is provided by NMR. In the fast-motion limit NMR also allows for a direct determination of the local polarization distribution function $W(p)$, the second moment of which is q_{EA}.

In the simplest case the local NMR frequency ω_i of the deuteron in an O—D···O bond is in the fast motion limit $[\omega_{AB}\tau \ll 1, \omega_{AB} = \omega_A - \omega_B]$ linearly related to the pseudospin polarization $\langle S_i^z \rangle = p_i$ of this bond:

$$\omega_i = \omega_0 + \omega_1 \langle S_i^z \rangle = \omega_0 + \omega_1 p_i \qquad (21)$$

O—D···O bonds with different local polarizations p_i will thus have different NMR frequencies resulting in an inhomogeneous broadening of the NMR line. The resulting frequency distribution,

$$f(\omega) = \frac{1}{N} \sum_i \delta(\omega - \omega_i) = [\delta(\omega - \omega_i)]_{AV} \qquad (22)$$

is related to the local polarization distribution $W(p)$ (Eq. (4)) via

$$f(\omega)d\omega = W(p)dp \qquad (23)$$

A measurement of $f(\omega)$ thus allows for a determination of $W(p)$ and its second moment q_{EA} if the relation between ω_i and p_i (Eq. (21)) is known from experiments on pure systems. q_{EA} in particular is obtained from

$$M_2[f(\omega)] = \int d\omega f(\omega)(\omega - \omega_0)^2 = \omega_1^2 q_{EA} \qquad (24)$$

Such experiments have indeed been performed [4.4], and the results can be well described by the random-bond–random-field Ising model [4.3, 4.4, 4.5]. The temperature dependence of q_{EA} (Fig. 4.13) in particular demonstrates that in contrast to magnetic spin glass, random fields are present in addition to random bonds. From the shape of the deuteron NMR spectrum (Fig. 4.14(a)) the temperature dependence of $W(p)$ can be determined (Fig. 4.14(b)). $W(p)$ here is symmetric around $p = 0$ and changes from a single-peaked to a double-peaked form with decreasing temperature.

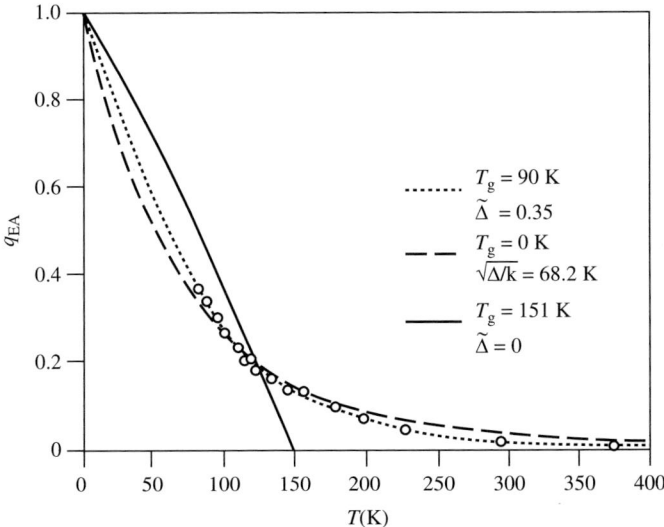

Fig. 4.13 Temperature dependence of q_{EA} as determined from the second moment of the $^{87}\text{Rb}\frac{1}{2} \to -\frac{1}{2}$ NMR spectra in $\text{Rb}_{0.56}(\text{ND}_4)_{0.44}\text{D}_2\text{PO}_4$. The dotted line represents the fit to the random-bond–random-field model with $\Delta/J^2 = 0.35, T_g = J/k = 90\,\text{K}$, whereas the dashed line represents the best fit for the pure random field model where $T_g = 0\,\text{K}$ and $\sqrt{\Delta}/k = 68\,\text{K}$. The solid line represents the fit to the pure random bond model with $T_g = 151\,\text{K}$ and $\sqrt{\Delta}/k = 0\,\text{K}$. Here $\tilde{\Delta}$ stands for Δ/J^2.

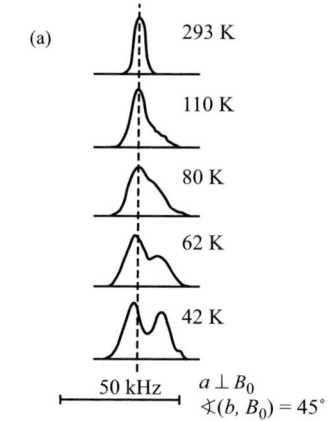

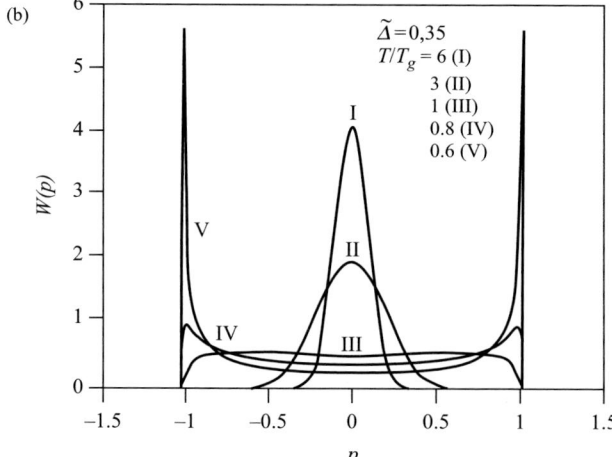

Fig. 4.14 (a) O—D···O deuteron NMR spectrum in $Rb_{0.56}(ND_4)_{0.44}D_2PO_4$ at various temperatures. (b) Temperature dependence of the local polarization distribution function $W(p)$ for $Rb_{0.56}(ND_4)_{0.44}D_2PO_4$, where $\tilde{\Delta} = \Delta/J^2 = 0.35$. Following [4.3–4.5].

At lower temperatures where the pseudospin fluctuations are no longer very fast as compared to the rigid lattice quadrupole splittings ω_{AB} we have to take into account the additional broadening due to the slowing down of the fluctuations by convoluting $W(p)$ with the well-known dynamic lineshape $I(\omega, p, \tau)$ for exchange in an asymmetric two-site potential. Equation (22) now becomes

$$f(\omega) = \int I(\omega, p, \tau) W(p) \, dp \qquad (25)$$

When the fluctuations become too slow, $\omega_{AB}\tau \geq 1$, the second moment of the lineshape is no longer proportional to q_{EA} but approaches a constant value determined by $I(\omega, p, \tau = \infty)$.

4.5.2 Determination of the Edwards–Anderson glass order parameter q in the slow-motion limit

In the slow-motion limit where $\omega_{AB}\tau \gg 1$ the nucleus 'sees' the instantaneous value of the Hamiltonian corresponding to $S^z = \pm 1$. The NMR spectrum of the O–D$\cdots$O bond deuteron will thus consist of two lines,

$$\omega_A = \omega_0 + \omega_1 \qquad (26a)$$

and

$$\omega_B = \omega_0 - \omega_1 \qquad (26b)$$

corresponding to the 'left' and 'right' position of the deuterons in the double-minimum hydrogen bond potentials. 2D 'exchange' NMR spectroscopy [4.20] can be here used to determine the glass order parameter q as well as the intrabond deuteron jump rate. The corresponding pulse sequence (Fig. 4.15) developed by Spiess *et al.* [4.21] selects the signal of the form

$$\begin{aligned}F(t_1, t_2, t_m) = &\exp(-t_m T_1) \exp[-(t_1 + t_2)/T_2] \\ &\times [a_{AA}(t_m) \cos(\omega_A t_1) \cos(\omega_A t_2) \\ &+ a_{BB}(t_m) \cos(\omega_B t_1) \cos(\omega_B t_2) \\ &+ a_{AB}(t_m) \cos(\omega_A t_1) \cos(\omega_B t_2) \\ &+ a_{BA}(t_m) \cos(\omega_B t_1) \cos(\omega_A t_2)] \end{aligned} \qquad (27)$$

After a Fourier transformation with respect to t_1 and t_2, the results are presented in the $\omega_1 - \omega_2$ plane. If no intrabond deuteron exchange, A $\leftrightarrow$ B, has taken place, the absorption peaks lie on the diagonal of the ω_1–ω_2 plane at (ω_A, ω_A) and (ω_B, ω_B). If intrabond deuteron transfer has taken place between the two physically non-equivalent sites A and B, we also find cross-peaks at (ω_A, ω_B) and (ω_B, ω_A). The intensities of the cross peaks (Fig. 4.16) are proportional to the fraction of the nuclei that have undergone exchange in the time t_m. Since

$$a_{AA}(\infty) = \frac{1}{N} \sum_i W_{Ai}^2 \qquad (28a)$$

$$a_{BB}(\infty) = \frac{1}{N} \sum_i W_{Bi}^2 \qquad (28b)$$

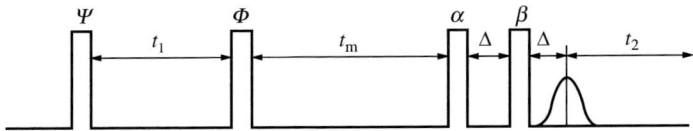

Fig. 4.15 2D exchange NMR pulse sequence [4.21].

and

$$a_{AB}(\infty) = a_{BA}(\infty) = \frac{1}{N}\sum_i W_{Ai}W_{Bi} \qquad (28c)$$

where W_{Ai} is the probability of finding the deuteron of the ith O—D$\cdots$O bond in site A, etc., we get for the ratio of the intensities of the cross peaks to the diagonal peaks in the limit $t_m \to \infty$ the expression

$$R(t_m \to \infty) = \frac{a_{AB}(\infty)}{a_{AA}(\infty)} = \frac{\sum_i W_{Ai}W_{Bi}}{\sum_i W_{Ai}^2} \qquad (29a)$$

Using $\langle S_i^z \rangle = W_{Ai} - W_{Bi}$ and

$$q = \frac{1}{N}\sum_i \langle S_i^z \rangle^2 = \frac{1}{N}\sum_i (W_{Ai} - W_{Bi})^2 \qquad (29b)$$

as well as

$$\frac{1}{N}\sum_i (W_{Ai} + W_{Bi})^2 = 1 \qquad (29c)$$

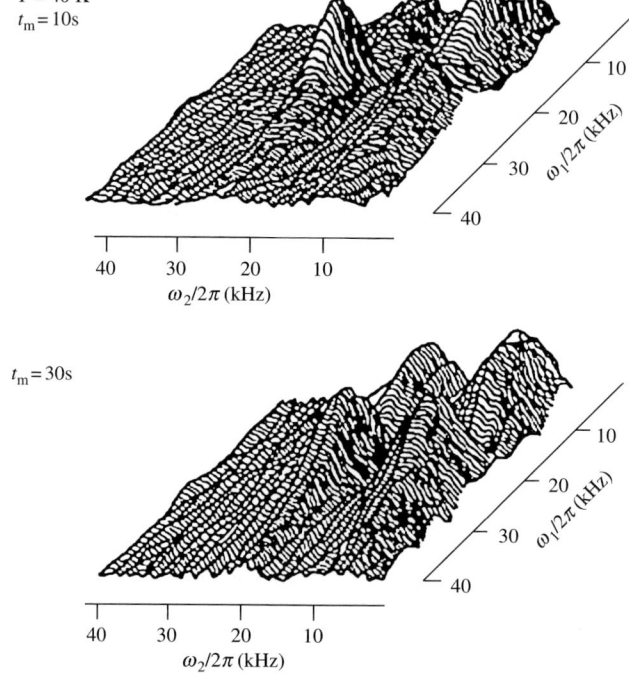

Fig. 4.16 2D O—D$\cdots$O intrabond deuteron 'exchange' NMR spectra of Rb$_{0.68}$(ND$_4$)$_{0.32}$ D$_2$AsO$_4$ for $\vec{a} \perp \vec{B}_0$, $\angle(\vec{b}, \vec{B}_0) = 45°$ at $T = 40$ K showing the gradual appearance of cross-peaks with increasing mixing time t_m.

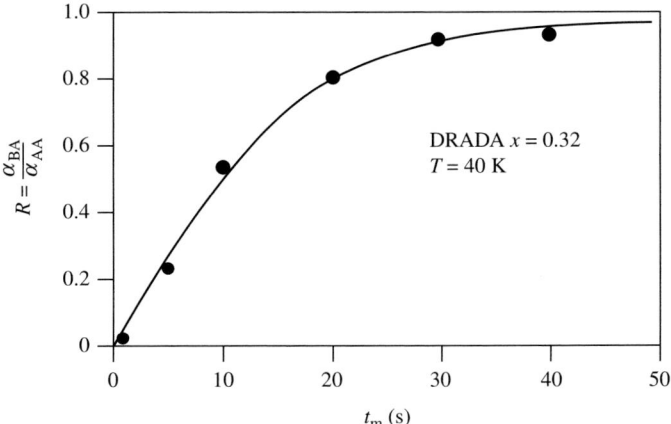

Fig. 4.17 Ratio of cross-peak to diagonal peak intensity R as a function of the mixing time t_m at $T = 40$ K. The same curve is obtained for different crystal orientations with different ω_{AB} splittings so that spin diffusion is excluded.

and $a_{AA} = a_{BB}$, Eq. (29a) can be rewritten as

$$R(t_m \to \infty) = \frac{1-q}{1+q} \tag{30}$$

A measurement of $R(t_m \to \infty)$ thus allows for a determination of q in the slow motion limit [4.20].

The time evolution of $R(t_m) = a_{BA}/a_{AA}$, on the other hand (Fig. 4.17), allows for a direct site-specific measurement of the deuteron intrahydrogen bond exchange rates, A ↔ B in the slow motion limit:

$$R(x) = \tanh x, \quad x = t_m/\tau \tag{31}$$

Here, t_m is the mixing time.

The observed intrahydrogen bond deuteron exchange times [4.20] vary from 14.4 s at $T = 45\,K$ to 235 s at $T = 24\,K$. The motion is thermally activated with $E_a = 12.8$ meV and $\tau_0 = 0.43\,s$, showing that phonon-assisted deuteron tunneling may be important.

4.6 NMR determination of order parameters in inhomogeneous ferroelectrics

For $0 \leq x \leq x_{FE}$ or $x_{AFE} \leq x \leq 1$ (see Fig. 4.2) where $|J_0| > J$, a long-range-ordered inhomogeneous ferroelectric or anti-ferroelectric phase exists below T_c. Here, long-range and glassy order should coexist below T_c. Above T_c, long-range order vanishes, whereas glassy order should persist. The glassy state is characterized above T_c by the Edwards-Andersson (EA) parameter q_{EA} and below T_c by $\tilde{q}_{EA} = q_{EA} - P^2$, i.e. by the difference between the EA order parameter and the square of spontaneous polarization P^2. Both $\tilde{q}_{EA}$ and q_{EA} have been determined [4.22] by ^{75}As NQR ($T < T_c$) and quadrupole perturbed NMR ($T > T_c$) in Rb$_{1-x}$(NH$_4$)$_x$H$_2$AsO$_4$ for $x = 0.01$ and $x = 0.02$ as well as by ^{31}P in NMR in RADP [4.23].

4.7 Theory of dipolar glasses: The random-bond–random-field Ising model

The above observations can be quantitatively described by the random-bond–random-field Ising model [4.7]

$$\mathcal{H} = -\frac{1}{2}\sum_{ij} J_{ij} S_i^z S_j^z - \sum_i f_i S_i^z - E \sum_i S_i^z \tag{32}$$

Here, the Ising pseudospin $S_i^z = \pm 1$ describes the left–right equilibrium positions of a deuteron in the ith O—D$\cdots$O bond in DRADP, and the random-bond interactions J_{ij} are assumed [4.24] to have a Gaussian probability distribution with a mean J_0/N and variance J^2/N. The second term in Eq. (32) represents the random local bias field [4.3] f_i that has an independent Gaussian distribution with mean zero and variance Δ. In magnetic spin glasses, $f_i = 0$. The third term in Eq. (32) represents the external field E.

Applying the replica formalism to model (32) one obtains in the replica-symmetric phase above T_f the following self-consistent equation for the glass order parameter q [4.3]:

$$q = \int Dz \tanh^2[J(q+\tilde{\Delta})^{1/2} z/T], \tag{33}$$

where $\tilde{\Delta} \equiv \Delta/J^2$ and $\int Dz \cdots$ is a Gaussian measure $(2\pi)^{-1/2} \int dz \exp(-z^2/2) \cdots$. For simplicity, here we have set $E = 0$, assuming $J_0 \neq 0$ and $|J_0| < J$. The freezing temperature $T_f = T_f(\Delta)$ is determined by the stability condition

$$T_f^2/J = \int Dz \cosh^{-4}[J(q+\tilde{\Delta})^{1/2} z/T_f] \tag{34}$$

which must be solved simultaneously with Eq. (33) at $T = T_f$.

For $T < T_f$ the replica-symmetric solution (33) is unstable and should be replaced by the Parisi [4.3] order-parameter function $q(\hat{x})$. Using the Parisi two-step approximation for $q(\hat{x})$, i.e. $q(\hat{x}) = q_0$ for $0 \leq \hat{x} < m$ and $q(\hat{x}) = q_1$ for $m \leq \hat{x} \leq 1$, one obtains the free energy in the form

$$\beta F = -(\beta J/2)^2[(1-q_1)^2 - m(q_1^2 - q_0^2)] - m^{-1}\int Dz \ln \int Dy Z^m(y,z) \tag{35}$$

where $\beta \equiv 1/T$ and $Z(y,z) = 2\cosh[\beta h(y,z)]$ with $h(y,z) = J[(q_1 - q_0)^{1/2} y + (q_0 + \tilde{\Delta})^{1/2} z]$.

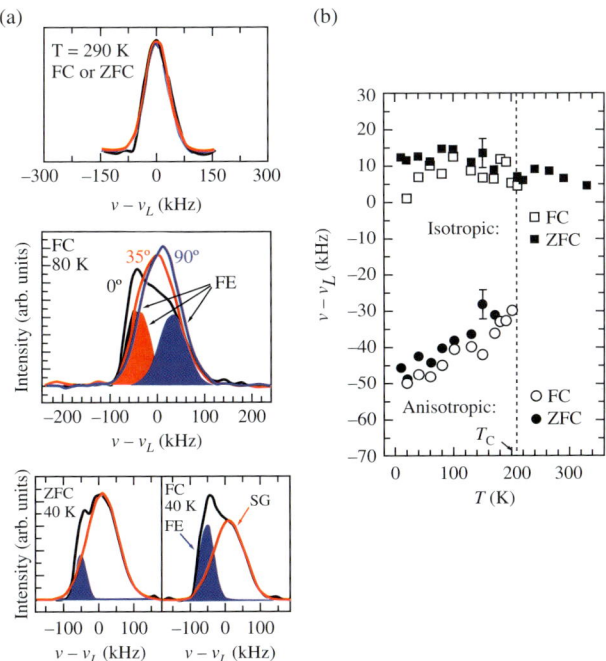

Plate 1 (Fig. 6.20) (a) Decomposition of the field-cooled and zero-field-cooled ^{207}Pb NMR spectra of a PMN single crystal, taken at $\vec{B}\|\vec{E}\|$ [111] at different temperatures, into an isotropic and anisotropic component. (b) Temperature dependence of the positions of isotropic and anisotropic ^{207}Pb NMR lines. The anisotropic line disappears above 210 K for both the FC ($E = 3\,\mathrm{kV/cm}$) and the ZFC spectra [6.29].

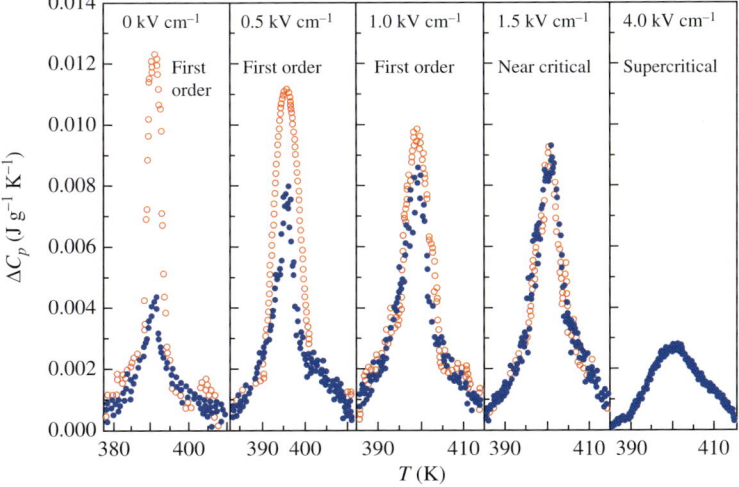

Plate 2 (Fig. 6.24) Temperature dependence of the excess heat capacity data obtained in PMN–PT $x = 0.295$ at various bias electric fields, E, shown at the top of each panel. Filled (blue) and open (red) circles represent data obtained in the ac and relaxation mode, respectively. The relaxation runs measure the total enthalpy changes including the latent heat, whereas the ac runs measure the continuous variations of the enthalpy. The nature of the paraelectric to tetragonal phase transition is denoted by 'first order', 'near critical' and 'supercritical' labels.

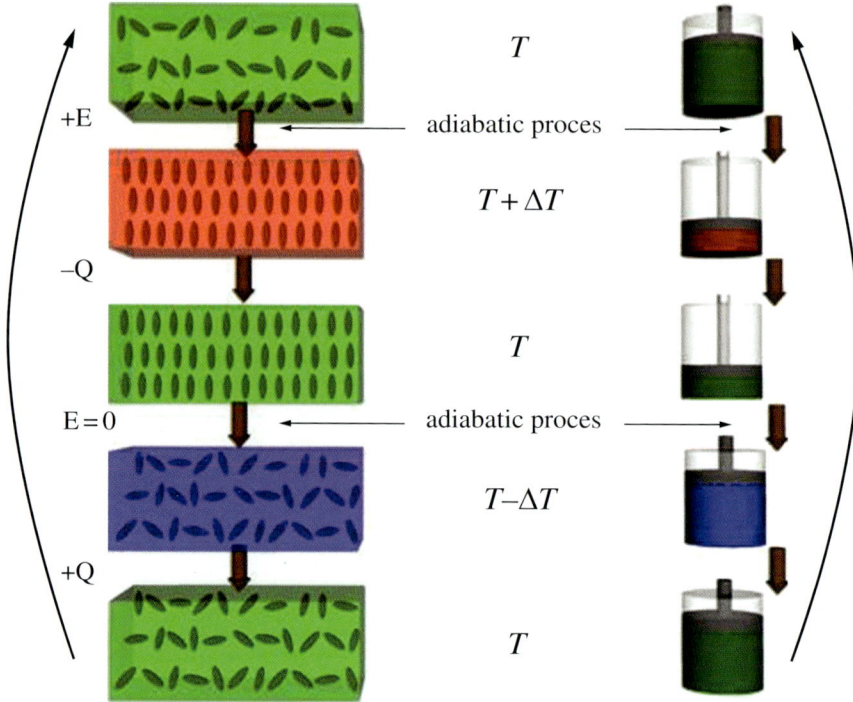

Plate 3 (Fig. 8.1) Analogy between EC cooling and a vapour-compression cooling cycle in a heat engine. Here, $-Q$ is the heat ejected from the EC module and $+Q$ is the heat absorbed (courtesy of Z. Kutnjak and B. Rožič).

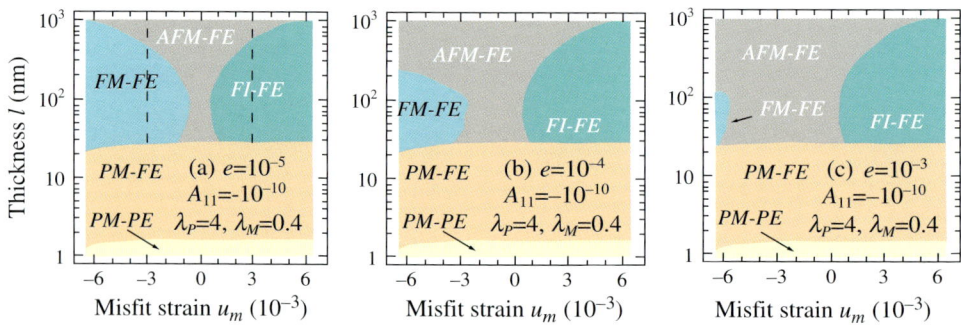

Plate 4 (Fig. 9.12) Phase diagrams of strained ferroic films: AFM–FE designates antiferromagnetic and ferroelectric phases, FM–FE are ferromagnetic and ferroelectric phases (secondary ferroic phase), FI–FE is a ferrimagnetic (week plane at $\theta = \pi/2$) and ferroelectric phase, PM–FE is a paramagnetic-ferroelectric phase, PM–PE is a paramagnetic-paraelectric phase. External fields are zero. A built-in electric field is absent ($g_{31}^e = 0$). Material parameters are listed in Fig. 9.10. Different values of A_{11} and c, λ_P and λ_M (in nanometres) are listed [9.34]. Reproduced with permission from M. D. Glinchuk, A. N. Morozovska, E. A. Eliseev, and R. Blinc: Misfit strain induced magnetoelectric coupling in thin ferroic films, J. Appl. Phys. 105, 084108 (2009).

The value of the free energy corresponding to a stable solution $q(\hat{x})$ is obtained by maximizing the functional (35) with respect to q_0, q_1, and m.

The Edwards–Anderson order parameter $q_{EA} = N^{-1}\sum_i \langle S_i\rangle^2$ is equal to q in the high-temperature phase and to $q(1) = q_1$ in the low-temperature frozen-in glassy phase [4.25, 4.26]. Following Parisi [4.26], the local ($J_0 = 0$) linear-response susceptibility in zero bias field is given by $\tilde{\chi}_{ZFC} = \beta(1 - q_{EA})$, i.e.

$$\tilde{\chi}_{ZFC} = \begin{cases} \beta(1-q) & for \quad T > T_f \\ \beta(1-q_1) & for \quad T > T_f. \end{cases} \tag{36}$$

In contrast, the local field-cooled susceptibility is $\tilde{\chi}_{FC} = \beta[1 - \int_0^1 d\hat{x}\, q(\hat{x})]$. Therefore,

$$\tilde{\chi}_{FC} = \begin{cases} \beta(1-q) = \tilde{\chi}_{ZFC} & for \quad T > T_f \\ \beta[1-q_1+m(q_1-q_0)] & for \quad T < T_f. \end{cases} \tag{37}$$

The above local susceptibilities are related to the bulk susceptibility [4.24] and the measured static dielectric constant ε via

$$\varepsilon_M = 1 + \chi_\infty + A\frac{J\tilde{\chi}_M}{1 - J_0\tilde{\chi}_M}, \quad M = \text{FC or ZFC} \tag{38}$$

where χ_∞ and A are dimensionless constants, and $\tilde{\chi}_M$ refers to either $\tilde{\chi}_{FC}$ or $\tilde{\chi}_{ZFC}$, depending on the measuring procedure.

The observed temperature variation of χ_{FC} and χ_{ZFC}, respectively, ε_{FC} and ε_{ZFC}, can be described both above and below T_f by Eqs. (36)–(38), where the temperature dependence of q for $T > T_f$ is determined from Eq. (33), and the values of the parameters q_0, q_1, and m for $T < T_f$ are obtained by maximizing the free energy. Here, $J = 119.6\,\text{K}$, $J_0 = -55\,\text{K}$, $\tilde{\Delta} \equiv \Delta/J^2 = 0.34$, $\chi_\infty = 7.9$, and $A = 168.2$. The value of J_0 has been obtained by a linear extrapolation between the known critical temperatures $T_c = 218\,\text{K}$ and $T_N = 237\,\text{K}$ for $x = 0$ and 1, respectively, i.e. $J_0 = (1-x)T_c - xT_N$. The values of the two crucial parameters J and $\tilde{\Delta}$ are rather close to the ones obtained from NMR data [4.4] on a sample with $x = 0.44$. Since Δ, J, and J_0 are concentration (i.e. x) dependent, the agreement between the values obtained by NMR and dielectric measurements is within the expected limits.

The above analysis further allows a determination of the transition point $T_f = 62.4\,\text{K}$ for DRADP with $x = 0.6$ on the $T_f(\Delta)$ instability line [4.3] (see Fig. 4.18), which characterizes the phase transition in deuteron glasses. It should be noted that this line, which represents an $E = 0$ cut of the $T_f(E, \Delta)$ surface (see Fig. 4.18), is analogous to the AT line [4.25] in spin glasses in a magnetic field. Such an instability surface with non-zero values of T_f exists only if random bonds are present in addition to random fields, i.e. $T_f = 0$ if only random fields are present. Since the random-field variance Δ depends on the relative concentration of ferroelectric and anti-ferroelectric components, the $T_f(\Delta)$ line [4.3] in deuteron glasses should be experimentally accessible by measuring χ_{FC} and χ_{ZFC} for DRADP and related systems at various values of x.

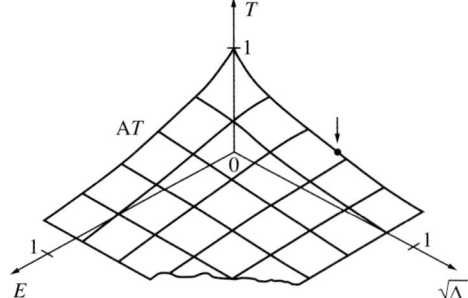

Fig. 4.18 Calculated instability surface $T_f(E, \Delta)$ for deuteron glasses separating the region where $\chi_{FC} = \chi_{ZFC}$ (above the surface) from the region where $\chi_{FC} > \chi_{ZFC}$ (below the surface). The arrow indicates the transition point for DRADP with $x = 0.6$. T, E, and $\sqrt{\Delta}$ are in units of J.

In addition to H-bonded systems, dipolar glass behaviour has been also found in disordered perovskites, e.g. Mn^{++}-doped $SrTiO_3$ [4.27] and other systems.

4.8 Conclusions

In contrast to magnetic spin glasses where we deal with random-bond-type interactions, proton and deuteron glasses are characterized by both random bonds and random fields. The Ising random-bond–random-field model (RBRF) gives a satisfactory description of H-bonded proton and deuteron glasses such as $Rb_{1-x}(ND_4)_xD_2PO_4$ (DRADP)-like systems. The variance of the random field Δ acts as an effective ordering field conjugate to the Edward–Anderson order parameter q_{EA} and can be determined by NMR. The local polarization distribution function $W(p)$ characterizes the proton respectively deuteron order and has been experimentally obtained. Using 2D deuteron exchange NMR spectroscopy the H-bond deuteron exchange time has been measured in the slow-motion limit in the glassy phase. A classical to quantum cross-over has been found in proton RADP type glasses where tunnelling of Takagi groups becomes rate determining as $T \to 0$.

5
Magnetoelectric ferroelectrics

5.1 Introduction

Magnetoelectric (ME) crystals show spontaneous magnetic and electric order. This means that they show simultaneously ferroelectric, ferromagnetic or anti-ferromagnetic properties. Whereas the existence of magnetization has been known since ancient times and ferroelectricity has been known for about 100 years, research in magnetoelectrics was revived relatively recently. As a matter of fact the possibility of a magnetoelectric effect was pointed out already in 1894 by Pierre Curie [5.1]. Between 1957–62 the linear magnetoelectric effect was proved theoretically by Dzyaloshinskii [5.2] for Cr_2O_3 and later experimentally by Astrov [5.3] and Rado [5.4].

An anti-ferromagnet thus acquires a magnetization $\vec{M}$ in an electric field $\vec{E}$ or a polarization $\vec{P}$ in a magnetic field $\vec{H}$:

$$P_i = \alpha_{ij} H_j \tag{1}$$

$$M_j = \alpha_{ij} E_i \tag{2}$$

Magnetoelectric materials allow for the possibility of manipulating the ferromagnetic or anti-ferromagnetic state by an electric field and the ferroelectric structure by magnetic interactions. As such they are of great potential interest as random access memories and spintronic devices [5.5–5.11].

In magnetoelectric systems the electric polarization and the ferromagnetic magnetization are simultaneously allowed [5.5–5.13]. This is usually not the case in normal crystals. The magnetization is an axial vector

$$\vec{M} \to -\vec{M}, \quad t \to -t \tag{3}$$

$$\vec{M} \to \vec{M}, \quad \vec{r} \to -\vec{r} \tag{4}$$

which is invariant against space reversal but not time reversal (Fig. 5.1).

The polarization, on the other hand, is a polar vector

$$\vec{P} \to -\vec{P}, \quad \vec{r} \to -\vec{r} \tag{5}$$

$$\vec{P} \to \vec{P}, \quad \vec{t} \to -\vec{t} \tag{6}$$

invariant against time reversal but not space reversal (Fig. 5.2).

122 Magnetoelectric ferroelectrics

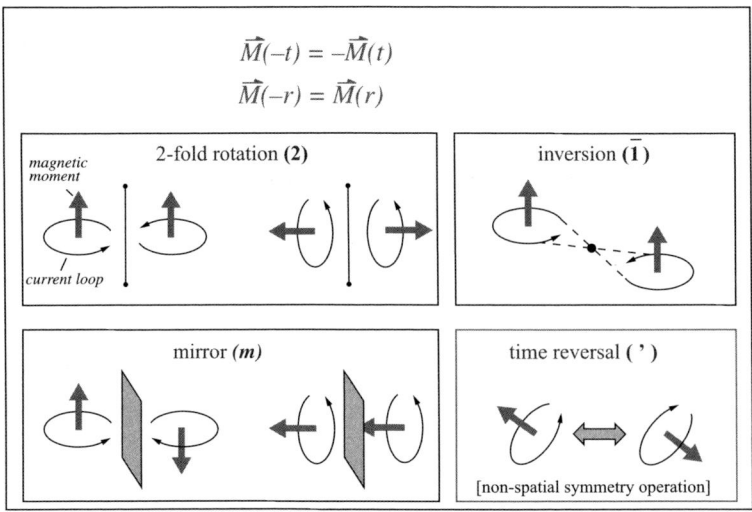

Fig. 5.1 Magnetic order: broken symmetry on time inversion (courtesy of D. Arčon).

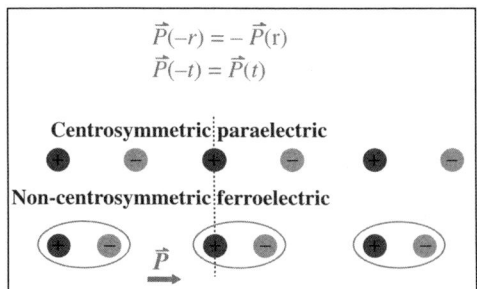

Fig. 5.2 Polar order: broken symmetry on space inversion (courtesy of D. Arčon).

A linear coupling term $P_i M_j$ is thus allowed by symmetry only in certain magnetic groups (see chapter on Nanoferroelectrics).

A systematic phenomenological description of the magnetoelectric effect is obtained by an expansion of the free energy density in powers of $\vec{E}$ and $\vec{H}$:

$$F(\vec{E}, \vec{H}) - F_0 = P_j E_i + M_i H_j - \frac{1}{2}\varepsilon_0 \varepsilon_{ij} E_i E_j - \frac{1}{2}\mu_0 \mu_{ij} H_i H_j$$
$$+ \alpha_{ij} E_i H_j - \frac{1}{2}\beta_{ijk} E_i H_j H_k - \frac{1}{2}\gamma_{ijk} H_i E_j E_k + \ldots \quad (7)$$

Here, $\vec{E}$ and $\vec{H}$ are the electric and magnetic fields and Einstein's summation symbols are used. In Eq. (7) the MKS system of units was used where ε_0 and μ_0 are the electric and magnetic permeabilities of vacuum.

Table 5.1 Introduction to magnetoelectrics.

Multiferroics	*Coexistence of two or more properties: (anti)ferroelectricity, (anti)ferromagnetism and (anti)ferroelasticity*
Old ME materials	$BiMnO_3, BiFeO_3, Ni_3B_7O_{13}I, Cr_2O_3, \ldots$
Problem	Magnetoelectric coupling is weak.

Differentiation yields

$$P_i = \alpha_{ij}H_j + \frac{1}{2}\beta_{ijk}H_jH_k \ldots \tag{8}$$

and

$$\mu_0 M_i = \alpha_{ji}E_j + \frac{1}{2}\gamma_{ijk}E_jE_k + \ldots \tag{9}$$

As mentioned above α_{ij} can be non-zero only in systems where the time- and space-reversal symmetries are broken. The linear magnetoelectric response, PM, is thus only allowed in time-asymmetric media. The quadratic effect, P^2M^2, is, however, always allowed. It arises from higher-order effects such as the strain coupling and the electrostrictive and magnetostrictive interaction (see section on Bi-relaxors). Piezoelectric and piezomagnetic effects may also be important. It should be noted that the phenomenological Landau-type approach is very useful to look for non-zero components of α but says nothing about the size or microscopic origin of the ME response. First-principles 'ab initio' calculations are necessary for elucidating the microscopic origin of these effects.

α_{ij} is bounded by the geometric mean of the diagonalized dielectric susceptibility ε_{ij} and magnetic susceptibility μ_{ij} tensors so that

$$\alpha_{ij}^2 \leq \varepsilon_0\mu_0\varepsilon_{ij}\mu_{ij} \tag{10}$$

If, however, the coupling becomes so strong that it produces a phase transition to a more stable state, then $\alpha_{ij}, \varepsilon_{ij}$ and μ_{ij} take on values appropriate for the new phase.

The temperature dependence of the magnetoelectric coefficient a_{ij} in a typical magnetoelectric Cr_2O_3 is shown in Fig. 5.3. It should be mentioned that Cr_2O_3 has a non-zero ME response at zero temperature only perpendicular to the trigonal [111] axis.

One should mention that the coexistence of electric and magnetic order is difficult from a microscopic point of view. Ferroelectricity in perovskites requires empty d-orbitals for the off-centre displacement of the cations, whereas ferromagnetism requires partially filled d-orbitals. In $BiFeO_3$ the 6s lone pair electrons on the Bi ion are responsible for ferroelectricity, while partially filled d-orbitals of the Fe ions lead to magnetic ordering.

The $BiFeO_3$ system belongs to the R3c space group with a rhombohedrally distorted perovskite structure. The magnetic phase shows a cycloidal spin structure

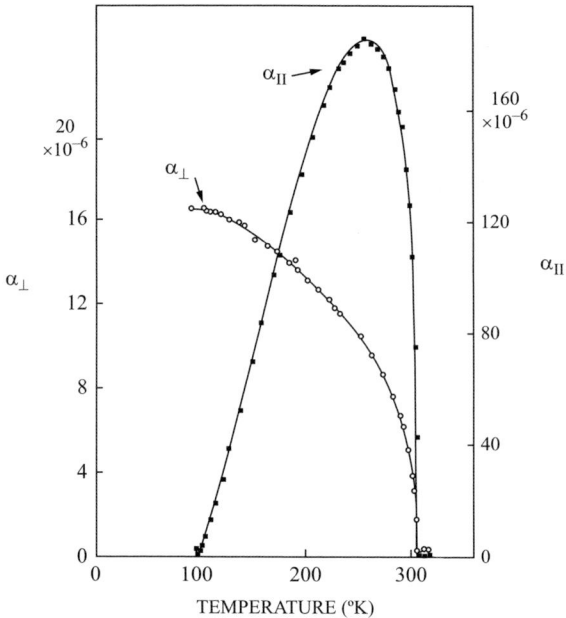

Fig. 5.3 Temperature dependence of the magnetoelectric coefficient in Cr_2O_3. The αs are dimensionless in the Gaussian units used [5.4].

Table 5.2 Why there are so few magnetoelectrics [5.14]?

Magnetic systems usually rely on $3d^1, 3d^2, 3d^3, \ldots$ transition-metal ions.
Ferroelectric systems require empty d orbitals (d^0 and lone pair s electrons).
These two requirements seem to be contradictory.

---

incommensurate with the $[110]_h$ direction of the hexagonal unit cell of the rhombohedral structure. The incommensurate spin structure can be suppressed by an electric or magnetic field or by chemical substitution.

In epitaxial thin films of $BiFeO_3$ the room temperature polarization is $P_r \sim 55\,\mu C\,cm^{-2}$ and the magnetization $M_s \sim 1$ emu cm^{-3}. In $HoMnO_3$, in particular, an electrically driven magnetic phase transition has been found. Unfortunately the change in polarization occurs at low temperatures only. As already mentioned, very few room-temperature single-phase magnetoelectrics are available. Among them $BiFeO_3$ shows the highest ferroelectric polarization with a ferroelectric Curie temperature of $T_C \approx 1100\,K$ and a Neel temperature of $T_N \approx 640\,K$. Both ferroelectricity and anti-ferromagnetism occur in these systems [5.15–5.19].

Multiferroic systems where magnetism and ferroelectricity are strongly coupled can be classified according to the mechanism which breaks the spatial inversion symmetry and allows ferroelectricity: (1) spin order, (2) charge order and (3) orbital order. $LuFe_2O_4$ is a prototype for the second group [5.20]. Recently, Fe_3O_4 has also been

Table 5.3 Ferroelectric multiferroics.

$6s^2$ lone-pair electrons of Bi or Pb [5.22]	Pb(A,B)O$_3$ (A = Fe, Mn, Co ...; B = W, Nb ...)	$T_c(\text{FE}) \approx 300\text{–}400\,\text{K}$
		$T_N(\text{M}) \approx 100\text{–}200\,\text{K}$
	BiFeO$_3$	$T_c(\text{FE}) \approx 1123\,\text{K}$
		$T_N(\text{M}) \approx 650\,\text{K}$
	BiMnO$_3$	$T_c(\text{FE}) \approx 773\,\text{K}$
		$T_N(\text{M}) \approx 110\,\text{K}$
Geometric Ferroelectrics [5.23]	Hexagonal RMnO$_3$ (R = Y, Ho, Er, Tm, Yb, Lu, Sc)	$T_c(\text{FE}) \approx 900\,\text{K}$
		$T_N(\text{M}) \approx 80\text{–}120\,\text{K}$
Exchange striction (collinear spin order) [5.24]	RMn$_2$O$_5$ (R = Tb, Dy, Y, Ho, Er, Tm, Yb)	$T_c(\text{FE}) \approx 37\,\text{K}$
		$T_N(\text{M}) \approx 39\,\text{K}$
Charge ordering [5.25]	LuFe$_2$O$_4$; (Pr, Ca)Mn$_2$O$_7$; some low-dimensional organic anti-ferromagnets	
Spiral magnetic ordering [5.26]	TbMnO$_3$	$T_c(\text{FE}) \approx 27\,\text{K}$
		$T_N(\text{M}) \approx 41\,K$
	Ni$_3$V$_2$O$_8$	$T_c(\text{FE}) \approx 9\,\text{K}$
		$T_N(\text{M}) \approx 13\,\text{K}$
	MnWO$_4$	$T_c(\text{FE}) \approx 9\,\text{K}$
		$T_N(\text{M}) \approx 13\,\text{K}$

suggested to be a ferrimagnet with ferroelectricity being induced by Fe^{2+}/Fe^{3+} charge ordering [5.21].

Recently, a new mechanism for magnetoelectric coupling has been discovered in multiferroic FeTe$_2$O$_5$Br [5.27] where ferroelectric order and incommensurately amplitude-modulated long-range magnetic order appear simultaneously. The ferroelectric order involves polarizable Te^{4+} lone-pair electrons and magnetostriction. The long-range-ordered magnetic structure has no spatial inversion centre, thus removing the symmetry restriction for the coexistence of ferroelectric and magnetic order.

A memory element combining ferroelectric and ferromagnetic properties would permit an electric write-in operation. Combined with a magnetic read-out operation it would eliminate the need for the destructive read-out (and reset) in today's ferromagnetic RAMs, thus making possible fatigue-free memories. It would also allow control of the magnetic properties via dielectric ones and vice versa. Magnetoelectric capacitors and transistors could be built.

It should be noted that the small values of the magnetic-field-induced polarization or electric-field-induced magnetization make technical applications difficult. At present, more than 80 single phase compounds showing the magnetoelectric effect have been found. The Curie or Neel temperatures of most compounds are far below room temperature. The largest magnetoelectric effects have been observed for LiCoPO$_4$, TbPO$_4$ and yttrium iron garnet (YIG) films ($\sim$30 psm^{-1}).

5.2 The quadratic ME effect in Pb (Fe$_{1/2}$Nb$_{1/2}$)O$_3$

The temperature dependence of the d.c. magnetic susceptibility χ_m of Pb(Fe$_{1/2}$Nb$_{1/2}$)O$_3$ – abbreviated as PFN – in the vicinity of the ferroelectric transition T_c [5.19] is shown in Fig. 5.4. Figure 5.5 shows the relation between $\Delta \chi_m$ and P^2. $\Delta \chi_m$ is plotted against $(T-T_c)$ and $(T-T_c)^{1/2}$. The plot of $\Delta \chi_m$ against $(T-T_c)$ is linear, demonstrating that we deal here with a quadratic P^2M^2 coupling. In the case of linear coupling $\Delta \chi_m$ should be proportional to $(T-T_c)^{1/2}$.

An alternative strategy for obtaining large magnetoelectric effects is to introduce indirect coupling via strain between a ferromagnet and a ferroelectric. The coupling can be achieved via composites, laminates or epitaxial multilayers. A very large coupling of 4800 mV cm^{-1} Oe^{-1} is obtained in terfenol –D/PZT. Other two-phase systems with large coupling constants are listed in [5.28].

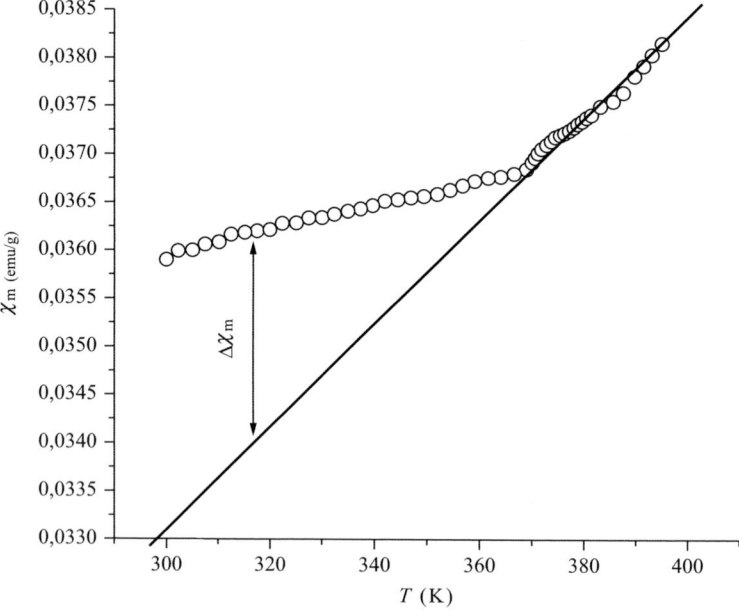

Fig. 5.4 Temperature dependence of the d.c. magnetic susceptibility of a fresh PFN polycrystalline sample in the vicinity of the paraelectric–ferroelectric transition T_C and its extrapolation from the paraelectric phase. The measurements were performed by a SQUID magnetometer in a magnetic field of 200 Gauss [5.18].

The quadratic ME effect in Pb (Fe$_{1/2}$Nb$_{1/2}$)O$_3$

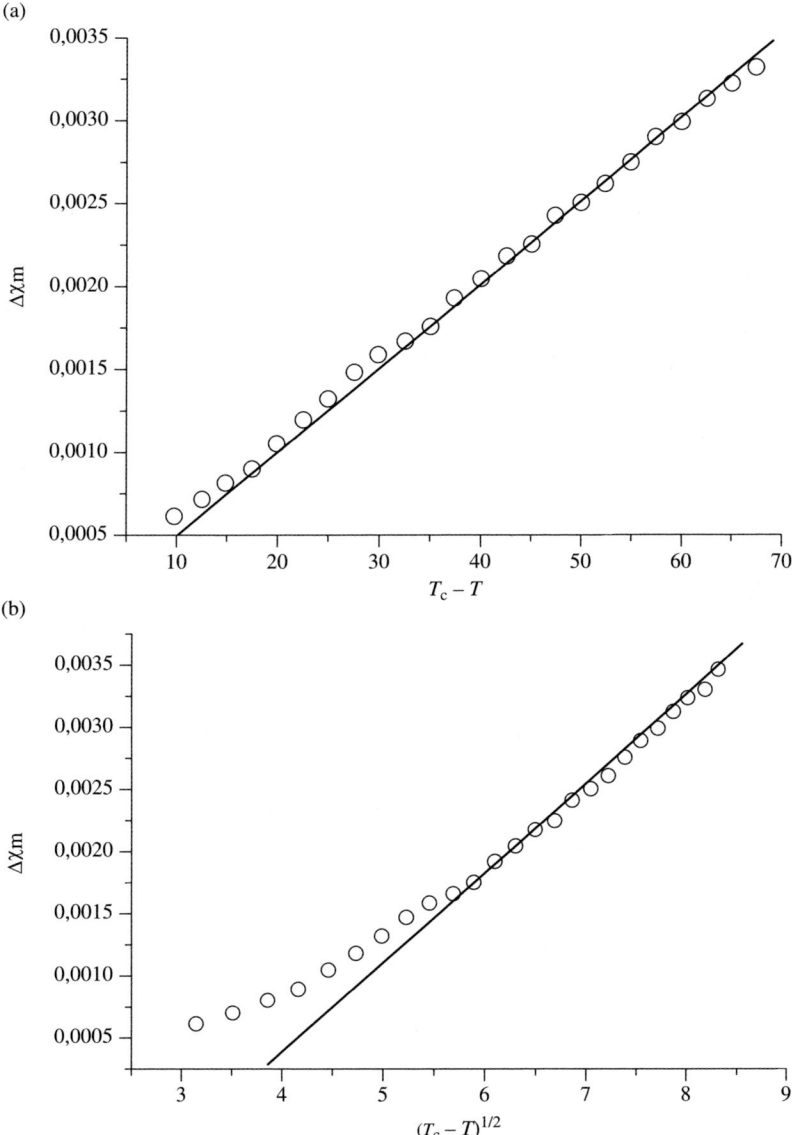

Fig. 5.5 The relation between $\Delta\chi_\mathrm{m}$ and P^2 below T_C. (a) Plot of $\Delta\chi_\mathrm{m}$ vs. $(T_\mathrm{c}-T)$ and (b) plot of $\Delta\chi_\mathrm{m}$ vs. $(T_\mathrm{c}-T)^{1/2}$ [5.18].

5.3 Ferroelectric polarization reversal by electric and magnetic fields

As mentioned, electric control of ferromagnetic domains and magnetic control of ferroelectric order are of great interest for the realization of a number of new devices. Control of magnetic behaviour with electric fields has been realized in a number of composite multiferroic thin films and heterostructures [5.29]. Magnetic control of ferroelectric properties has been achieved in, e.g., $TbMnO_3, Ni_3V_2O_8, CoCr_2O_4$ and $Ca_3Co_{2-x}Mn_xO_6$ where ferroelectricity is induced by magnetic structures breaking inversion symmetry [5.30–5.35].

Ferroelectric polarization reversal can be finely tuned by using both magnetic and electric fields in $CuCrO_2$ [5.34]. This is a triangular lattice anti-ferromagnet with the space group $R\bar{3}m$. It exhibits two magnetic transitions at $T_{N1} \approx 23.6\,K$ and $T_{N2} \approx 24.2\,K$. Ferroelectric polarization appears below T_{N1} where the onset of the out-of-plane 120° spin-chiral structure breaks the inversion symmetry [5.35]. Figure 5.6(a) shows the P–E hysteresis loops in various magnetic fields at 5 K, whereas Fig. 5.6(b) shows the P–H relation at various electric fields.

Figures 5.7 and 5.8 show the dielectric relaxation time as a function of applied magnetic field and the Vogel–Fulcher-type fitting of the dielectric relaxation frequency obtained on a $(PbFe_{0.67}W_{0.33}O_3)_{0.2}(PbZr_{0.53}Ti_{0.47}O_3)_{0.8}$ ('0.2PFW/0.8PZT') sample [5.36].

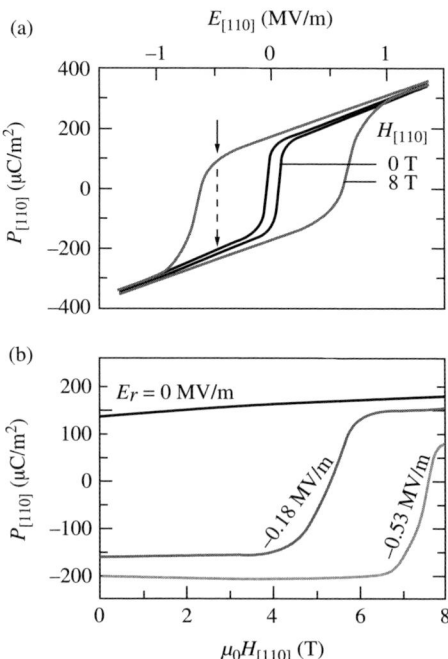

Fig. 5.6 (a) P–E hysteresis loops in various H at 5 K for $P \parallel H \parallel$ [110] in $CuCrO_2$. (b) P–H relation at various E at 5 K in $CuCrO_2$. T stands for Tesla. Following ref. [5.35].

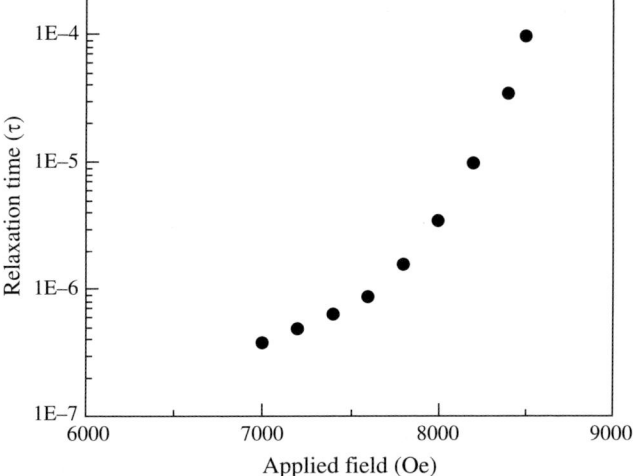

Fig. 5.7 Dielectric relaxation time at room T as a function of applied magnetic field in 0.2PFW/0.8PZT [5.36].

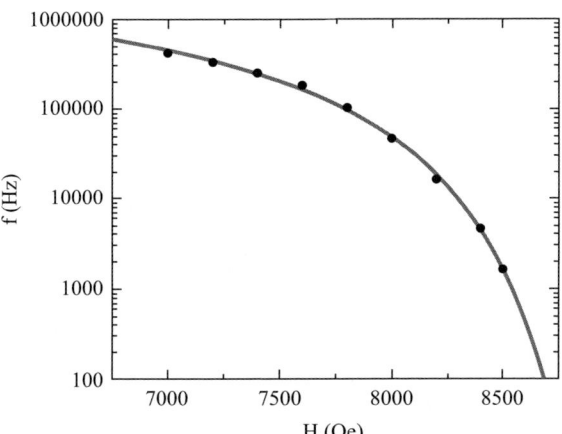

Fig. 5.8 Vogel–Fulcher-type fitting of the dielectric relaxation frequency of 0.2PFW/0.8PZT at room T [5.36].

5.4 The modified Vogel–Fulcher relation in external fields and the polar nanocluster size

Let us now try to understand the basic physical mechanism leading to the relation, describing the observed magnetic field modified Vogel–Fulcher (VF) law. The relaxation mechanism in relaxor ferroelectrics is growth, percolation and reorientation of polar nanoregions (PNRs) [5.37].

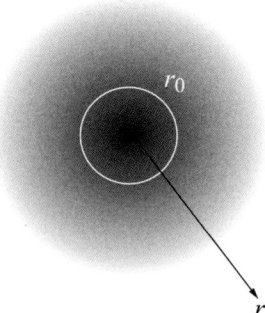

Fig. 5.9 Simplified model of a PNR showing an assumed polarization and electric-field distribution. The actual distribution is determined by compositional fluctuations.

One can assume that the local electric field $E_{\rm loc}$ inside the polarization cloud $P(r)$ falls off at large distances with distance as $1/r^3$ for a three-dimensional cluster (Fig. 5.9):

$$E_{\rm loc}(r) = E_0 \left(\frac{r_0}{r}\right)^3 \tag{11}$$

The size of the PNR is determined by the local electric field $E_{\rm loc}(r)$ acting on a dipole near its boundary. A dipole moment μ at a distance r has the potential energy $V_{\rm pot} = -\mu E_{\rm loc}(r)$. If the potential energy will be much larger than the thermal energy, $|V_{\rm pot}| > kT$, the dipole moment μ is strongly correlated with the PNR. The correlation radius $r_{\rm c}$ is obtained from

$$\mu E_{\rm loc}(r_{\rm c}) \approx kT \tag{12}$$

The correlation volume $v_{\rm c}$ then grows as $\sim 1/T$, $v_{\rm c} \propto 1/T$. This is so as $E_{\rm loc} \propto \frac{1}{r^3}$.

The correlation radius increases with decreasing temperature as

$$r_{\rm c} \propto 1/T^{1/3} \tag{13}$$

The volume fraction of PNRs increases until the percolation limit is reached at $T_{\rm p}$. It is $T_{\rm p}$ which here corresponds to the Vogel–Fulcher temperature T_0.

The local electric field in a PNR is modified in the presence of magnetoelectric (ME) coupling. Even though there is no linear ME coupling in PWF-PZT materials and thin films, a quadratic ME effect P^2M^2 is induced via electrostrictive and magnetostrictive strains.

The local electric field within each PNR now acquires an additional contribution proportional to M^2. This results in an increase in the size of the PNR and a resulting change in the relaxation time.

The main result is that the VF temperature becomes magnetic-field dependent.

In the distribution of τ values the longest relaxation time τ now diverges on a line of percolation critical points: $T = T_{\rm p}(H)$. It behaves as

$$\tau = \tau_0 \exp\left[\frac{U}{T - T_{\rm p}(H)}\right] \tag{14}$$

At constant temperature it varies with the magnetic field and behaves as

$$\tau = \tau_0 \exp\left[\frac{U}{H_p^2 - H^2}\right] \quad (15)$$

thus explaining the experimental result described in the preceding section. Here,

$$H_p^2 = \frac{T - T_0}{T_0} H_0^2 \quad (16)$$

where H_0^2 is a constant that depends on the electrostrictive and magnetostrictive terms as shown in the last section.

It should be noted that in a magnetically disordered relaxor ferroelectric and in bi-relaxors, the non-linear magnetoelectric effect leads not only to a magnetic- but also to an electric-field dependence of the freezing temperature. The dielectric relaxation time thus obeys a modified Vogel–Fulcher relaxation and diverges on a critical surface in the E, H, T space.

The longest dielectric relaxation time τ is thus obtained as:

$$\tau = \tau_0 \exp\left[\frac{U}{T - T_p(E, H)}\right], \quad T > T_p. \quad (17)$$

It is understood that $\tau \longrightarrow \infty$ for $T \leq T_p$. Here

$$T_p(E, H) = T_0 \left[1 + (\varepsilon_0 \chi_e)^3 K_1 E^2 - \varepsilon_0 \chi_e \chi_m^2 K_2 H^2\right] \quad (18)$$

where K_1 and K_2 are constants [5.37].

If there is no applied external electric or magnetic field, $T_p = T_0$ and the classical Vogel-Fulcher law is recovered. For $E = 0, H \neq 0$ we obtain as before

$$\tau = \tau_0 \exp\left[\frac{U_m}{H_c^2 - H^2}\right] \quad (19)$$

whereas we find for $E \neq 0, H = 0$

$$\tau = \tau_0 \exp\left[\frac{U_e}{E_c^2 - E^2}\right] \quad (20)$$

5.5 Theory of bi-relaxors

The multiferroic concept [5.5–5.7,5.21,5.38,5.39] should, in principle, also be applicable to compositionally disordered systems, such as dipolar glasses and relaxor ferroelectrics on the one hand, and spin glasses – or relaxor ferromagnets – on the other. Several examples of disordered systems of the above type have so far been described, for example, ME relaxors [5.40, 5.41], multiferroic relaxors [5.42], and ME multiglasses [5.43]. Here, a theoretical model of a system, which possesses both relaxor ferroelectric and relaxor ferromagnetic properties, to be referred to as bi-relaxor, will be described. In analogy to relaxor ferroelectrics [5.44], a relaxor ferromagnet [5.45, 5.46] is expected to have no long-range magnetic order, but exhibits a high value of the quasi-static magnetic response in a broad temperature range, strong frequency dispersion, as well as a Vogel–Fulcher (VF)-type freezing of the dynamic magnetic response. Therefore, in bi-relaxors these magnetic properties should coexist with the corresponding dielectric features. From a mesoscopic point of view, bi-relaxors are characterized by the existence of polar nanoregions (PNRs) and magnetic nanoregions (MNRs), which appear as a result of strong compositional fluctuations (Fig. 5.10). In principle, the above description applies to both single-phase and composite systems.

The possibility of direct magnetoelectric coupling between the PNRs and MNRs in bi-relaxors will not be discussed here. Since long-range order (LRO) is absent in a normal relaxor state by definition and the average spatial-inversion and time-reversal symmetry are not broken, the occurrence of a linear ME effect is not allowed. Strain modulation of the PNR–PNR and MNR–MNR interactions may, however, gives rise to indirect higher-order coupling between electric and magnetic degrees of freedom [5.7]. In the absence of piezoelectric and piezomagnetic coupling, the strain-mediated PNR–MNR interactions will be dominated by the electrostriction and magnetostriction coefficients, which are related to the strain (stress) derivatives of the PNR–PNR and MNR–MNR interaction parameters. The resulting PNR–MNR interaction term in the Hamiltonian is bilinear in the pseudospin coordinates of both the PNRs and MNRs, and its sign depends on the relative signs of the electrostriction and magnetostriction coefficients. A positive sign will result in a positive fourth-order ME effect, i.e. the dielectric constant will increase upon application of a static magnetic field. This opens up the possibility of a magnetic-field-induced phase transition from a relaxor state to a LRO ferroelectric state. The value of the critical magnetic field H_c can be

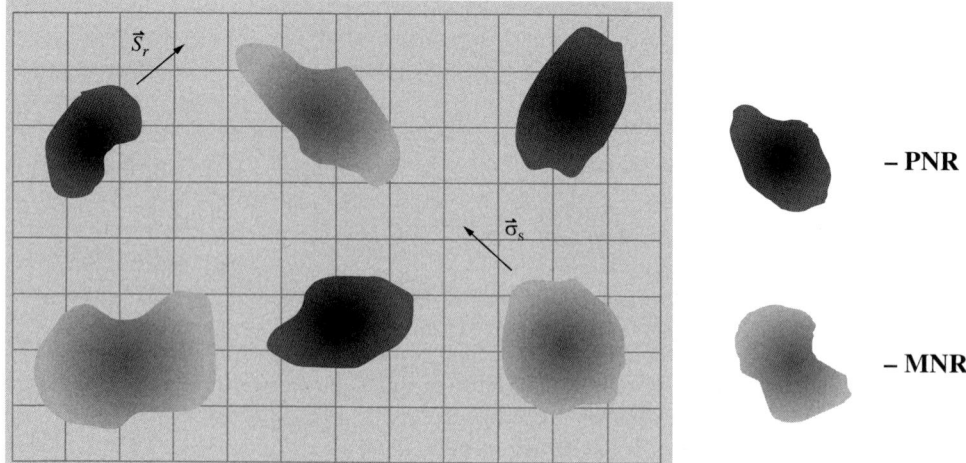

Fig. 5.10 Reorientable polar (PNR) and magnetic (MNR) nanoclusters in bi-relaxors.

estimated from the values of the striction coefficients. On the other hand, a negative sign means that the value of the dielectric constant should decrease with the magnetic field. This implies, for example, that if the system is originally in a long-range-ordered ferroelectric state, the field $H > H_c$ will tend to destroy this order, leading to a normal bi-relaxor state.

A similar reasoning can be applied to the free-energy density of bi-relaxors. As discussed in the preceding section the local electric field inside a PNR, which determines its correlation radius R_c and volume v_c [5.47], in a bi-relaxor acquires an extra contribution due to the magnetic field H, which either enhances or reduces v_c, depending on the coupling sign. The percolation temperature T_p, at which an infinite cluster of PNRs is formed, is also shifted by the field H. Thus, the VF temperature $T_0(H) = T_p$ becomes a function of the field, and the VF relaxation time diverges on a line of percolation critical points $T = T_0(H)$ in the H, T plane. This effect agrees with recent observations of magnetic-field-controlled relaxation of the dielectric polarization [5.36].

5.5.1 Spherical model of bi-relaxors

Following the spherical random-bond–random-field (SRBRF) model of relaxor ferroelectrics [5.48, 5.49] we assume that the PNRs are coupled through infinitely ranged Gaussian random interactions in the presence of random local electric fields, and an analogous assumption is made for the MNRs [5.37, 5.50–5.55].

As a straightforward generalization of the SRBRF model [5.55] we introduce two sets of pseudospins, $S_{ri}, r = 1, 2, \ldots, N_e$ and $i = x, y, z$ for PNRs, and $\sigma_{sj}, s = 1, 2, \ldots, N_m$ for the MNRs. The spherical conditions for the pseudospins are chosen as

$$\sum_{r=1}^{N_e} (\vec{S}_r)^2 = N_e, \quad \sum_{s=1}^{N_m} (\vec{\sigma}_s)^2 = N_m. \tag{21}$$

The Hamiltonian can be symbolically written as the sum

$$\mathcal{H}_0 = \mathcal{H}_e + \mathcal{H}_m \tag{22}$$

where $\mathcal{H}_e$ and $\mathcal{H}_m$ are the electric and the magnetic term

$$\mathcal{H}_e = -\frac{1}{2} \sum_{r \neq r'} J_{rr'} \vec{S}_r \cdot \vec{S}_{r'} - \sum_r \vec{f}_r \cdot \vec{S}_r - g_e \sum_r \vec{E} \cdot \vec{S}_r \tag{23a}$$

$$\mathcal{H}_m = -\frac{1}{2} \sum_{s \neq s'} K_{ss'} \vec{\sigma}_s \cdot \vec{\sigma}_{s'} - \sum_s \vec{h}_s \cdot \vec{\sigma}_s - g_m \sum_s \vec{B} \cdot \vec{\sigma}_s \tag{23b}$$

respectively. In these expressions, $J_{rr'}$ and $K_{ss'}$ are the PNR–PNR and MNR–MNR coupling parameters, respectively, and $\vec{f}_r$ and $\vec{h}_s$ are the corresponding random fields, while $\vec{E}$ and $\vec{B} = \mu_0 \vec{H}$ are external fields. Finally, g_e and g_m represent the average dipole moments of the PNRs and MNRs, respectively.

The random variables $J_{rr'}$ and $K_{ss'}$ obey two independent Gaussian distributions with mean values $[J_{rr'}]_{av} = J_0/N_e$ and $[K_{ss'}]_{av} = K_0/N_m$, and with variances

134   Magnetoelectric ferroelectrics

$[J_{rr'}^2]_{\text{av}}^{\text{c}} = J^2/N_{\text{e}}$ and $[K_{ss'}^2]_{\text{av}}^{\text{c}} = K^2/N_{\text{m}}$, respectively. The random fields have zero mean values and the respective variances are $[f_{ri}f_{r'j}]_{\text{av}} = \delta_{rr'}\delta_{ij}\Delta_{\text{e}}$ and $[h_{si}h_{s'j}]_{\text{av}} = \delta_{ss'}\delta_{ij}\Delta_{\text{m}}$. We choose $\Delta_{\text{m}} = 0$ and write $\Delta_{\text{e}} \equiv \Delta$.

Next, we assume that the PNR–PNR and MNR–MNR interactions depend on the components of the lattice strain (or stress) tensor u_{ij} (or X_{ij}). Expanding $J_{rr'}(\underline{X})$ to linear order in stresses we obtain

$$J'_{rr'}(\underline{X}) = J_{rr'}(0) + \sum_{ij} J_{rr',ij}{}^{(1)} X_{ij} + \cdots \tag{24}$$

with $J_{rr',ij}^{(1)} = \partial J_{rr'}(\underline{X})/\partial X_{ij}$, and similarly for $K_{ss'}(\underline{X})$. Since by assumption $J_{rr'}$ in Eq. (23a) is isotropic, we will limit ourselves to the isotropic part of the second term in Eq. (24). The random average of $J_{rr',ii}^{(1)}$ is parametrized by first introducing the stress derivatives of the average coupling parameter J_0, namely, $J_{0,iii}^{(1)} = \partial J_0/\partial X_{ii}$. Thus, the derivative of J_0 is simply related to the derivative of the field-cooled static dielectric susceptibility, which according to the SRBRF model is given by

$$\left(\chi_{\text{e}}^{-1}\right)_{ij} = \frac{1}{k\theta_{\text{e}}}\left(\frac{kT}{1-q_{\text{e}}} - J_0\right)\delta_{ij} \tag{25}$$

This result has been derived earlier by use of the replica formalism[5.48, 5.49], and contains the spherical glass order parameter q_{e} of the PNR subsystem, which depends on the parameters J, J_0, and Δ. The parameter $\theta_{\text{e}} = g_{\text{e}}^2/(v_{\text{e}}ke_0)$ plays the role of an effective Curie constant. It should be noted, however, that in relaxors a Curie–Weiss-type behaviour holds only approximately in the asymptotic regime $T \gg J/k$. An analogous expression is obtained for the magnetic case.

We can then apply the thermodynamic Maxwell relations between the electro-(magneto-)striction constants $Q_{\text{e},ijkl}$ and $Q_{\text{m},ijkl}$, and the stress derivatives of χ_{e}^{-1} [5.49–5.51] and χ_{m}^{-1}, respectively,

$$Q_{\text{e},ijkl} = -\frac{1}{2e_0}\left(\frac{\partial\left(\chi_{\text{e}}^{-1}\right)_{kl}}{\partial X_{ij}}\right)_T; Q_{\text{m},ijkl} = -\frac{\mu_0}{2}\left(\frac{\partial\left(\chi_{\text{m}}^{-1}\right)_{kl}}{\partial X_{ij}}\right)_T \tag{26}$$

where the limit $\vec{P} \to 0, \vec{M} \to 0$ is understood. The average of $J_{rr',ij}^{(1)}$ in Eq. (24) and similarly of $K_{rr',ij}^{(1)}$ then leads to the relations

$$J_{0,k} \cong 2e_0k\theta_{\text{e}}Q_{\text{e},ki}, K_{0,k} \cong 2\mu_0^{-1}k\theta_{\text{m}}Q_{\text{m},ki} \tag{27}$$

with $k, i = 1, 2, 3$, using the Voigt notation. In view of the presumed isotropy of J_0 the above expressions must be i independent. Thus, we can perform the averages over i, and introduce hydrostatic electro-(magneto-)striction coefficients $Q_{\text{e},h}, Q_{\text{m},h}$ defined by $Q_h = Q_{k1} + Q_{k2} + Q_{k3}$, which are k-independent. It follows that $J_{0,k}$ and $K_{0,k}$ are both k independent.

In deriving relations (27) we have ignored the effects of stress deformations on the random distribution widths J and K, since they are $O(1/N_e^2)$ and $O(1/N_m^2)$, respectively. The stress derivatives of θ_e, θ_m have been neglected assuming rigid PNRs and MNRs. Also, it should be noted that q_e, q_m do not depend on J_0, K_0 and hence on $\underline{X}$ for $\vec{P} = 0, \vec{M} = 0$ [5.48].

After adding the elastic energy $\mathcal{H}_{\text{elastic}} = (V/2) \sum_{ij=1}^{3} C_{ij}^{-1} X_i X_j$, where C_{ij}^{-1} is the elastic compliance tensor, the total Hamiltonian is minimized, thus eliminating terms linear in X. This generates a fourth-order PNR–MNR magnetoelectric coupling term of the following structure:

$$\mathcal{H}_{\text{me}} = -\frac{\Lambda_h}{2VN_eN_m} \sum_{r \neq r'} \sum_{s \neq s'} \left(\vec{S}_r \cdot \vec{S}_{r'}\right)(\vec{\sigma}_s \cdot \vec{\sigma}_{s'}) \tag{28}$$

The coupling parameter Λ_h is given by

$$\Lambda_h = 2\varepsilon_0 \mu_0^{-1} k\theta_e k\theta_m C_h Q_{e,h} Q_{m,h} \tag{29}$$

where $C_h = (1/9) \sum_{ij=1}^{3} C_{ij}$ is the bulk modulus.

The prefactor $(VN_eN_m)^{-1}$ in Eq. (28) ensures proper scaling of $\mathcal{H}_{\text{me}}$ with the system's size (for short-range interactions, the prefactor should be $O(l/V)$ only), Λ_h being intensive.

Stress mediated fourth-order PNR–PNR and MNR–MNR coupling terms are not given here since they represent a correction of the anharmonic part of $\mathcal{H}_0$, which has been ignored from the outset.

5.5.2 Static dielectric properties under constant magnetic field

In the following, we consider a system where the relaxor dielectric peak is well separated from the magnetic one and lies at a higher temperature, and no LRO is present. This corresponds to the case $J > K$ and $J_0 > K_0$, as well as $K_0 < K$ and $J_0 < J$. Furthermore, we assume that a constant magnetic field $\vec{B} = \mu_0 \vec{H}$ has been applied to the system. In discussing the effects of $\mathcal{H}_{\text{me}}$ on the time evolution of PNR pseudospins $\vec{S}_i$, we can effectively replace the magnetic part of the coupling by its average value $\sum_{s \neq s'} [\langle \vec{\sigma}_s \cdot \vec{\sigma}_{s'} \rangle]_{\text{av}}$. In the spirit of perturbation theory, the averages can be evaluated with respect to the initial Hamiltonian $\mathcal{H}_0$. This can be done exactly using the eigenstates and eigenvalues of the random matrix $K_{ss'}$ [5.52]. The result to order $O(1/V)$ is

$$\frac{1}{VN_m} \sum_{k \neq s'} [\langle \vec{\sigma}_s \cdot \vec{\sigma}_{s'} \rangle]_{\text{av}} = (v_m/g_m^2) \sum_i \chi_{m,ii}^2 B_i^2 \tag{30}$$

The dimensionless magnetic susceptibility $\chi_{m,ij}$ is defined through the phenomenological relation $M_i = \sum_j \chi_{m,ij} H_j$ and can be explicitly obtained from the magnetic version of Eq. (25). Since all averages have the same value, namely, $[\langle \sigma_{s,i} \rangle]_{\text{av}} \equiv <\sigma>$, we can write $M_i = (g_m/v_m) <\sigma>$, where $v_m \equiv V/N_m$ represents the effective average MNR volume, which is essentially a measure of the MNR concentration $c_m = 1/V_m$. The corresponding quantities for the PNRs are $v_e \equiv V/N_e$ and $c_e = 1/v_e$.

In the isotropic case the magnetic susceptibility tensor is isotropic, i.e. $\chi_{m,ij} = \delta_{ij}\chi_m$, where $\chi_m = \mu - 1$. Similarly, the dielectric susceptibility is $\chi_{e,ij} = \delta_{ij}\chi_e$ with $\chi_e = \varepsilon - 1$.

Finally, the coupling term (28) can be rewritten as

$$\mathcal{H}_{\mathrm{me}}(H) = -J_1(H) \frac{1}{2N_e} \sum_{r \neq r'} \vec{S}_r \cdot \vec{S}_{r'} \tag{31}$$

where $J_1(H)$ represents the shift of the mean PNR interaction J_0, i.e.

$$J_0(H) = J_0 + J_1(H) \tag{32}$$

for which we obtain with use of Eq. (29)

$$J_1(H) = \frac{\mu_0(\mu-1)^2}{k\theta_m} \Lambda_h H^2 \tag{33}$$

The sign of $J_1(H)$ depends on the signs of $Q_{e,h}$ and $Q_{em,h}$, and can be either positive or negative. Thus, if both signs are equal, $J_1(H)$ will be positive, implying that the magnetic field enhances the average PNR interaction. In this case, there is a possibility of a magnetic-field-induced phase transition from a relaxor state into a long-range-ordered ferroelectric state. In general, LRO exists if $J_0(H) \geq J_{0c} \equiv \sqrt{J^2 + \Delta}$ and $T \leq T_c(H)$, where

$$T_c(H) = J_0(H) \frac{J_0(H)^2 - J_{0c}^2}{J_0(H)^2 - J^2} \tag{34}$$

For $J_0 < J_{0c}$, a critical value of the magnetic field H_c exists such that for $H > H_c$ LRO is possible. Introducing $\delta J_0 \equiv J_{0c} - J_0$ and assuming $\delta J_0 > 0$ (i.e. no LRO in zero field), we find

$$H_c^2 = \frac{|\delta J_0|}{2e_0(\mu-1)^2 k\theta_e C_h |Q_{e,h} Q_{m,h}|} \tag{35}$$

If, however, the system in zero magnetic field is in a ferroelectric state corresponding to the case $\delta J_0 < 0$, and the sign of $J_1(H)$ is negative, then $J_0(H)$ will decrease with H until the critical field H_c is reached. At that point a phase transition from a long-range-ordered ferroelectric into a relaxor ferroelectric state occurs, and LRO is not possible for all $H > H_c$.

To estimate the possible value of H_c from Eq. (35) we use the following representative parameter values: $\mu = 5$, $C_h = 10^{11}\,\mathrm{J\,m^{-3}}$, $\theta_e = 10^5$ K, and $\delta J_0/k = 33$ K. Then, for $Q_{e,h}$ in the range 10^{-3}–$10^{-2}\,\mathrm{m^4 A^{-2} s^{-2}}$ and $Q_{m,h}$ between 10^{-15} and $10^{-14}\,\mathrm{m^2 A^{-2}}$, we find

$$0.43\,\mathrm{T} < \mu_0 H_c < 4.3\,\mathrm{T} \tag{36}$$

Thus, H_c lies in the experimentally accessible regime.

In Fig. 5.11 the static zero-field cooled dielectric susceptibility $\chi_{e,s}$, Eq. (25) is plotted as a function of temperature for several values of magnetic field H. The

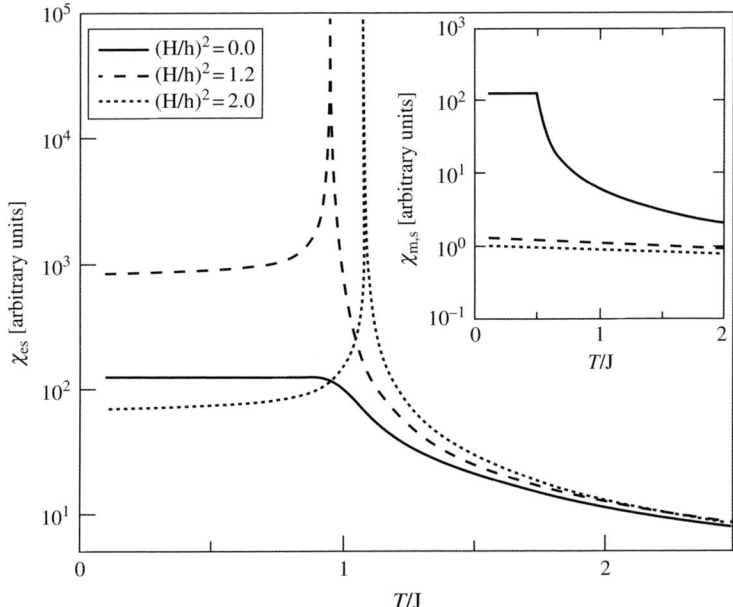

Fig. 5.11 Temperature dependence of the static zero-field-cooled dielectric susceptibility of a bi-relaxor for several values of magnetic field H/h, as indicated, where $h^2 = 0.9H_c^2$. Inset: Field-cooled static magnetic susceptibility $\chi_{m,s}$ for the same values of H/h.

parameter values are $J_0 = 0.9J$ or $\delta J_0/J = 0.1, \Delta/J^2 = 0.001, k\theta_e/J = 12.5$, and $J_1(H)/J = 0.1(H/h)^2$, where the random field fluctuation is $h^2 = 0.9H_c^2$. As expected, for $H > H_c$, i.e. $(H/h)^2 > 10/9$, the system develops LRO and $\chi_{e,s}$ shows a Curie–Weiss-type divergence at $T_c(H)$. The inset shows the magnetic-field-cooled susceptibility $\chi_{m,s}$, which monotonically decreases with H at fixed temperature. The cusp for $H = 0$ is due to the absence of magnetic random fields ($\Delta_m = 0$). The remaining parameters are $K/J = K_0/J_0 = 0.5$ and $\theta_m = \theta_e/4$.

5.5.3 Dynamic dielectric response

We now focus on the dynamic dielectric response of the bi-relaxor in the presence of a static magnetic field. The Langevin equations of motion for the PNR pseudospins are [5.52]

$$\tau_e \frac{\partial S_{ri}}{\partial t} = -\frac{\partial(\beta\mathcal{H}_S)}{\partial S_{ri}} - z_e(t)S_{ri}(t) + \xi_{ri}(t) \qquad (37)$$

where $\mathcal{H}_S = \mathcal{H}_e + \mathcal{H}_{me}(H)$ and $z_e(t)$ is a Lagrange multiplier enforcing the spherical condition (21) for PNRs. The Langevin forces $\xi_{ri}(t)$ formally obey the Einstein relations $\langle \xi_{ri}(t)\xi_{r'j}(t')\rangle_{av} = 2\tau_e\delta_{rr'}\delta_{ij}\delta(t-t')$. In the asymptotic regime $t \gg \tau_e$ [5.52], we obtain the following result for the complex linear dynamic dielectric susceptibility

$$\chi_e(\omega) = k\theta_e \frac{z(\omega) - r(\omega) - \beta J_0(H)}{\beta[J^2 + J_0(H)^2] - 2z(\omega)J_0(H)} \tag{38}$$

where $\chi_e(\omega)_{ij} = \chi_e(\omega)\delta_{ij}$ and we use the notation $z \equiv Z_e$, etc., with $z(\omega) = z - i\omega\tau/2$ and $r(\omega) = \sqrt{z(\omega)^2 - \beta^2 J^2}$.

In principle, the parameter z can be calculated from the static spherical condition (21) [5.52]. Alternatively, one can determine z from the static susceptibility $\chi_{e,s} = \chi_e(0)$, which is known from replica theory [5.48, 5.49], by inverting Eq. (38).

The relaxation time $\tau \equiv \tau_e$ is regarded as a phenomenological parameter. However, one may introduce a statistical distribution of $\ln\tau$ by writing

$$\langle \chi_e(\omega) \rangle_{av} = \int_0^\infty d(\ln\tau) f(\ln\tau) \chi_e(\omega) \tag{39}$$

and relate $\ln\tau$ to the dielectric response.

Empirically, the response of relaxors and dipolar glasses is characterized by a broad probability distribution of relaxation times. An approximate numerical scheme has been developed [5.53] to extract $f(\ln\tau)$ from the experimentally known response $\langle\chi_e(\omega)\rangle_{av}$. The longest relaxation time appearing in $f(\ln\tau)$ has been found to diverge according to the Vogel–Fulcher (VF) law

$$\tau = \tau \exp[U/(T - T_0)] \text{ for } T > T_0 \tag{40}$$

and $\tau = \infty$ for $T \leq T_0$, where T_0 is the VF or freezing temperature. The function $f(\ln\tau)$ can be expressed in terms of two probability distributions, i.e. $g(U)$ and $w(T_f)$ describing the distributions of barrier heights and freezing temperatures T_f, respectively. The result is:

$$f(\ln\tau) = \int_0^\infty dT_f w(T_f)(T - T_f) g\left[(T - T_f)\ln\tau\right]. \tag{41}$$

For the purpose of illustration we choose $g(U)$ as a Fröhlich box, $g(x) = 1/U_{max}$ for $0 < U < U_{max}$, and a triangular distribution $w(T_f) = (2/T_0)(1 - T_f/T_0)$ for $0 < T_f < T_0$ with parameter values $U_{max}/J = 11.0$, and $T_0/J = 1.0$. The calculated frequency dependence of the real and imaginary parts of $\langle\chi_e(\omega)\rangle_{av}$ is plotted in Fig. 5.12 at a fixed temperature $kT/J = 1.025$ for the same set of parameter values as in Fig. 5.11. However, the magnetic field H is now varied in a broader range. For $\delta J_0 > 0$ the maxima of the imaginary part of permittivity move towards higher frequencies on increasing H, but this effect seems rather weak.

We can also calculate the temperature dependence of the quasi-static zero-field-cooled susceptibility by choosing a very small value for the frequency. This is shown in Fig. 5.13 for $\omega\tau_0 = 10^{-8}$ and the same set of parameters and fields as in Fig. 5.11. At high temperatures the real part of the response agrees with the field-cooled response in Fig. 5.11, but for $T \to 0$ the zero-field-cooled response tends to zero, as required.

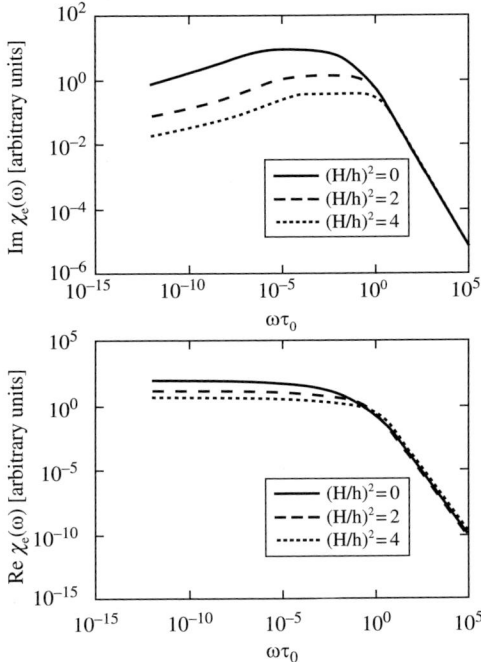

Fig. 5.12 Frequency dependence of the real (lower) and imaginary part (upper) of the dynamic dielectric susceptibility for several values of magnetic field H/h, calculated at fixed temperature $T/J = 1.025$.

5.5.4 Relaxation of dielectric polarization in magnetic field

We now discuss the effects of a magnetic field on the relaxation of PNRs due to the presence of a fourth-order magnetoelectric coupling between the PNRs and MNRs in a bi-relaxor. Our aim is to find a general expression for the internal electric field inside a PNR, which is modified by the presence of a magnetic field. We can then apply the mechanism of growth and percolation of PNRs [5.47] to derive a modified Vogel–Fulcher relaxation time.

Rather than follow the Hamiltonian approach used in the preceding sections, we start by writing down the phenomenological Landau-type free-energy density functional of a bi-relaxor in the absence of stress coupling:

$$\mathcal{F}_0 = \frac{1}{2e_0}(\chi_e^{-1})_{ij} P_i P_j + \frac{\mu_0}{2}(\chi_m^{-1})_{ij} M_i M_j + \frac{1}{4} b_e P^4 + \frac{1}{4} b_m M^4 + \cdots - E_i P_i - B_j M_j. \quad (42)$$

The inverse susceptibility tensors are explicitly given by Eq. (25). The anharmonic P^4 and M^4 terms have been added to formally ensure thermodynamic stability. Note that here are no direct ME terms in the above free energy.

Following the approach used in the section on The Modified Vogel–Fulcher Relation in External Fields and the Polar Nanocluster Size, we assume that $(\chi_e)^{-1}$ and $(\chi_m)^{-1}$

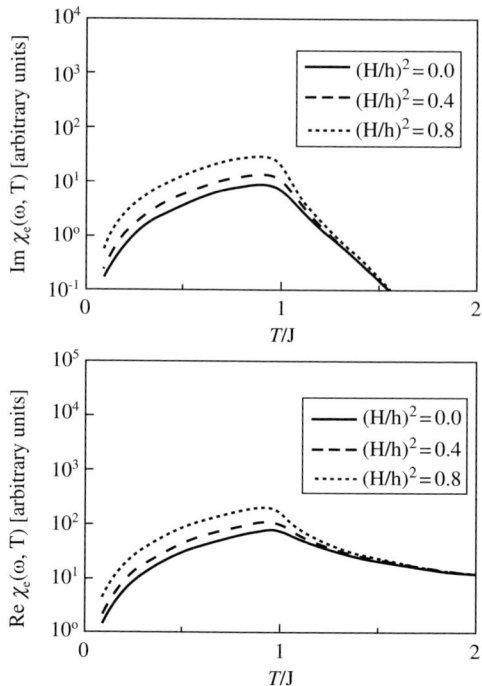

Fig. 5.13 Real (lower) and imaginary part (upper) of the quasi-static zero-field-cooled dielectric susceptibility for the same values of H/h as in Fig. 5.10, evaluated at a low frequency so that $\omega\tau_0 = 10^{-8}$.

are functions of the stress field $\underline{X}$ and expand them both to linear order in $\underline{X}$. We can again apply the Maxwell relations (26), and after including the elastic energy density $(1/2)\sum_{ij=1}^{3}(C^{-1})_{ij}X_iX_j$ and minimizing F_0 with respect to $\underline{X}$, we generate a new $\sim P^2M^2$ term, namely,

$$\mathcal{F}_1 = -\frac{1}{2}\lambda_{ij}P_i^2M_j^2 \qquad (43)$$

where the indirect fourth-order ME coupling constant is

$$\lambda_{ij} = 2C_{kl}Q_{e,ki}Q_{m,lj} \qquad (44)$$

By averaging λ_{ij} over all i,j we obtain its average value $\lambda_h = 2C_hQ_{e,h}Q_{m,h}$, and by comparing with Eq. (29) we see that $\lambda_h = \Lambda_h\mu_0/(e_0k\theta_e k\theta_m)$. This establishes a connection between the fourth-order coupling in the Hamiltonian and the corresponding coefficient in the free-energy density. The value of λ_h can be estimated by using the same parameter values as in Eq. 36, yielding $|\lambda_h|$ roughly between 2×10^{-7} and 2×10^{-5} m^3 s^{-1}V A^{-3}.

The macroscopic electric field is obtained from the equilibrium condition $\partial(F_0 + F_1)/\partial P_i = 0$ and is given by

$$E_i = \frac{1}{e_0}\chi_e^{-1}P_i(1 - e_0\chi_e\lambda_{ij}M_j^2) \qquad (45)$$

Since PNRs are mesoscopic objects, we expect that this relation is also valid inside the polarization cloud of a PNR. Following the example of relaxor ferroelectrics [5.47], we assume that the polarization inside an isolated PNR falls off with distance as $\vec{P}(r) = \vec{P}_0(r_0/r)^3$, and that the local field $\vec{E}(r)$ is proportional to $\vec{P}(r)$. In the present case, $\vec{P}(r)$ acquires an additional contribution due to ME coupling and $\vec{E}(r)$ becomes

$$E_{\text{loc},i} = \frac{\phi}{3e_0}P_i(1 - e_0\chi_e\lambda_{ij}M_j^2) \qquad (46)$$

where ϕ is the local field correction factor. The extra term in parentheses can be interpreted as the contribution due to stress field fluctuations, which are induced by the magnetic field via magnetostriction. Since magnetic LRO is absent and the temperature is far above the magnetic relaxor peak, the system is effectively in a (super)paramagnetic regime. Thus, we can write $M_j^2 = \chi_m^2 H_j^2$ and for an isotropic case we also have $\lambda_{ij} = \lambda_h \delta_{ij}$.

From this point on we can follow the reasoning used in Ref. [5.47] and consider the electrostatic energy $U_{\text{dip}} = -\vec{\mu} \cdot \vec{E}_{\text{loc}}$ of a virtual electric dipole $\vec{\mu}$ at distance r from the PNR centre. The correlation radius r_c is determined by the condition that $|U_{\text{dip}}|$ should be balanced against the thermal fluctuation energy $\sim kT$. Thus, the correlation volume behaves as $v_c = v_0 T_d^*(H)/T$, where v_0 is the 'core' volume of PNR at some reference temperature $T_d^*(H)$. In the present case, we can write $T_d^*(H) = T_d^*(0)[1 - e_0\chi_e\lambda_h(\chi_m)^2 H^2]$. As the temperature is lowered, v_c increases and so does the volume fraction η occupied by PNRs. As η reaches the percolation threshold η_p an infinite cluster of PNRs is formed. This occurs at the percolation temperature $T_p = 4\pi n T_d^*(H)/3\eta_p$, where n is the volume concentration of PNRs and η_p is around ~ 0.3 in $d = 3$ dimensions and $\eta_p \sim 0.6$ for $d = 2$ [5.54]. The dielectric relaxation time τ then behaves as

$$\tau = \tau_0 \exp\left(\frac{U}{T - T_p(H)}\right) \qquad (47)$$

which has the same general form as the VF equation (40) with $T_0 = T_p(H)$. It follows that the relaxation time diverges on a line of percolation critical points $T = T_p(H)$ in the H, T plane, where

$$T_p(H) = T_0[1 - \text{sgn}(\lambda_h)(H/H_0)^2] \qquad (48)$$

with $H_0^2 = 1/(|\lambda_h|(\chi_m)^2)$, or explicitly,

$$H_0^2 = \frac{1}{2e_0\chi_e\chi_m^2 C_h |Q_{e,h}Q_{m,h}|}, \qquad (49)$$

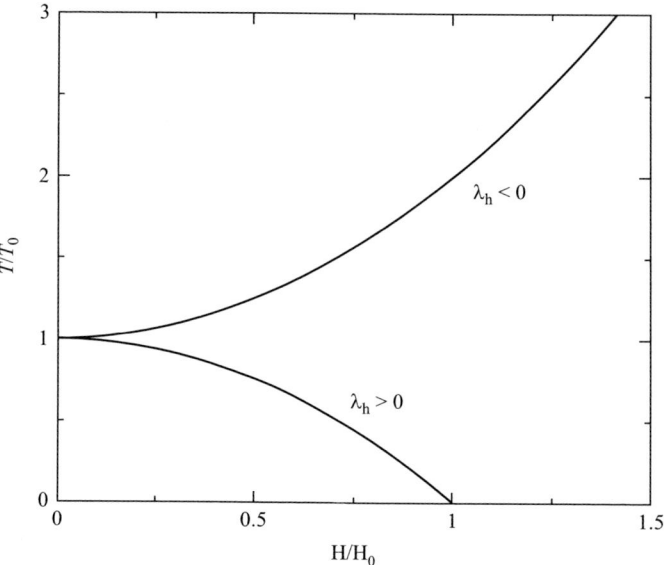

Fig. 5.14 Lines of percolation critical points in the H, T plane corresponding to positive and negative values of the coupling parameter λ_h. The Vogel–Fulcher relaxation time diverges on approaching the corresponding line from above.

The sign of λ_h in Eq. (48) depends on the signs of $Q_{e,h}$ and $Q_{m,h}$, i.e. $\mathrm{sgn}(\lambda_h) = +1$ if the two signs are equal, and -1 if they are opposite. The two cases are illustrated schematically in Fig. 5.14.

Equation (49) is consistent with the approximate relation (35) since in view of Eq. (25) one can write $\chi_e \cong k\theta_e/|J - J_0|$.

Let us consider the case $\mathrm{sgn}(\lambda_h) = -1$. The percolation temperature then increases with magnetic field, and at fixed temperature the relaxation time diverges according to

$$\tau = \tau_0 \exp\left(\frac{U_1}{H_p^2 - H^2}\right) \quad (50)$$

where $U_1 = UH_0^2/T_0$ and the critical field H_p is given by

$$H_p(T)^2 = \frac{T - T_0}{T_0} H_0^2 \quad (51)$$

We can estimate the value of H_0 by assuming $\chi_e \sim 3 \times 10^3$ near $T \approx J/k$ and $\chi_m \cong 4$, and by using the same values of $C_h, Q_{e,h}$, and $Q_{m,h}$ as in Eq. (36). We find $0.43\,\mathrm{T} < \mu_0 H_0 < 4.3\,\mathrm{T}$, in agreement with Eq. (36). The value of the critical field H_{rmp} at any temperature can then be obtained from Eq. (51) if the zero field value of the VF temperature T_0 is known.

A mesoscopic spherical random-bond–random-field (SRBRF) model of a bi-relaxor is thus formulated, which shows both relaxor ferroelectric and relaxor ferromagnetic behaviour. The underlying physical scenario is based on the simultaneous presence of polar nanoregions (PNRs) and magnetic nanoregions (MNRs). The PNR–PNR and MNR–MNR random inter-actions are modulated by the lattice stress tensor, thus giving rise to fourth-order PNR–MNR coupling terms. Using thermodynamic Maxwell relations the strength of the PNR–MNR interaction can be expressed in terms of the electrostriction and magnetostriction coefficients and elastic constants. Depending on the relative signs of these coefficients, a long-range-ordered ferroelectric state can be induced by the magnetic field larger than a critical field H_c. Estimates for the possible range of values of H_c are given.

The probability distribution of the dipole moments of PNRs and MNRs is at present unknown. As shown earlier for a relaxor ferroelectric, [5.49], the choice of an asymmetric Gaussian distribution leads to the spherical condition for the associated pseudospin variables. The same approach can be applied to the corresponding magnetic degrees of freedom for which the spherical condition has been originally introduced, mainly because it leads to an exactly solvable model. The infinite range interactions of the Sherrington–Kirckpatrick-type are a reminder of the long-range nature of the PNR–PNR and MNR–MNR interactions. Their principal advantage is that the number of physical parameters is kept at a minimum of two for each subsystem, to which one adds only the strength of the random electric fields. One should, of course, be aware of the possible limitations of this approach and more realistic interaction models need to be considered.

The dynamics of the PNR subsystem has been studied by means of the Langevin equations of motion, which had been used earlier both for Ising and spherical spin glasses, as well as for relaxor ferroelectrics. Quenched randomness is taken into account on two levels, namely, first by performing random averages over the eigenvalues of the random interaction matrix and over random fields, and secondly by assuming a broad probability distribution of relaxation times in the spirit of dynamic heterogeneity that is assumed to be applicable to relaxors and bi-relaxors alike.

Using a standard phenomenological Landau-type free-energy density functional of a bi-relaxor the polarization–magnetization or magnetoelectric (ME) coupling term of $\sim P^2 M^2$ type has been derived. It has been shown that the local internal field of a PNR acquires an extra component due to the ME coupling. According to the model of growth and percolation of PNRs in relaxors, the correlation radius r_c and volume v_c of PNRs is reduced (enhanced) by the magnetic field, again depending on the sign of the ME coupling. As the volume fraction of PNRs reaches the percolation threshold η_p, an infinite cluster of PNRs appears and its reorientation time diverges asymptotically according to the Vogel–Fulcher (VF) relation. The VF freezing occurs on a line of percolation critical points $T = T_p(H)$ in the H, T plane, where $T_p(H)$ corresponds to the VF temperature in a magnetic field [5.37].

6
Relaxor ferroelectrics

6.1 Introduction

Relaxor ferroelectrics (abbreviated relaxors) [6.1, 6.2] are site- and charge-disordered crystals, usually solid solutions. They are characterized by the presence of:

a) a broad distribution of relaxation times;
b) polar nanoclusters [6.3];
c) a highly polarizable neutral matrix.

The mobile polar nanoclusters [6.3] are due to compositionally or frustration-induced charge disorder that in turn induces random fields [6.4]. The clusters exist as islands (Fig. 6.1(a)) in a highly polarizable neutral matrix. In $PbMg_{1/3}Nb_{2/3}O_3$ (PMN) the polar clusters are above 400 K typically 2–3 nm in size and grow to 10 nm at ∼5 K. They occupy ∼20% of the crystal volume [6.4]. They are thus much smaller than typical FE domains. At a given temperature (T) the clusters are roughly of the same size (Fig. 6.1(a)) [6.4–6.6].

The important point is that they are embedded in a highly polarizable matrix [6.5, 6.6] with a high dielectric constant such as that provided by ABO_3 perovskite oxides (Fig. 6.2(a)). It is this matrix that determines whether the mobile polar clusters will at lower T grow in size to percolate the whole sample and form a ferroelectric or whether they will stay small enough to form a relaxor state down to $T \to 0$. Burns and Dacol [6.3] were the first to suggest from measurements of the optic index of refraction that randomly oriented polar nanoclusters appear in PMN below $T_B = 620$ K. Another signature of polar nanoclusters is the presence of elastic diffuse scattering which is temperature dependent but persists down to 10 K, i.e. down to the lowest temperatures studied. In relaxor ferroelectrics the nanoclusters percolate at the percolation temperature T_p

$$T_B \gg T_p. \qquad (1)$$

The Burns temperature T_B seems to depend on the method of measurement.

The size of the nanoclusters should be smaller than 500 Å so that they cannot be seen on the profile of the X-ray diffraction lines.

In addition to the presence of polar nanoclusters [6.8], the main differences between relaxors and classical ferroelectrics are:

(i) In relaxors the static zero-field-cooled dielectric constant monotonically increases with decreasing temperature and slowly saturates at low temperature. There is a splitting between the field-cooled and the zero-field-cooled

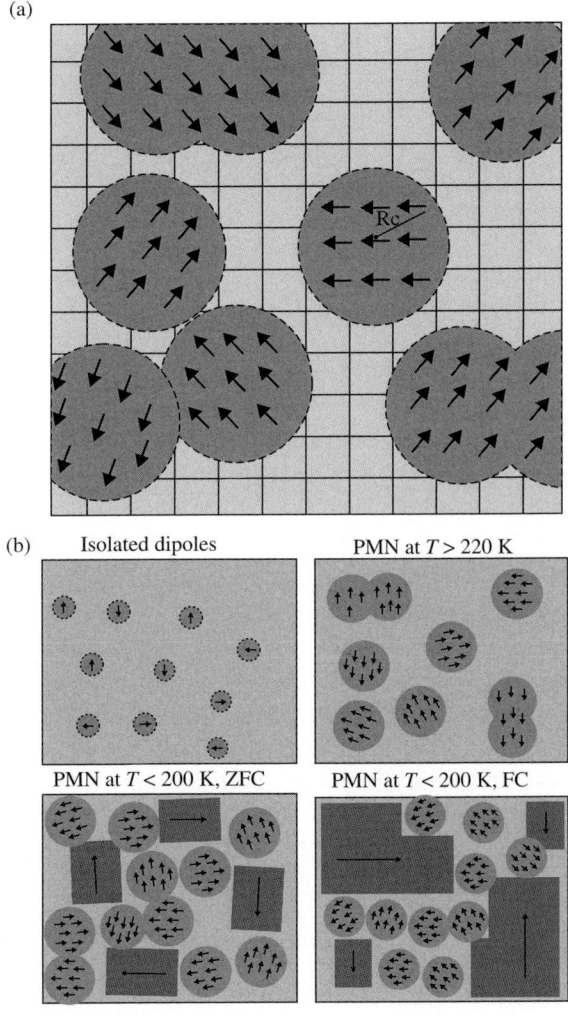

Fig. 6.1 (a) Randomly distributed nanoclusters in a polarizable host lattice. At low temperature the nanodomains grow and may coalesce into FE domains. (b) Schematic representation of the transition from the nanodomain (cubic) state to the macrodomain FE (trigonal) state [6.5, 6.6].

linear dielectric constant below the freezing temperature (Fig. 6.2(b)). This effect is dynamic and is absent in classical ferroelectrics. Another characteristic feature is the presence of a remanent polarization in a field-cooled zero-field-heated experiment (Fig. 6.2(b)).

(ii) Relaxors are characterized by a strong frequency dispersion in the radiofrequency region below the freezing transition (Fig. 6.3). The frequency dispersion is due to the slowing down of the dynamics and freezing of the fluctuations of

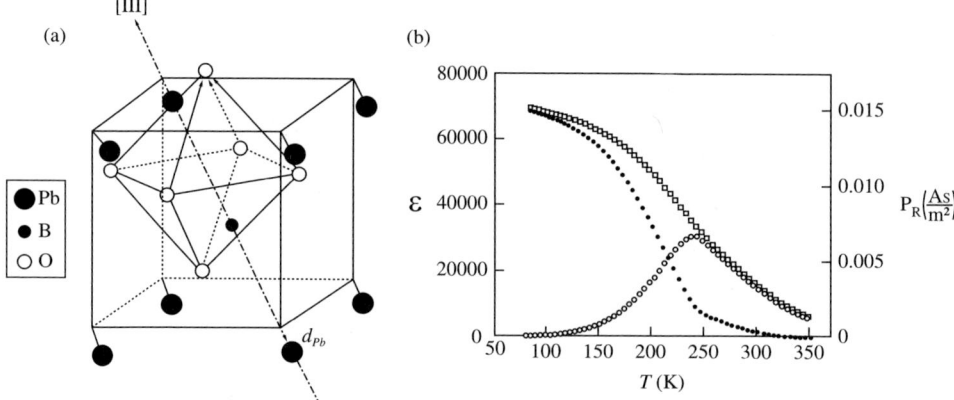

Fig. 6.2 (a) Displacements of the ions along the $\langle 111 \rangle$ direction for the rhombohedral distortion in PMN [6.2, 6.6]. (b) Temperature dependence of the field-cooled (□) and zero-field-cooled (○) quasi-static dielectric constant ε' of PMN measured perpendicularly to the [001] plane. The remanent polarization P_R after field cooling and zero field heating (●) is also shown [6.7].

polar clusters. Classical ferroelectrics, on the other hand, show a sharp, narrow dielectric response, which is frequency independent in the radiofrequency region.

(iii) The key to the understanding of relaxors is their local structure. At any temperature between 1000 K and 4 K the average macroscopic symmetry of PMN is cubic (m3̄m) [6.2]. This is true even on the micrometric scale. On the nanometric scale, on the other hand, the local structure is trigonal(3m) and different from the average structure.

(iv) Classical ferroelectrics show a change in crystal symmetry on cooling through the paraelectric–ferroelectric transition T_C, whereas there is no macroscopic symmetry change on cooling through the freezing transition in relaxors. The ferroelectric transition is thus a structural transition, whereas there is no structural transition in a relaxor.

(v) There is no specific-heat anomaly in relaxors around the maximum of the dielectric constant in zero electric field. On cooling in an electric field higher than the critical value E_c, a sharp heat-capacity anomaly occurs demonstrating the occurrence of a phase transition and the onset of long-range order.

(vi) As shown by recent neutron-scattering measurements, in addition to the zone-centre soft TO mode, soft modes exist in PMN in zero electric field at the R and M points of the Brillouin zone that disappear when PMN becomes a well-defined ferroelectric by doping. This shows that competing ferroelectric and anti-ferroelectric interactions are responsible for the glassy state.

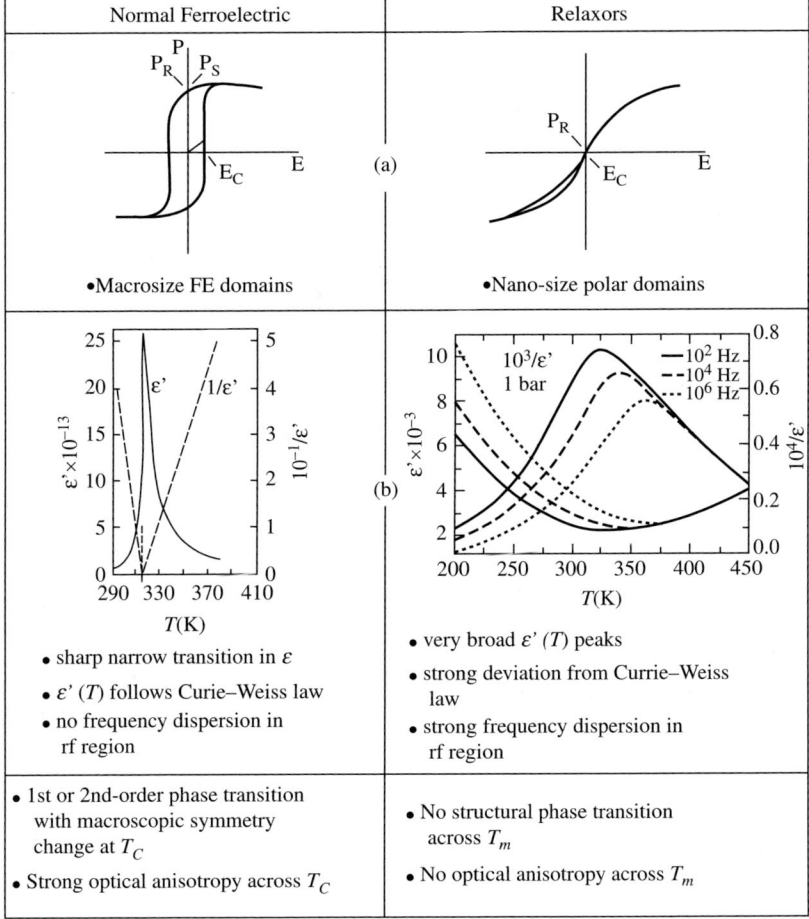

Fig. 6.3 Contrast between the properties of normal ferroelectrics and relaxors.

(vii) The temperature dependence of the dielectric response of a classical ferroelectric above T_C obeys a Curie–Weiss law $\varepsilon = C/(T-T_C)$, whereas there is a strong deviation from this law in relaxors.

(viii) Another important point is that in relaxors [6.5, 6.9] a ferroelectric state can be induced by an electric field, $E > E_c$, whereas this is not possible in dipolar glasses.

(ix) In contrast to classical ferroelectrics, there is no macroscopic spontaneous polarization in relaxors below T_C (Fig. 6.3). Rather, we deal with a polarization distribution function $W(p)$ with zero average value [6.8].

(x) The heat conductivity of relaxors is glass-like, as described later [6.13].

Relaxors thus provide a conceptual link between classical ferroelectrics and dipolar glasses.

6.2 Specific heat of relaxors

The cubic high-temperature paraelectric phase of PMN (space group $Pm\bar{3}m$) transforms below the Burns temperature, $T \approx 620\,\text{K}$, to the relaxor phase, which is a heterostructure composed of two phases, i.e. ferroelectric nanoregions (PNR) with a space group $R3m$, formed in a paraelectric matrix ($Pm\bar{3}m$). At zero electric field no anomaly in the heat capacity has been observed [6.9] in the temperature range 170–250 K where a broad temperature-dependent dielectric maximum is found [6.10, 6.11]. If there were a phase transition in this range at zero electric field, an anomaly in the specific heat would be observed even if the transition is smeared by random field interactions [6.8, 6.12]. Since there is no specific-heat anomaly in relaxors around the maximum of the dielectric constant no long-range order is formed and the ferroelectric 'domain state model' smeared by random fields should be dismissed [6.3].

There are two other things to be mentioned:

(i) There is a weak and broad excess specific heat ΔC_P in PMN below the Burns temperature that coincides with the formation and growth of the polar nanoregions. The excess entropy ΔS is here about 20% of $\Delta S = R\ln 8$, i.e. of the expected value for the ordering of the Pb ionic shifts between the eight equivalent $<111>_\text{cub}$ orientations. The profile analysis of the powder neutron diffraction data [6.13] of PMN revealed that the maximum volume ratio of the PNR to the whole crystal is also about 20%. Figure 6.4. shows the temperature dependence of the excess entropy $\Delta S/R\ln 8$ and the volume fraction of the polar nanoclusters V_PNR/V in PMN. According to this model, PMN is a glass even at very low temperatures with a residual entropy of $\approx 0.8 R\ln 8$.

(ii) On cooling in an electric field higher than the critical value $E_\text{c} \approx 1.7\,\text{kV/cm}$ a ferroelectric transition takes place in PMN. A heat-capacity anomaly occurs at $T_\text{FC} = 223\,\text{K}$ upon field cooling in a field of $E = 2.3\,\text{kV/cm}$ [6.13]. This is in sharp contrast to the zero-field case where no anomaly is observed in the range 170–250 K. For zero field heating after field cooling (ZFHaFC) the heat-capacity anomaly takes place around 210 K (Fig. 6.5) and reaches $\sim 6\,\text{J/(mol\,K)}$,

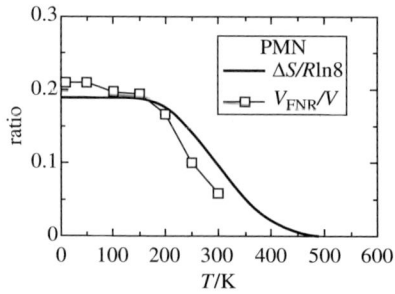

Fig. 6.4 Temperature dependence of the ratio of the excess entropy to $R\ln 8$–$\Delta S/R\ln 8$ – and the volume ratio of the PNR to the whole crystal –V_PNR/V– of PMN.

$\vec{S}_i$ has a large number of equilibrium orientations lying on a spherical shell of radius $|\vec{S}_i|$. Here, N is the number of reorientable nanoclusters that is subject to the spherical condition

$$\sum_{i=1}^{N} \vec{S}_i^2 = 3N \quad (3)$$

The pseudospin SRBRF [6.8, 6.14, 6.15] Hamiltonian on a rigid lattice in the presence of an external field $\vec{E}$ is thus given by

$$\mathcal{H}_S = -\frac{1}{2} \sum_{i,j} J_{ij}^{(b)} \vec{S}_i \vec{S}_j - \sum_i \vec{h}_i \vec{S}_i - g\vec{E}_i \sum_i \vec{S}_i \quad (4)$$

Here, g is an effective dipole moment corrected by the appropriate local field factor and $\vec{E}$ is the applied field. The bare interactions $J_{ij}^{(b)}$ are randomly frustrated and infinitely ranged. They are described by a Gaussian distribution with mean value $J_0^{(b)}/N$ and variance $(J^{(b)})^2/N$. It should be noted that in relaxors $J_0 \leq J$, whereas in dipolar glasses $J_0 << J$. The random fields $\vec{h}_i$ are similarly normally distributed but with zero mean value and non-zero second cumulant

$$[h_{i\mu} h_{j\mu}]_{\text{av}}^C = \delta_{ij} \delta_{\mu\nu} \Delta \quad (5)$$

6.4 Pseudospin phonon coupling

The total Hamiltonian of the coupled SRBRF – phonon system can be written as:

$$\mathcal{H} = \mathcal{H}_S + \mathcal{H}_L + \mathcal{H}_{SL} \quad (6)$$

Here, $\mathcal{H}_L$ represents the lattice Hamiltonian

$$\mathcal{H}_L = \frac{1}{2} \sum_{kp} \left[\omega_{\vec{k}p}^2 Q_{\vec{k}p} Q_{-\vec{k}p} + P_{\vec{k}p} P_{-\vec{k}p} \right]. \quad (7)$$

The pseudo-spin polar phonon coupling is

$$\mathcal{H}_{SL} = \sum_{kp} Q_{-\vec{k}p} \vec{\gamma}_{\vec{k}p} \vec{S}_{\vec{k}} \quad (8)$$

The displaced phonon coordinates are determined by the equilibrium condition $\delta H / \delta Q_{-\vec{k}p} = 0$.

The parameters $J_0^{(b)}$ and $J^{(b)}$, respectively, are now replaced by J_0 and $J : J_0^{(b)} \to J_0$, $J^{(b)} \to J$. After a lengthy calculation one finds

$$J_0 = J_0^{(b)} + \sum_{\gamma} \frac{|\vec{\gamma}_{0p}|^2}{\omega_{0p}^2} - \frac{1}{N} \sum_{kp} \frac{|\vec{\gamma}_{\vec{k}p}|^2}{\omega_{\vec{k}p}^2} \quad (9)$$

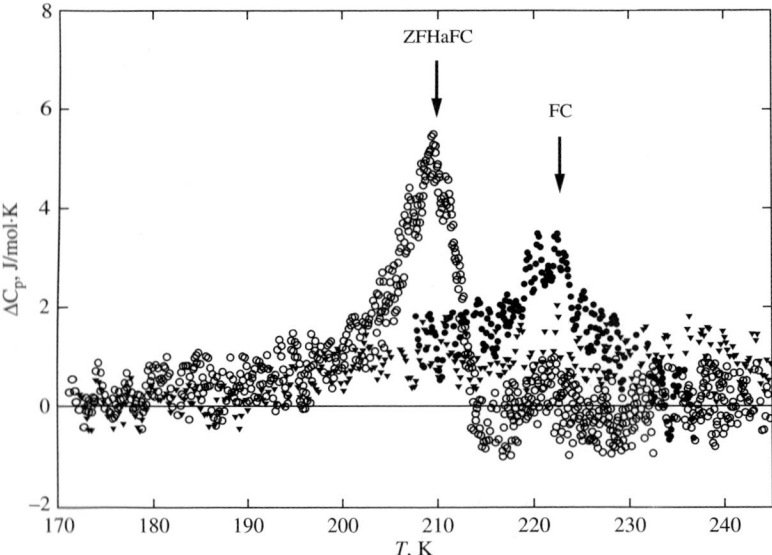

Fig. 6.5 Temperature dependence of the anomalous heat capacity of PMN upon field cooling (FC) and zero field heating after field cooling (ZFHaFC) [6.13].

i.e. ~5% of the lattice heat capacity. The narrow ΔC_P peak confirms the first-order nature of this ferroelectric relaxor transition. The small value of ΔS at the field-induced phase transition shows that no significant change in V_{PNR} occurs at this transition.

6.3 The rigid spherical random-bond–random-field (SRBRF) model

Relaxors can be, to a first approximation, described by the rigid-spherical random-bond–random-field (SRBRF) model [6.8]. The model is based on randomly competing ferroelectric (FE) and anti-ferroelectric (AFE) interactions between reorientable polar nanoclusters of different sizes, orientations and subject to different random fields (Fig. 6.1). It has been already described in part in the chapter on bi-relaxors. The various polar units in the clusters may also vary in orientation and size. Since the polar clusters – designated as pseudospins – which are the basic reorientable dipoles in the structure, vary in their size as well as orientation, the system cannot be described by an ordering field of fixed length as in Ising systems. Thus, we have to introduce a dimensionless continuous pseudospin order parameter field $\vec{S}$ related to the dipole moment of the i th polar cluster

$$-N^{1/2} \leq S_{i\mu} \leq N^{1/2}, \mu = x, y, z \tag{2}$$

and

$$J^2 = (J^{(b)})^2 + \frac{1}{N}\sum_k \left[\sum_p \frac{|\gamma_{\vec{k}p}|^2}{\omega_{\vec{k}p}^2}\right] - \left[\frac{1}{N}\sum_{kp}\frac{|\gamma_{\vec{k}p}|^2}{\omega_{\vec{k}p}^2}\right]^2 \qquad (10)$$

Here, p is the branch index and $\vec{k}$ is the wavevector.

6.5 The SRBRF phase diagram

The temperature-coupling parameter J_0 phase diagram for a relaxor according to the SRBRF model [6.8] is shown in Fig. 6.6.

The polarization properties are given by the polarization distribution function $W_{(p)}$ [6.8]. The polarization is determined by the first moment of $W_{(p)}$

$$P = \int_{-\infty}^{+\infty} W_{(p)} p\, dp \qquad (11)$$

whereas the second moment determines the Edwards–Anderson order parameter

$$q = \int_{-\infty}^{+\infty} W_{(p)} p^2\, dp \qquad (12)$$

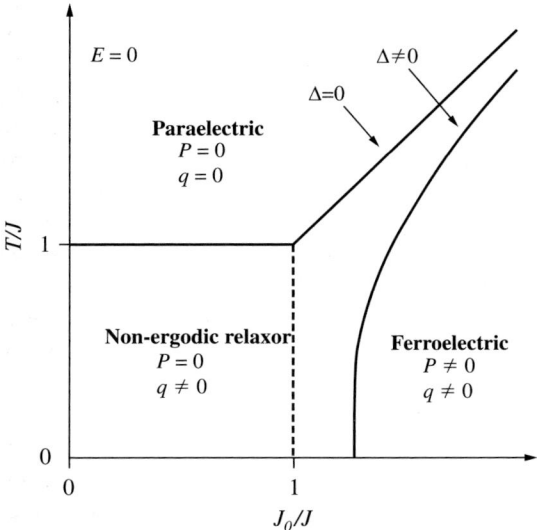

Fig. 6.6 Generalized temperature-coupling parameter J_0 phase diagram deduced from the spherical random-bond–random-field model of a relaxor [6.16].

This is equivalent to

$$P_\mu = \frac{1}{N}\sum_{i=1}^{N}\langle S_{i\mu}\rangle \qquad (13)$$

and

$$q_\mu = \frac{1}{N}\sum_{i=1}^{N}\langle S_{i\mu}\rangle^2 \qquad (14)$$

Within the model there are three phases: (i) the paraelectric phase ($P = q = 0$), (ii) the spherical glass phase ($P = 0$ and $q \neq 0$) and (iii) the inhomogeneous ferroelectric phase ($P \neq 0, q \neq 0$).

Let us first discuss the results in the absence of the biasing field, i.e. for $\vec{E} = 0$.

a) For $E = 0$ and

$$J_0 < (J^2 + \Delta)^{1/2} \qquad (15)$$

one gets a spherical glass without long-range order with $P = 0$ and $q \neq 0$.

b) For $E = 0$ and

$$J_0 > (J^2 + \Delta)^{1/2} \qquad (16)$$

one gets below a transition temperature T_c an inhomogeneous ferroelectric phase with a non-zero polarization $P \neq 0, q \neq 0$. The critical transition temperature is given by $P^2 = 0$, i.e. by $T_\text{c} = J_0/k = T_\text{m}$, where T_m is the maximum in the non-linear susceptibility. The FE phase can only exist for $J_0 > J_{0\text{c}}$ where $J_{0\text{c}} = \sqrt{J^2 + \Delta}$.

Assuming $P \neq 0$ we get

$$q = 1 - T/J_0 \qquad (17)$$

As mentioned before T_c is obtained from $P^2 = 0$. For $T > T_\text{c}$ we have $P = 0$. At $T \to 0$ the polarization is given by

$$P^2(0) = 1 - (J^2 + \Delta)/J_0^2 \qquad (18)$$

This implies that a critical value

$$\Delta_\text{c} = J_0^2 - J^2 \qquad (19)$$

exists such that for $\Delta > \Delta_\text{c}$ there is no long-range order and $P = 0$ at any temperature. The value of q at the critical temperature is

$$q(T_\text{c}) = \Delta/\Delta_\text{c} \qquad (20)$$

For $J_0 < J_{0\text{c}}$ there is no long-range order and $P = 0$. It should be noted that J_0 is electric field dependent. We have $J_0(E) = J_0(0) + \alpha E^2$.

In the presence of an external field $E \neq 0$, or a non-zero random field, $\Delta \neq 0$, there is no sharp phase transition. For $\Delta \neq 0$ and $J_0 < J_{0c}$ there is no sharp freezing transition into the low T spherical glass (SG) state, i.e. there is no replica symmetry breaking in the SRBRF model. This is different from the case of a spin glass where replica symmetry breaking exists as there are no random fields.

6.6 Linear and non-linear dielectric response

Let us first treat the linear dielectric response in the SRBRF model [6.17]. The evaluation of the average free energy yields [6.8] that if $J_0 < \sqrt{J^2 + \Delta}$ a spherical glass without long-range order is formed, whereas a ferroelectric state becomes stable if $J_0 > \sqrt{J^2 + \Delta}$. The resulting coupled equations for the Edwards–Anderson spherical glass order parameter q and the polarization P are:

$$q = \beta^2 J^2 (q + \Delta/J^2)(1-q)^2 + P^2 \qquad (21)$$

$$P = \beta(1-q)(J_0 P + gE) \qquad (22)$$

Here, $\beta = 1/(kT)$ and g the average cluster dipole moment. Numerical solutions of Eqs. (21) and (22) yield $q(E,T)$ and $P(E,T)$ from which the derivative $\varepsilon_1(E,T) = 1 + \partial P/\partial E$ can be obtained. A comparison between the experimentally obtained temperature dependence of the static dielectric constant ε_S in PMN measured perpendicularly to the (001) plane (open squares) and the fit to the SRBRF model (solid line) is presented in Fig. 6.7. The fit is rather good and shows that in zero bias field $J_0 = 193 \pm 4\,\text{K}$ is indeed smaller, but rather close to $J = 219 \pm 4\,\text{K}$.

Since $J_0 = J_0(E)$, the condition $J_0 > \sqrt{J^2 + \Delta}$ can be satisfied in relaxors at much lower bias fields than in dipolar glasses, where typically $J_0 \ll J$ [6.18].

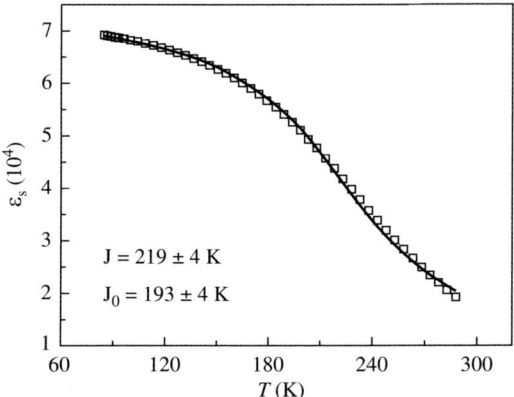

Fig. 6.7 Comparison between the experimentally obtained temperature dependence of the static field-cooled dielectric constant ε_S in PMN (open squares) and the fit to the SRBRF model (solid line) for $J = 219 \pm 4\,\text{K}$, $J_0 = 193 \pm 4\,\text{K}$, and $\Delta/J^2 = 0.001$.

154 Relaxor ferroelectrics

Recently, the third-harmonic non-linear permittivity has been investigated in relaxor systems PMN and(Pb,La)(Zr,Ti)O_3 ceramics (PLZT) with La/Zr/Ti ratio 9/65/35 and in $Sr_{0.61}Ba_{0.39}Nb_2O_6$ [6.19 and references therein].

In a system with average cubic symmetry the relation between the biasing field $E_\mu (\mu = x, y, z)$ and the polarization P_μ can be for small amplitudes of E represented as a power series in E:

$$P = \chi_1 E - \chi_{122} E_1 (E_2^2 + E_3^2) - \chi_{111} E_1^3 + \cdots \tag{23}$$

where $\varepsilon_1 = 1 + \chi_1/\varepsilon_0$ in MKS units, etc.

The inverse relation is

$$E = a_1 P_1 + a_{122} P_1 (P_2^2 + P_3^2) + a_{111} P_1^3 + \cdots \tag{24}$$

Now we get $a_1 = 1/\chi_1, a_{122} = \chi_{122}/\chi_1^4, a_{111} = \chi_{111}/\chi_1^4$. Here, χ_1 is the linear susceptibility and χ_{122}, χ_{111}, etc. are third-order non-linear susceptibilities [6.15]; see also [6.20].

For $\vec{E} \parallel [111]$ and $P_1 = P_2 = P_3$ the above equations simplify to

$$P = \chi_1 E - \chi_3 E^3 \tag{25a}$$

where

$$\chi_1 = \left(\frac{\partial P}{\partial E}\right)_{E=0} \text{ and } \chi_3 = -\frac{1}{6}\left(\frac{\partial^3 P}{\partial E^3}\right)_{E=0} \tag{25b}$$

This can be inverted to

$$E = a_1 P + a_3 P^3 \tag{26}$$

where $a_1 = \chi_1^{-1}$ and $a_3 = \frac{\chi_3}{\chi_1^4}$.

The phenomenological theory predicts that in the FE phase χ_3 diverges as

$$\chi_3 \sim |T - T_c|^{-3\gamma - 2\beta} \tag{27a}$$

so that a_3 should behave as

$$a_3^{\text{FE}} \sim |T - T_c|^{\gamma - 2\beta}, \tag{27b}$$

In general $\gamma - 2\beta > 0$ for a cubic system, and $\gamma = 2\beta$ in the mean-field approximation. Thus, in a random-field frustrated ferroelectric a_3^{FE} should not diverge at T_c, but should fall off to a constant value of zero at T_c. In contrast, for a dipolar glass (DG) one expects a divergent a_3

$$a_3^{\text{DG}} \sim |T - T_c|^{-\gamma_3} \tag{28}$$

where the mean-field value of the exponent is $\gamma_3 = 1$.

According to the rigid SRBRF model in the spherical glass for $J_0 = 0$ and $\Delta = 0$, a_3 diverges at $T_f = J$ according to $a_3 \sim |T - T_f|^{-1}$. For $\Delta \neq 0$ the divergence of a_3 does not occur. However for $\Delta/J^2 \ll 1$, a_3 shows a sharp peak near T_f. At $\Delta/J^2 > 0.1$ the peak starts to broaden and completely disappears at larger values of Δ/J^2 (Fig. 6.8).

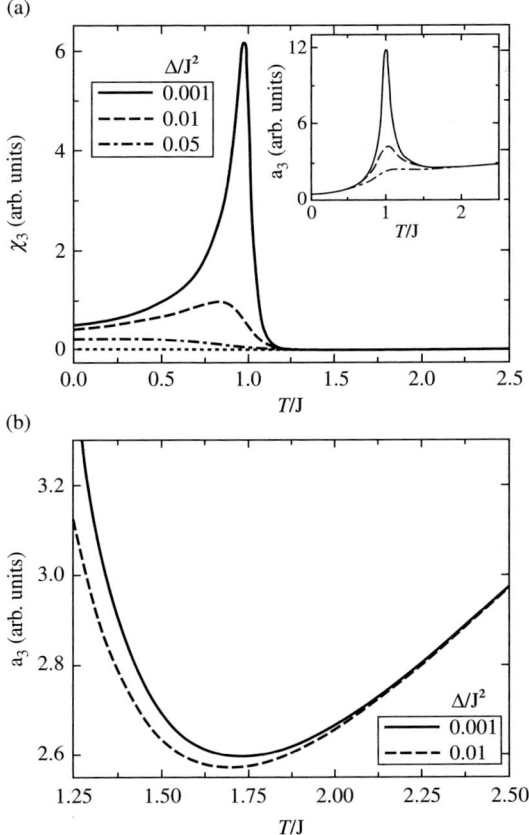

Fig. 6.8 (a) Calculated third-order non-linear response χ_3 in the SRBRF model as a function of temperature for three values of Δ/J^2, as indicated. Inset: Scaled non-linear response $a_3 = \chi_3/\chi_1^4$. (b) Zoomed view of a_3, illustrating the cross-over between paraelectric and relaxor-like behaviour [6.15].

It has been also suggested that $a_3(T)$ could be extracted from the dynamic (Fig. 6.9) linear and non-linear susceptibilities:

$$a_3'(T,\omega) = \frac{\overline{\chi_3'(\omega)}}{\overline{\chi_1'(3\omega)}\,\overline{\chi_1'(\omega)}^3} \qquad (29)$$

Here, the bar represents an average over the relaxation times or equivalently Vogel–Fulcher temperatures T_0.

6.6.1 The difference between relaxors and ferroelectrics

The difference between the linear and non-linear susceptibilities of relaxors and ferroelectrics is illustrated in Figs. 6.10(a)–(c).

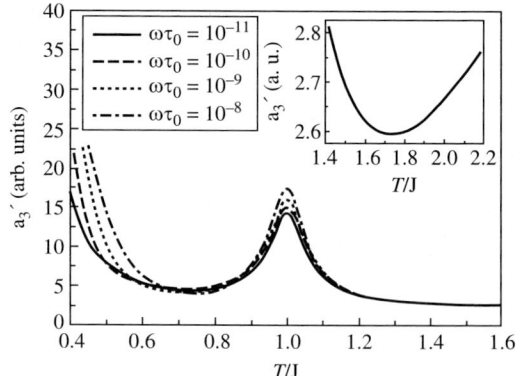

Fig. 6.9 Temperature dependence of the scaled dynamic non-linear susceptibility $a'_3(T,\omega) = \bar{\chi}'_3(\omega)/\bar{\chi}'_1(3\omega)\bar{\chi}'_1(\omega)^3$ of a relaxor according to the rigid SRBRF model. The inset shows the cross-over from the decreasing paraelectric-like to the increasing glass-like behaviour in the quasi-static regime above the peak [6.15].

Figure 6.10(a) shows the static linear and non-linear susceptibilities of a classical ferroelectric TGS. Both the linear (ε_1) and non-linear (ε_3) dielectric constants show a peak at T_C. The static non-linearity coefficient $a_3 = \varepsilon_3/\varepsilon_1^4$ shows a cusp at T_C.

In a relaxor, on the other hand (Fig. 6.10(b)), the static field-cooled dielectric constant (ε_1) smoothly and monotonously increases with decreasing T (see also Fig. 6.7). The same is true for the static non-linear dielectric constant ε_3. It shows the same monotonous behaviour as ε_1 without any anomaly at the freezing temperature T_f. The static field-cooled non-linearity coefficient a_3 smoothly decreases with decreasing T as in the case of classical ferroelectrics.

The situation is quite different for relaxors in the dynamic case (Fig. 6.10(c)). Here, ε_1 and ε_3 show broad dispersive peaks. ε_1 and ε_3 first increase and then decrease with decreasing T. The dynamic a_3 as measured at 3ω first decreases and then increases with decreasing T but shows no anomaly at T_f. A classical ferroelectric like TGS behaves similarly but shows a small peak at T_C that is absent in relaxors.

The most pronounced difference between the non-linear behaviour of ferroelectrics and relaxors is thus the temperature dependence of the static non-linear dielectric constant ε_3.

There are essentially two different ways to measure the third-order non-linear dielectric response χ_3, respectively, ε_3 and the non-linearity coefficient a_3. One is to measure the third-harmonic response at 3ω simultaneously with the linear response at ω at zero-bias electric field. These results can be explained within the rigid SRBRF model as shown in Figs. 6.8 and 6.9.

The second technique exploits the electric-field dependence of the linear dielectric constant. Here, ε_3^{FC} is obtained from

$$\varepsilon_3^{FC} = \varepsilon_0 \left[\varepsilon_{\text{eff}}(E_2) - \varepsilon_{\text{eff}}(E_1)\right]/\left(E_2^2 - E_1^2\right) \qquad (30)$$

Linear and non-linear dielectric response 157

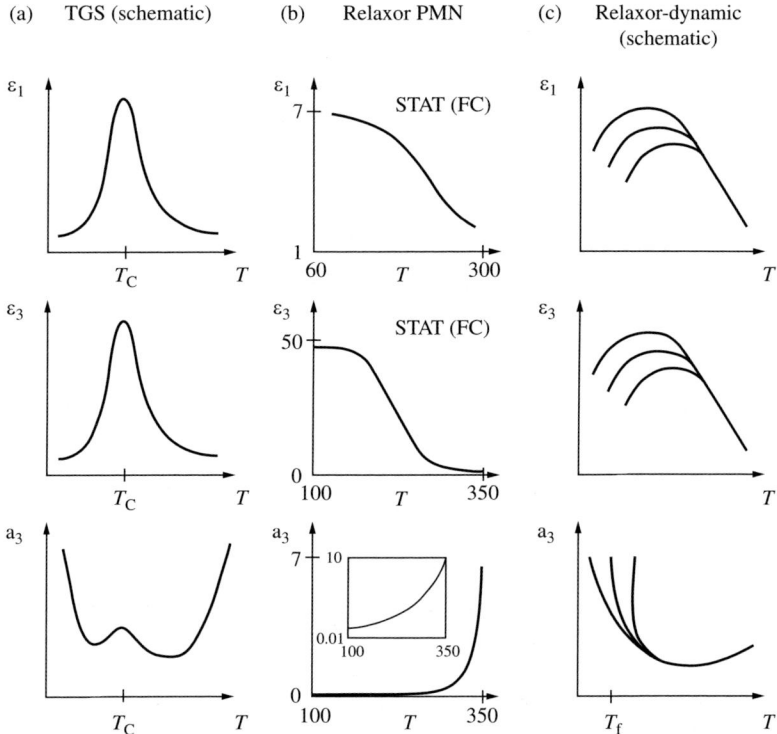

Fig. 6.10 (a) Temperature dependence of the linear (ε_1) and non-linear (ε_3) dielectric constant in ferroelectric TGS. The temperature dependence of the non-linearity coefficient (a_3) that shows a small peak at T_C is also shown (schematical). (b) Temperature dependence of the linear (ε_1) and non-linear (ε_3) dielectric constant in the relaxor PMN for the static field-cooled case. The temperature dependence of the non-linearity coefficient (a_3) is also shown. The inset shows the same a_3 data plotted on a semilog scale. (c) Temperature dependence of the linear (ε_1) and non-linear (ε_3) susceptibilities of a relaxor in the dynamic case. The non-linearity coefficient (a_3) is also schematically shown. No peak is seen at T_f in ε_1 or ε_3.

for $E_2, E_1 \ll E_c$. The thus obtained a_3 (designated as $\hat{a}_3$ below) monotonically decreases with decreasing temperature and saturates at low temperature (Fig. 6.10(b)).

This cannot be described by the rigid SRBRF model. It can be, however, described by the SRBRF model if the electric-field dependence of the interaction $J_0 = J_0(E)$ between the polar nanoclusters is taken into account:

$$J_0(E) = J_0 + J_1 \left(E^2/E_c^2 - 1\right) \tag{31}$$

The non-linear permittivity can be calculated as $\varepsilon_3 = (1/2)\frac{\partial^2 \chi_s(E)}{\partial E^2}\big|_{E=0}$ where

158 Relaxor ferroelectrics

$$\chi_s(E) = \frac{g^2}{\bar{v}_c} \frac{z - \sqrt{z^2 - \beta^2 J^2} - \beta J_0}{\beta(J^2 - J_0^2) - zJ_0^2} \tag{32}$$

Here, $J_0 = J_0(E), z = z(E)$ is the Lagrange multiplier enforcing the spherical condition $\sum_{i=1}^{N} \left(\vec{S}_i\right)^2 = 3N$, $\bar{v}_c$ the average volume of the cluster and g the average dipole moment of the cluster. Since both z and J_0 are functions of E^2 only, there will be two separate contributions to ε_3:

$$\varepsilon_3 = \varepsilon_3^I + \varepsilon_3^{II} = \frac{1}{2}\left(\frac{\partial \chi_s}{\partial z}\frac{\partial z}{\partial E^2} + \frac{\partial \chi_s}{\partial J_0}\frac{\partial J_0}{\partial E^2}\right) \tag{33}$$

Since $\varepsilon_3^I << \varepsilon_3^{II}$ we find

$$\varepsilon_3 \approx \varepsilon_3^{II} = \left(J_1/E_c^2\right)\frac{\partial \chi_s}{\partial J_0} \tag{34}$$

It should be noted that the two different measuring techniques explore different parts of the E–T phase diagrams and that the measurement of ε_3 by varying the bias field always gives the field-cooled ε_3 and the field-cooled a_3. For PLZT ceramics we find $J_1 = 0.04 J_0, J/k = 261\,\text{K}$ and $J_0/k = 214\,\text{K}_1$. Similar values are obtained for PMN.

The corresponding temperature dependences of $|\varepsilon_3|$ and $|a_3|$ obtained from the SRBRF model with a field-dependent interaction between nanoclusters are shown for PMN in Fig. 6.11. The agreement with the experimental results (Fig. 6.10(b)) is satisfactory [6.19].

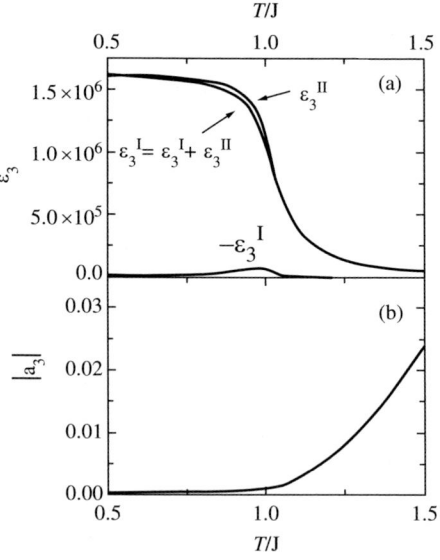

Fig. 6.11 Calculated temperature dependence of ε_3^I and ε_3^{II} (a), and $a_3 = -\varepsilon_3/\varepsilon_1^4$ (b), obtained from Eqs. (32)–(34). The estimated value of the average cluster dipole moment here is $gE_c/J = 0.022$.

6.7 Ferroelectrics in random fields

This corresponds to the case $J \ll J_0$ and we will set $J = 0$ without much loss of generality. We can distinguish between the paraelectric (PE) phase $T > T_c$ and the FE phase $T \leq T_c$, where we find $T_c = J_0 \left(1 - \Delta/J_0^2\right)$. For $T > T_c$, we get

$$q_> = \frac{1}{2\beta^2 \Delta} \left(1 + 2\beta^2 \Delta - \sqrt{1 + 4\beta^2 \Delta}\right) \tag{35}$$

The linear susceptibility χ_1 diverges as $\sim (T-T_c)^{-1}$.

The non-linear susceptibility χ_3 diverges as $\sim (T-T_c)^{-4}$; however, the scaled non-linear response a_3 is given by

$$a_{3>} = \frac{T}{(1-q)^2 \left[1 + 2\beta^2 \Delta (1-q)\right]} \tag{36}$$

and remains finite at $T \to T_c$.

For $T \leq T_c$ one has

$$P^2 = 1 - T/J_0 - \Delta/J_0^2 \tag{37}$$

The scaled non-linear response $a_{3<}$ can be calculated numerically after deriving a set of the appropriate field derivatives of $P(E, T)$. The result for a_3 with $\Delta/J_0^2 = 0.01$ at temperatures both above and below T_c is shown in Fig. 6.12. Also shown is the non-linear response $a_3(E)$ at $T > T_c$ and various values of the external field E. Obviously, a_3 remains finite as $T \to T_c$, but makes a jump at T_c in accordance with mean-field theory.

6.8 PbMg$_{1/3}$Nb$_{2/3}$O$_3$(PMN) and related perovskite relaxors: Phase diagrams, neutron scattering, Raman spectra and heat conductivity

PMN is the first and most investigated perovskite relaxor [6.2]. It is chemically a well-defined compound and not a solid solution. The charge disorder is produced by the differences in valence between Nb^{+5} and Mg^{2+} (+5 versus 2+) at the B lattice sites as well as the corresponding changes in the ionic radii (0.64 Å versus 0.72 Å). The compositional randomness of Mg$_{2/3}$ and Nb$_{1/3}$ at the B sites produces random Pb^{2+} ion displacements from the A sites. The early literature described PMN – and ABO$_3$ relaxors in general – as ferroelectrics with diffuse phase transitions [6.1]. The broad diffuse peak in the dielectric constant was seen as a result of compositional fluctuations leading to microregions with different ferroelectric transition temperatures T_C. The fact that there is no macroscopic symmetry change, respectively phase transition, on going across the dielectric maximum $\varepsilon'(T)$ speaks against this model.

The corresponding field-induced relaxor–ferroelectric transition [6.9] for a PLZT 9/65/35 ceramic sample is shown in Fig. 6.13. The relaxor–ferroelectric transition occurs well below the maxima in the dielectric response T_m. The results were obtained

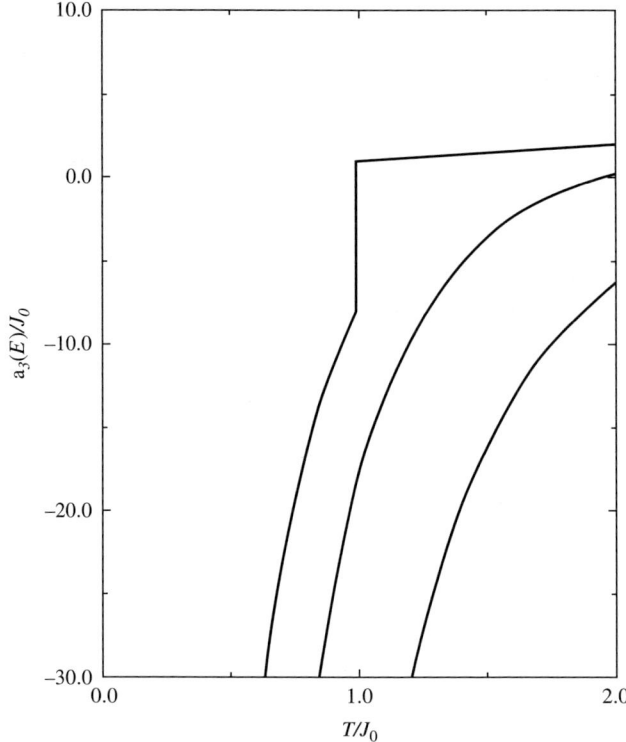

Fig. 6.12 Temperature and field dependence of the non-linear response in the FE phase ($J = 0$) and fixed $\Delta/J_0^2 = 0.01$. Top to bottom: $E/J_0 = 0, 0.2, 0.5$.

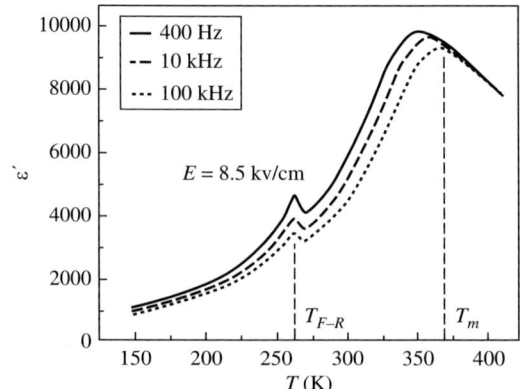

Fig. 6.13 Field-induced relaxor–ferroelectric transition at T_{F-R} for PLZT 9/65/35 [6.9].

for zero field heating (ZFH) after field cooling (FC) in a field of 8.5 kV/m. Field cooling here transforms the relaxor phase to a ferroelectric phase. The critical field is 3 kV/cm. The ferroelectric to relaxor transition here is characterized by a sharp cusp in ε'.

The electric field–temperature (E–T) phase diagram of the PLZT 9/65/35 sample is shown in Fig. 6.14.

The ferroelectric phase is approached in Fig. 6.14(a) by varying the temperature T at constant bias field E. In Fig. 6.14(b), on the other hand, it is approached by varying the field E at constant temperature. There is a significant difference between these two cases [6.9]. Here, FC means field cooled, FH field heated and ZFH zero field heated. The phase diagram in Fig. 6.14 contains the ferroelectric, the ergodic relaxor and the non-ergodic relaxor state. In the ergodic relaxor state every fluctuating polar cluster is statistically representative of the whole sample, whereas this is not true in the non-ergodic relaxor state. The ergodic–non-ergodic transition takes place at the temperature where the longest relaxation time of the system diverges. Here, the polar clusters freeze out in random orientations.

This is also considered as the static freezing temperature that can be determined from the Vogel–Fulcher relation [6.21] and the Kutnjak plot [6.21, 6.22].

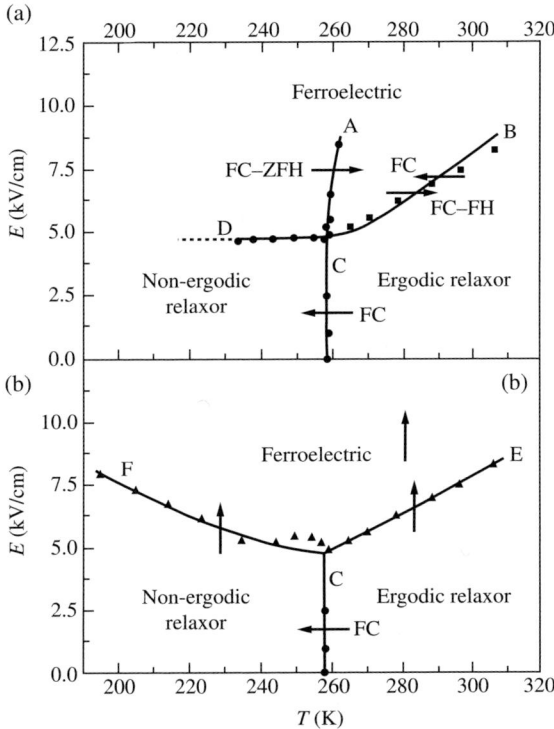

Fig. 6.14 Electric field–temperature phase diagram of PLZT 9/65/35 where the FE phase is approached by varying T at constant E (a) and by varying E at constant T (b) [6.9].

As already pointed out the relaxor phase in compositionally disordered perovskites is characterized by the appearance of nanosized polar clusters embedded in a non-polar matrix. Evidence for that is also provided by the waterfall effect (Fig. 6.15). Optical phonons become overdamped and seem to fall into the acoustic branch when their wavelength becomes comparable to the size of polar nanoclusters.

The low-frequency mode in $Pb(Zn_{1/3}Nb_{2/3})O_3$ – abbreviated as PZN – cannot be observed by neutron scattering (Fig. 6.15(b)) because of the waterfall effect [6.24].

Such a behaviour has been observed for $PbMg_{1/3}Nb_{2/3}O_3(PMN)$, $PbSc_{1/2}Ta_{1/2}O_3$ (PST), $Pb_{1-x}La_x(Zr_yTi_{1-y})_{1-x/4}O_3(PLZT)$ [6.3], Ca-doped $KTa_{1-x}Nb_xO_3$ [6.2] and others.

Another signature of relaxors is the strong elastic diffuse scattering near the zone centre that occurs as the result of the formation of polar nanoregions below the Burns temperature T_B. Neutron-scattering measurements in zero electric field have shown [6.25] soft modes in PMN at the R and M points of the Brillouin zone. These modes disappear when PMN becomes a well-defined ferroelectric by PT doping. The softening of these modes around 400 K occurs at nearly the same temperature as the minimum in the frequency of the zone-centre TO mode so that competing antiferroelectric and ferroelectric interactions seem to be responsible for the glassy state in PMN [6.25].

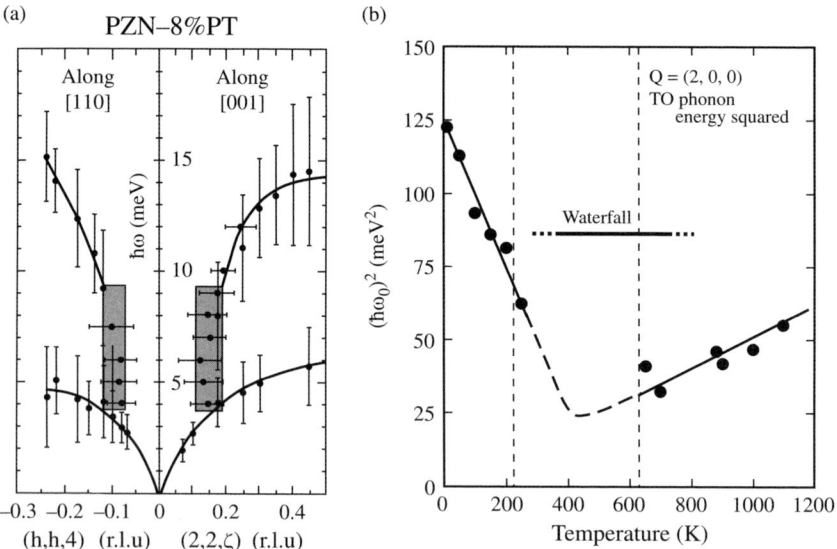

Fig. 6.15 (a) Waterfall effect. Phonon-dispersion relations for PZN-8 showing the unusual behaviour of the soft optic phonon as a function of wavevector [6.23]. For small enough wavevectors the optic phonon becomes overdamped and seems to drop into the acoustic branch. The behaviour seems to be due to scattering of phonons on the nanosized polar domains when the phonon wavelength becomes of the order of the size of polar clusters. (b) Low-frequency mode in PZN by neutron scattering [6.24].

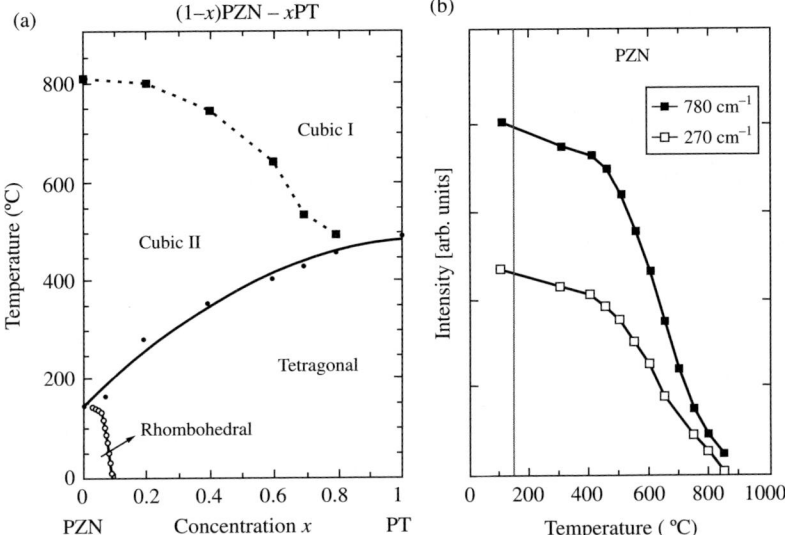

Fig. 6.16 (a) Phase diagram of PZT–PT. In the cubic II phase Raman modes are observed, proving the existence of polar nanoclusters that break the local symmetry. (b) Temperature dependence of the intensity of the Raman spectra in the cubic phase of $Pb(Zn_{1/3}Nb_{2/3})O_3$ (PZN) [6.26].

Another significant feature are the Raman spectra of relaxors. In perovskite-type classical ferroelectrics there are no Raman-active modes in the cubic phase because of symmetry restrictions. In relaxors, on the other hand, Raman-active modes exist even in the cubic phase. The intensity of these modes vanishes around 800°C, i.e. around the Burns temperature (Fig. 6.16). This demonstrates that the Raman-active modes in the cubic phase of relaxors are due to polar nanoclusters where the symmetry is ferroelectric [6.26].

Another point characterizing relaxors is the heat conductivity, which is glass-like [6.13] and different from the one in classical ferroelectrics.

6.9 Effect of pressure

The pressure dependence of the relaxor–FE phase boundary is due to the different shapes of the phonon branches and the pressure dependence of the phonon frequencies (expressions (9) and (10)). It should be noted that $J_0(p) = J_0(0) - \alpha p$.

The effect of pressure on the dielectric response of ferroelectric $KTa_{0.98}Nb_{0.02}O_3$ (KTN) is shown [6.27] in Fig. 6.17. At 1 bar the response is frequency independent, as expected for a ferroelectric. Above 3 kbar a ferroelectric to relaxor cross-over takes place. The system is for $p \geq 4$ kbar a relaxor. The maximum in the dielectric response T_m vanishes at $T_m \to 0$ with a finite slope [6.28]. This important result is also characteristic of other relaxors. The dielectric dispersion disappears at $p = 9.2$ kbar [6.27].

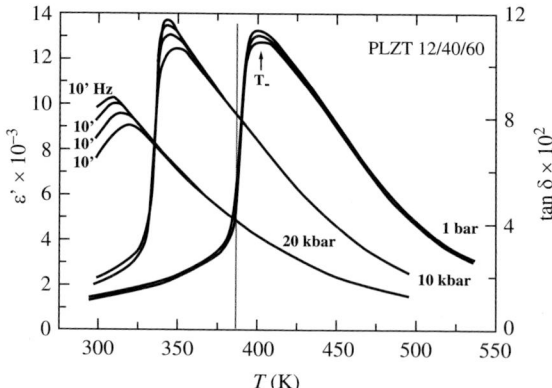

Fig. 6.17 The dielectric response of PLZT 12/40/60 showing the pressure-induced FE–R cross-over [6.27].

This seems to be due to the freeze out of the dynamics of polar clusters and to the vanishing of their dipole moments.

The above pressure response of KTN can be qualitatively understood in terms of the hopping motion of the off-centre Nb^{+5} ion among eight equivalent $<111>$ sites. The energy barrier here is proportional to the off-centre distance of the Nb^{+5} ion. The barrier decreases with increasing pressure. At a sufficiently high pressure the Nb^{+5} ion moves to the centrosymmetric position, i.e. to the B site of the cubic perovskite lattice. The corresponding dipole moment and the dielectric dispersion thus vanish. The same model also explains the effect of the electric field. The biasing field adds a linear term to the potential versus distance relation (Fig. 6.18) thus making one minimum deeper and the other shallower. At a sufficiently high electric field, the ion hopping and corresponding cluster reorientations freeze out completely and the dielectric polar cluster dispersion is quenched.

From a lattice point of view, one can explain the effect of pressure by the change of the phonon frequencies. The interactions between the polar clusters are phonon mediated and the increase of the frequencies with pressure makes the system stiffer and reduces the tendency to form a ferroelectric state.

Evidence for polar clusters around Nb^{+5} ions is also obtained from NMR [6.29, 6.30], Brillouin- and Raman-scattering data (Fig. 6.18) [6.30–6.33]. On fast enough time scales the system shows some static long-range order, whereas at slow enough time scales the response is glass-like and dynamic. Van der Klink [6.30] estimated that the polar clusters and the polarization cloud contain about 100 unit cells.

The main effect of hydrostatic pressure is thus to move the off-centre Nb^{5+} ions towards the centre and to reduce the barrier height between the potential minima. This shifts the freezing of the Nb ions and allows for Nb hopping motion. At the same time it provides for reorientations of the polar clusters.

Hydrostatic pressure thus introduces a cross-over from a ferroelectric to a relaxor state (Figs. 6.18 and 6.19), whereas the electric field induces a relaxor to ferroelectric transition.

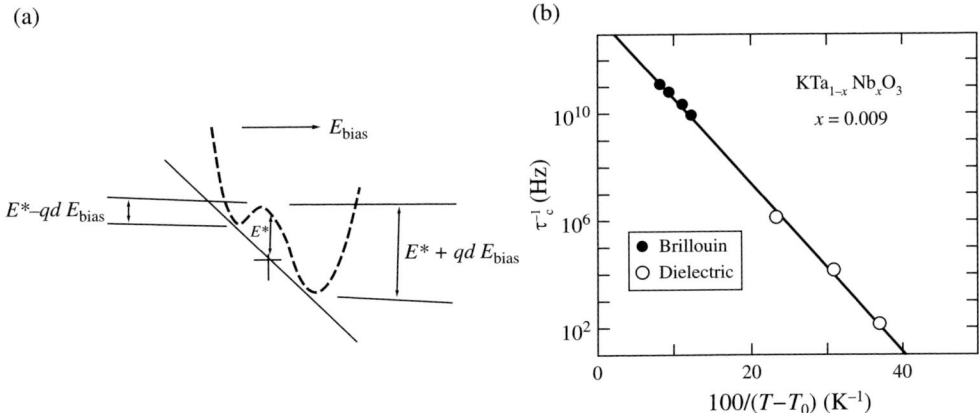

Fig. 6.18 (a) Double-well potential model to describe the hopping of Nb ions in KTN between adjacent equilibrium sites. The influence of a dc bias on the potential is also depicted. (b) Vogel–Fulcher representation of the temperature dependence of the cut-off frequency of the distribution of relaxation times for a KTN (0.9 atomic % Nb) crystal [6.27, 6.28].

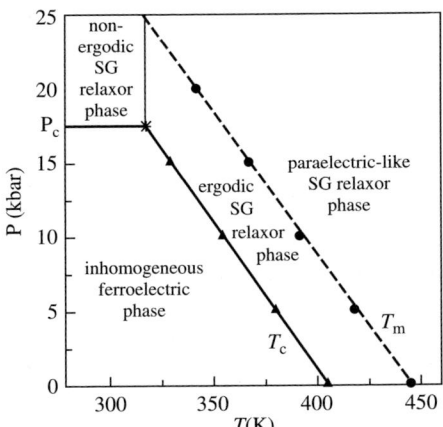

Fig. 6.19 Temperature–pressure phase diagram of the relaxor ferroelectric PLZT 6/65/35. The solid and dashed lines are evaluated from the static and dynamic coupled SRBRF-phonon models, respectively, whereas the solid triangles and solid circles are experimental points obtained by Samara [6.27]. The vertical dotted line corresponds to the freezing temperature T_f, where the static non-linear permittivity has a peak.

6.10 NMR lineshapes and relaxation times in relaxor PMN: Evidence of polar clusters

In the fast-motion limit the NMR nucleus does not 'see' the instantaneous value of the ordering field $\vec{S}$ but its time average $\vec{p}_i = \langle \vec{S}_i(t) \rangle$, which can be replaced by the thermodynamic average [6.8, 6.29]; see also [6.34].

$$\vec{p}_i = \left\langle \vec{S}_i \right\rangle \tag{38}$$

Here, p_i is the local polarization.

The NMR frequency of a quadrupole or chemical shift perturbed nucleus at a given site depends on the local polarization

$$\nu_i = \nu(p_i) \tag{39}$$

This relation can be expanded in a power series

$$\nu(p_i) = \nu_0 + \vec{\alpha} \cdot \vec{p}_i + \vec{p}_i \underline{\beta} \vec{p}_i + \cdots \tag{40}$$

where the coefficients $\vec{\alpha}$ and $\underline{\beta}$ depend on the orientation of the external magnetic field with respect to the principal axes of the electric-field gradient (EFG) or chemical-shift tensors.

The inhomogeneous NMR lineshape

$$f(\nu) = \frac{1}{N} \sum_i \delta(\nu - \nu_i) \tag{41}$$

can be now related to the polarization distribution function $W(\vec{p})$ as:

$$f(\nu) = \int W(\vec{p}) \delta(\nu - \nu_0 - \vec{\alpha} \cdot \vec{p} - \vec{p} \cdot \underline{\beta} \cdot \vec{p} + \cdots) d^3 p \tag{42}$$

For a system where all polar clusters are roughly of the same size at a given T, the lineshape is Gaussian

$$f(\nu) = \frac{1}{\sqrt{2\pi q \alpha^2}} \exp\left[-\frac{(\nu - \nu_0)^2}{2 q \alpha^2}\right] \tag{43}$$

In the linear case, where $|\alpha| \gg \|\underline{\beta}\|$, the second moment is proportional to the Edwards–Anderson order parameter q

$$M_2 = q\alpha^2 \tag{44}$$

For an ion jumping in an asymmetric double well in a polar cluster the average spin-lattice relaxation rate will depend on the magnitude of the local polarization p and bond asymmetry A as [6.35]

$$\frac{1}{T_1} \propto J(\omega) \propto \int_{-\infty}^{+\infty} \rho(A) dA \cdot (1 - p^2) \frac{\tau_C}{1 + \omega^2 \tau_C^2} \tag{45}$$

Here, $\rho(A)$ stands for the distribution of the asymmetries A of the double-well potentials in the polar nanoclusters and $1 - p^2$ is related to the bond asymmetry A as $p = \frac{A}{\sqrt{A^2+\Gamma^2}} \tanh\left(\sqrt{A^2 + \Gamma^2}/2k_B T\right)$ where Γ is the tunnelling matrix element. If $\rho(A)$ and/or $W(p)$ are not delta functions we expect a variation of T_1^{-1} over the spectrum $f(\nu)$.

At any given frequency we have here a distribution of spin-lattice relaxation rates

$$F\left[T_1^{-1}, \nu\right] = \int W(p)\delta\left[\nu - \nu(p)\right]\delta\left[T_1^{-1} - T_1^{-1}(p)\right] d^3p \tag{46}$$

The magnetization recovery will be thus generally non-exponential

$$M(t) - M_0 = M_0 \int_0^\infty W(T_1^{-1}) \exp(-t/T_1) dT_1^{-1} \tag{47}$$

and will be given by the Laplace transform of the relaxation time distribution function $W(T_1^{-1})$. Here, $W(T_1^{-1})$ is $F(T_1^{-1}, \nu)$ integrated over the frequency interval of interest.

The local structure and nano-cluster dynamics have been checked [6.29] by FC and ZFC Pb^{207} NMR in a PbMg$_{1/3}$Nb$_{2/3}$O$_3$ (PMN) single crystal.

The method is based on the fact that in the pseudocubic spherical shell-type matrix into which the polar clusters are embedded the Pb^{207} chemical-shift tensor is isotropic:

$$\underline{\sigma} = \sigma_0 \underline{1} \tag{48}$$

When the Pb^{207} nuclei are displaced from their high-symmetry cubic perovskite sites, the Pb chemical-shift tensors become anisotropic and can be written as a sum of a scalar part σ_0 and a traceless second-rank tensor part $\underline{\sigma}_a$:

$$\underline{\sigma} = \sigma_0 \underline{1} + \underline{\sigma}_a \tag{49}$$

The eigenvalues of the traceless part are $-\sigma_a/2, -\sigma_a/2, \sigma_a$. The largest principal axis of $\underline{\sigma}_a$ should be parallel to the direction of the average shift of the Pb nucleus, i.e. to the direction of the polarization $\vec{p}_i$ of a given polar cluster i. In the present case this is the [111] direction.

The Pb^{207} spectrum above 290 K is isotropic and of a Gaussian lineshape [6.29]. Two- dimensional (2D) separation of interactions experiments show that the spectra are frequency distributions and are composed of a large number of individual Pb^{207} lines with different chemical shifts. This is incompatible with the assumption that the Pb ions sit at their high-symmetry cubic sites as in this case all Pb ions would be equivalent and only a single sharp line is expected. The fact that we see a Gaussian distribution demonstrates that we deal with a spherical glass where all Pb nuclei are displaced but there is no preferential frozen-out orientation or magnitude of displacement. If the shifts of the Pb ions varied only in orientation, a powder-like pattern rather than the Gaussian lineshape would be observed. Due to the short-range-order correlations among the displacements clusters are formed that fluctuate in time, orientation and magnitude of the dipole moment. The orientational bias here is spherically symmetric.

Below 209 K an anisotropic Pb^{207} line suddenly appears in addition to the isotropic one in the FC as well as ZFC spectra (Fig. 6.20). Its angular dependence in the external magnetic field follows the $(3\cos^2\vartheta - 1)$ law. Here, ϑ determines the orientation of the eigenframe of $\underline{\sigma}$ with respect to the crystal fixed frame. The anisotropic line requires a shift of the Pb ions in a given preferred direction, i.e. [111].

The ratio of the intensities of the isotropic and anisotropic lines is about 5:1 in the ZFC spectra. In the FC spectra the anisotropic line is much stronger. The angular dependence of the anisotropic lines is the same in the FC and ZFC spectra. The anisotropic line corresponds to polar clusters frozen out on the NMR time scale and oriented along [111]. The isotropic line corresponds, on the other hand, to the glassy spherical matrix, which does not respond to an electric field. It also contains time-averaged clusters that reorient fast on the NMR time scale.

Below 210 K the anisotropic line suddenly increases in intensity corresponding to Pb shifts along the [111] direction. The increase in the intensity is much stronger for

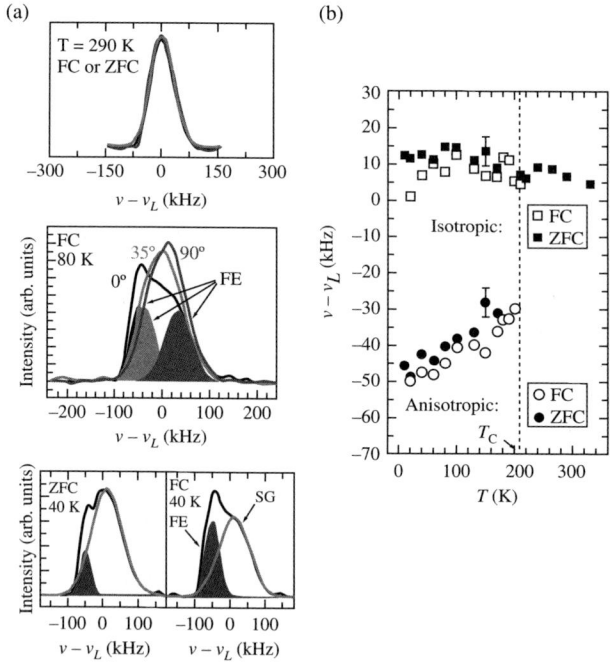

Fig. 6.20 (a) Decomposition of the field-cooled and zero-field-cooled ^{207}Pb NMR spectra of a PMN single crystal, taken at $\vec{B}\|\vec{E}\|$ [111] at different temperatures, into an isotropic and anisotropic component. (b) Temperature dependence of the positions of isotropic and anisotropic ^{207}Pb NMR lines. The anisotropic line disappears above 210 K for both the FC ($E = 3\,\text{kV/cm}$) and the ZFC spectra [6.29]. See Plate 1 in the colour plate section.

electric fields larger than a critical value, $E > E_c$. At the same temperature a peak in the dielectric constant appears for $E > E_c$. This means that the sudden increase in the intensity of the [111] oriented polar clusters signals the transition to a more ordered state on the NMR time scale. The state becomes ferroelectric for $E > E_c$.

The width of the anisotropic line at $\approx 80\,\text{K}$ is $\approx 40\,\text{kHz}$, whereas the width of the isotropic line reaches $100\,\text{kHz}$. This is much more than what can be accounted for by nuclear dipolar interactions. It shows that we deal in the case of the isotropic line with a frequency distribution due to differently oriented chemical-shift tensors.

In the absence of the electric field the concentration of the polar clusters in PMN is below the threshold for a percolation-type ferroelectric transition, $p_{\text{thres}} = 0.3$. This is changed in the presence of an electric field $E > E_c$. PMN is thus an incipient ferroelectric, which performs an orientational percolation transition for $p_c \to p_{\text{thres}}$.

The electric-field-induced transition to the ferroelectric state here is thus due to a sudden increase in the concentrations of nanoclusters with a polarization and Pb shift in the [111] direction.

To check on the nanoclusters dynamics the Pb^{207} spin-spin relaxation time (T_2) was measured for $\vec{B}\|\vec{E}\|[111]$. Here, T_2 measures the spectral density of the low-frequency fluctuations:

$$\frac{1}{T_2} \propto J_0(\omega) \propto \frac{\tau}{1+\omega^2\tau^2},\ \omega \to 0 \tag{50a}$$

$$\Delta\nu_{\text{dyn}} = (\pi T_2)^{-1} \approx 10\,\text{kHz} \tag{50b}$$

The dynamic NMR width $\Delta\nu_{\text{dyn}}$ is always small compared to 'static' glassy NMR width $\sim 100\,\text{kHz}$ induced by the inhomogeneous nature of the spectrum. Cluster dynamics can be therefore seen only by T_2 measurements and not by 1D lineshape data. On cooling from $400\,\text{K}$ T_2 decreases with decreasing T to reach a BPP-type minimum as $T_2 \propto 1/\tau$. The minimum, where $\omega\tau \approx 1$, occurs around $255\,\text{K}$, i.e. it clearly matches the freezing temperature $T_M = 250\,\text{K}$, where a maximum in the dielectric susceptibility is found. Here, ω is of the order of the NMR linewidth, i.e. it represents the Larmor frequency in the local field. The same mechanism, i.e. fluctuations in the nanocluster polarization, thus determines the dielectric losses and the T_2 value. Below the minimum $\omega\tau \gg 1$, and $T_2 \propto \tau$ increases with decreasing T as τ increases.

The sudden increase in the intensity of the anisotropic Pb^{207} NMR line near $210\,\text{K}$ is thus due to the slowing down and freeze-out of the polar nanocluster dynamics on the NMR time scale.

It should also be mentioned that at low T below the BPP minimum the field-cooled T_2 depends on the frequency off-set, i.e. on the position of the nucleus in the sample. This means that PMN is, at low T, a dynamically heterogeneous system.

At $T > 250\,\text{K}$, T_2 only weakly depends on the frequency off-set, which is in the present case $-40\,\text{kHz}$ (FC) and $+20\,\text{kHz}$ (FC and ZFC). This is different below $250\,\text{K}$ (Fig. 6.21). It clearly shows that the system became inhomogeneous and that different motions occur in different parts of the sample. We have a part where the motion is frozen out on the NMR ($10^7\,\text{Hz}$) – though not $T_2(10^3\,\text{Hz})$ – time scale and another part where the motion is still fast on the NMR time scale. The first part corresponds

170 Relaxor ferroelectrics

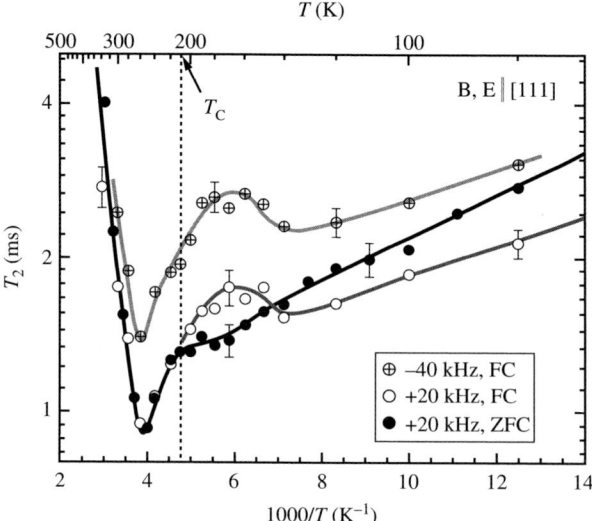

Fig. 6.21 Temperature dependence of the spin-spin relaxation time T_2. The minimum at 250 K corresponds to the well-known freezing of polar clusters. At about 140 K there is a second shallow minimum in the FC ($E = 3\,\mathrm{kV/cm}$) data, which could be related to the motion of domain walls in the ferroelectric state [6.29].

to 'frozen-out' polar nanoclusters and the second part to 'reorienting' nanoclusters. A second, rather flat T_2 minimum occurs in the FC data around 140–150 K with a frequency off-set – 40 kHz. It appears also in the ZFC data but is weaker. A new reorientation process thus takes place that may be due to the motion of the domain walls of [111] clusters at the percolation transition.

The NMR results thus separately show the anisotropic Pb^{207} NMR spectra of frozen ferroelectric-like polar clusters and the isotropic spectra of the dynamic spherical glass matrix into which the polar frozen out clusters are embedded. They also show that the nanocluster dynamics is different in different parts of the crystal [6.29].

The above conclusions are supported by recent neutron spin-echo results [6.29]. Here, a static nanocluster component appears in addition to the dynamic one already around 400 K. Since the neutron spin-echo time scale is in the ns region, whereas the NMR is in the kHz range, the two sets of data agree rather well. We deal with a motional effect. The anisotropic clusters represent the instantaneous and the isotropic matrix the time-averaged picture.

6.11 Electric-field-induced critical end points in PMN–PT relaxors and giant electrostriction

It has been recently shown [6.36] that by applying a sufficiently strong electric field $E \geq E_C$, the first-order paraelectric to ferroelectric phase transition in the PMN–PT system terminates in a line of critical (CL) points [6.36–6.38] of the liquid-vapour

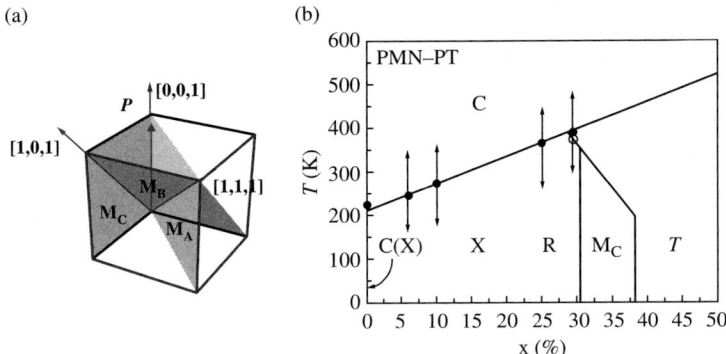

Fig. 6.22 (a) The polarization vector rotates from the $[001]_C$ direction in the tetragonal (P4mm) phase to the $[111]_C$ direction in the rhombohedral (R3m) phase via the $[110]_C$ direction in the orthorhombic ($mm2$) phase. The polarization vector is within the $(010)_C$, $(1^-01)_C$, and $(11^-0)_C$ planes in the monoclinic M_C, M_B, and M_A phases, respectively. (b) The composition-temperature (x–T) phase diagram of the PMN–PT system. Double arrows ↕ show the compositions measured in this study [6.55].

type above which supercritical behaviour is observed. On approaching this critical line ($E \to E_{CL}$) the piezoelectric coefficients as well as the static dielectric constant and the heat capacity exhibit maximum values. The electric field necessary for polarization rotations and the energy barriers involved significantly decrease as $E \to E_{CL}$, thus revealing a new driving mechanism for the giant electromechanical response of relaxors.

As already mentioned, classical relaxors [6.1, 6.2] are perovskite solid solutions characterized by site and charge disorder [6.7, 6.37–6.40]. They show no symmetry-breaking transition on cooling but exhibit broad and strongly frequency-dependent peaks in the dielectric and electromechanical responses as well as glassy-type freezing [6.7, 6.40–6.42]. In contrast to dipolar glasses [6.12], a ferroelectric phase can be induced in relaxors [6.41] by applying a sufficiently strong electric field [6.12, 6.19, 6.34, 6.42–6.44]. This is due to the fact that relaxors consist of small, randomly oriented polar nanoregions, which, in view of their relatively large dipole moments, couple to the electric field more efficiently than individual dipoles in dipolar glasses [6.20, 6.45–6.48].

The coupling to the electric field is still enhanced if the disorder is reduced by changing the composition, e.g., by adding ferroelectric PbTiO$_3$ (PT) to PMN [6.47]. The PMN$_{1-x}$-PT$_x$ system is for $x > 0.05$ at low temperatures ferroelectric and rhombohedral up to some specific PT content [6.47, 6.49]. At still higher PT concentrations, it undergoes a morphotropic phase transition [1.35, 6.50, 6.51] and becomes tetragonal. A giant piezoelectric response is observed near this transition. At higher temperatures, the system becomes paraelectric and cubic for all PT concentrations (Fig. 6.22).

The PMN–PT system exhibits a number of different phases [6.47]. Pure PMN exhibits in addition to the electric field induced ferroelectric state ($E > E_C$) also an ergodic relaxor state at high and a non-ergodic relaxor state at low temperatures ($E < E_C$). For sufficiently high PT concentrations, ferroelectric tetragonal (T), rhombohedral (R), monoclinic (M), and orthorhombic (O) phases occur at low and a paraelectric cubic (C) phase at high temperatures. The spontaneous polarization directions are [001] in the tetragonal (P4mm) phase, [111] in the rhombohedral (R3m) phase, and [110] in the orthorhombic (mm2) phase. There are possibly three different monoclinic phases ($M_A, M_B,$ and M_C) [6.51–6.54] between the rhombohedral and tetragonal phases (Fig. 6.22(b)).

The phase transition between the R and T ferroelectric phases observed by many authors [6.7, 6.40, 6.45, 6.50, 6.51] becomes degenerate near the morphotropic phase boundary. Schmidt and coworkers [6.51, 6.52] concluded that the R–T phase transition occurs as a rotation of the polarization through M phases. The piezoelectric effect is due to the coupling between the strain and the polarization. When the polarization is along the cube diagonal, the lattice strain is small, but when the polarization is along the cube axis, the lattice strain is large, resulting in a large piezoelectric effect (Fig. 6.22(a)).

The combinations of stress and electric field are particularly effective to drive the system through the phase transformations [6.55–6.58].

Cohen and coworkers[6.53–6.55, 6.58] have studied the polarization rotation mechanism under electric field using first-principles calculations. The polarization rotation mechanism has been directly verified by X-ray and polarized-light microscopy studies [6.51–6.54, 6.57, 6.58].

Very recently, a study of the phase diagram of PMN for electric bias fields applied along [111], [110], and [100] was published by Zhao et al. [6.58] Whereas they confirmed the existence of the critical point [6.36] when the bias field was applied along [111], they did not observe a critical point in the [110] and [100] directions up to the highest applied field of 7.5 kV/cm. In contrast, we have shown [6.37, 6.38] the existence of critical points also for the bias field along [110] in PMN–PT. It should also be noted that all electric-field-induced transitions occur in PMN at much higher fields than in the PMN–PT system.

The macroscopic symmetry of the PMN–PT system has been studied by optical microscopy in the absence of electric biasing fields by Shuvaeva et al. [6.55]. Here, we relate the observations of the line of critical points in the electric-field–temperature (E–T) phase diagram of PMN–PT to the polarization rotation mechanism and the piezoelectric response for fields applied along the [110] and [111] directions.

PMN_{1-x}–PT_x single crystals with $x = 0, 0.10, 0.15, 0.25,$ and 0.295 have been investigated [6.37]. In all these cases, electric-field-induced critical points have been found. The necessary electric fields strongly decreased on going from $x = 0$ to $x = 0.295$. It is this last case on which we shall focus. The electric bias field was applied along the [111] and [110] directions. The field-cooled (FC) and zero-field-cooled quasi-static dielectric (ε) measurements, the measurements of the pyroelectric current, and the polarization (P) determinations were performed by electrometer charge accumulation measurements as described by Levstik et al. [6.7, 6.40] The phase sequence was determined by polarized light microscopy. The ac dielectric spectroscopy measurements were performed between 1 mHz and 1 GHz. The piezoelectric and elastic coefficients were determined by the resonance response of a platelet [6.59] poled and excited along its thickness. The radius of the circular platelet was at least ten times larger than the thickness d. High-resolution ac and relaxation calorimetry [6.36, 6.37] in electric fields were used to determine the enthalpy, latent heat, and the critical behaviour. The difference between the a_C and the relaxation heat capacity data gives the latent heat which disappears when $E \to E_{\mathrm{CL}}$.

6.11.1 Landau theory

Let us now first discuss in the simplest possible way the electric-field-induced critical end point – designated in the following as the critical point – in the phase diagram of PMN–PT within the framework of the Landau theory [6.38].

The thermodynamic potential density can be in the isotropic approximation in the vicinity of the phase transition expanded in powers of the order parameter P:

$$f = f_0 + \frac{a}{2}P^2 + \frac{b}{4}P^4 + \frac{c}{6}P^6 - PE \tag{51}$$

where E is the normalized external electric field. Here, we truncate the expansion at the sixth order in P.

We assume for the sake of simplicity that we deal with a homogeneous medium and a scalar order parameter. We also assume that the polarization and the bias field point in the same direction. The idea is to determine the coordinates of the critical point and the evolution of the Widom line in the supercritical region [6.38]. The treatment of the anisotropic case that is essential for the complete phase diagram close to the morphotropic phase boundaries is non-trivial.

The coefficient a in expression (51) varies with temperature as $a = a_0(T-T_0)/T_0 = a_0\tau$ and changes its sign at the phase transition.

For second-order phase transitions, one has $b > 0$, whereas $b < 0$ for first order phase transitions. c is assumed to be always positive, $c > 0$.

For second order phase transitions, the critical exponents can be defined for the dielectric susceptibility $\chi(0)$, the order parameter P, and the specific heat at constant pressure C_p as

$$f_{PP}^{-1} = \chi(0) \propto |\tau|^{-\gamma} \tag{52}$$

$$P_0(\tau < 0) \propto |\tau|^{\beta} \tag{53}$$

$$C_P = -T\left(\frac{\partial^2 f}{\partial T^2}\right)_{P_0} \propto |\tau|^{-\alpha} \tag{54}$$

Here, $\partial E/\partial P = (\partial^2 f/\partial P^2)_{P_0} = f_{PP} = \chi(0)^{-1}$.

It is well known that the classical values of γ, β, and α in the Landau theory are $\gamma = 1, \beta = 1/2$, and $\alpha = 0$. For $b = 0$, a tricritical point exists, which separates the line of second-order transitions from the line of first-order transitions. Here, $\beta = 1/4$.

Let us now look for the critical point where the line of first-order phase transitions terminates. Above the critical point, the anomalies in the response functions become rounded and non-critical, with $\alpha < 0$. The difference between the two phases disappears in analogy to the liquid-vapor transition in water.

For a first order ($b < 0$) phase transition for $E = 0$, the $(f - f_0)$ versus P surface has two minima at $P \neq 0$ in addition to the one at $P = 0$. The two additional minima move downward with decreasing T and reach the line $f - f_0 = 0$ at a positive τ ($\tau = (T-T_0)/T_0$) value, i.e. for $T_C > T_0$. For $T < T_C$ these minima have $f - f_0 < 0$ so that a discontinuous (first-order) transition takes place from $P = 0$ into one of the two minima with $P \neq 0$. The first order transition takes place when $f - f_0 = 0$. Solving these equations, we find the transition temperature for $E = 0$ as

$$a_C = \frac{3}{16}\frac{b^2}{c} \tag{55}$$

The temperature (i.e. a) dependence of the order parameter for different applied electric fields and the relation between E and P are obtained from

$$\frac{\partial f}{\partial P} = aP + bP^3 + cP^5 - E = 0, \quad b < 0 \tag{56}$$

$$E = aP + bP^3 + cP^5 \tag{57}$$

The discontinuity in the temperature dependence of the order parameter P is maximal for $E = 0$ and vanishes for (a_{CE}, E_{CE}), i.e. at the critical point. Let us now determine the T and E coordinates of this point (Fig. 6.23).

Taking the derivative of expression (56) with respect to a, we obtain

$$\left(a + 3bP^2 + 5cP^4\right)\frac{\partial P}{\partial a} + P = 0 \tag{58}$$

At the transition point, $\partial P/\partial a$ diverges so that

$$\left(a + 3bP^2 + 5cP^4\right) = 0 \tag{59}$$

At (a_{CE}, E_{CE}), Eq. (59) has a double root so that $9b^2 - 20ca_{CE} = 0$. We thus find the coordinates of the critical point as

$$a_{CE} = \frac{9b^2}{20c} \tag{60}$$

$$P_{CE}^2 = \frac{-3b}{10c} \tag{61}$$

$$E_{CE} = \frac{6b^2}{25c}\sqrt{\frac{-3b}{10c}} \tag{62}$$

Beyond (a_{CE}, E_{CE}), the susceptibility χ shows a rounded peak along the Widom line, where $\partial\chi/\partial T = 0$ and also $(\partial\chi^{-1}/\partial T) = 0$.

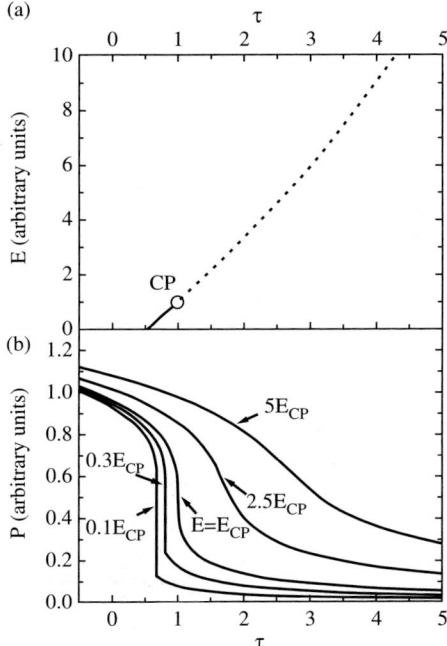

Fig. 6.23 (a) The schematical E–T phase diagram for the PMN–PT system. Here, $\tau = (T-T_0)/T_0$. The solid line represents a first-order transition line. CP stands for the critical end point, which terminates a line of first-order ferroelectric transitions. E_{CP} is much larger than E_C. The dotted line represents the Widom line, a locus of supercritical anomalies emanating from the critical point typically observed in the supercritical region. (b) The calculated temperature dependence of the order parameter for different bias fields E above and below the critical one (E_{CP}).

This is in sharp contrast to the tricritical point beyond which χ diverges at the second-order transition line. Beyond (a_{CE}, E_{CE}), we thus find with the help of Eq. (57) the coordinates of the Widom line for a given a as

$$\chi^{-1} = \frac{\partial^2 f}{\partial P^2} = a + 3bP^2 + 5cP^4 \tag{63}$$

$$\frac{d}{da}(\chi^{-1}) = 1 + (6bP + 20cP^3)\frac{dP}{da} = 0 \tag{64}$$

These relations thus lead to

$$a - 3bP^2 - 15cP^4 = 0 \tag{65}$$

$$P_a^2 = \frac{-b}{10c} + \frac{1}{30c}\sqrt{9b^2 + 60ac} \tag{66}$$

Figure 6.23(a) shows the schematical E–T phase diagram for the PMN–PT system, based on the above expressions. The theoretical curve agrees rather well with the experimental results [6.36].

6.11.2 Experimental data

Let us first discuss the [111] case for $PMN_{1-x}PT_x$ with $x = 0.295$. For $E = 0$, the high-temperature phase is cubic and the low-T phase rhombohedral. If we apply a bias field along [111], the symmetry including the field is rhombohedral both at low and high temperatures.

Therefore, the two phases can merge at high fields without a boundary, i.e. there exists a critical end point.

For the [110] case, the situation is similar. According to Lu et al. [6.60] the high-temperature phase is cubic and the low-T phase rhombohedral, so that critical points exist in agreement with our observations.

If, however, the system were monoclinic at low temperatures and tetragonal at high temperatures and the field were applied along [100], there should be some boundary between the two different symmetries.

Let us now look into the behaviour of pure PMN at $E \parallel [111]$. It is known that below a certain temperature, a sufficiently strong electric field $E \geq E_C$ can induce a first-order transition from a relaxor state to a ferroelectric state. At still higher electric fields, a critical point exists, $E = E_{CL} = 4\,\text{kV/cm}$ (here $E_{CL} = E_{CP}$), above which supercritical behaviour is found [6.36]. In the following, we wish to relate the critical response of PMN to its piezoelectric response.

The temperature dependence of the quasi-static FC polarization P of a PMN single crystal poled along [111] has been obtained [6.36–3.38] for different electric bias fields. The results can now be compared with the calculated temperature dependence of the order parameter for different electric fields above and below the critical one (Fig. 6.23(b)). For a zero field, the system is in a relaxor state with no spontaneous polarization. For small fields, $E < E_C < E_{CL}$, there is a small induced polarization that continuously increases with decreasing temperature. For $E_C < E < E_{CL}$, there is a jump in the polarization at the first order transition from the relaxor to the ferroelectric state. For $E \geq E_{CL}$, the jump disappears and P continuously varies with temperature in the supercritical regime, in agreement with the theoretical predictions (Fig. 6.23). The excess heat-capacity data (Fig. 6.24) also show with $E \to E_{CL}$ a transition from a first order behaviour to a near-critical and finally to a supercritical behaviour. It should be stressed that it is the continuous supercritical evolution along the Widom line (Fig. 6.25) that distinguishes the critical end point from a tricritical point where a line of first-order transition meets a line of second-order transition. Whereas in the supercritical case the anomalies in the response functions become non-critically smeared out and finally disappear [6.59–6.61], there are sharp critical anomalies at the line of second-order transition in the tricritical case. This is demonstrated in Fig. 6.24 where the T dependences of the excess heat capacity in PMN-$(PT)_x$ with $x = 0.295$ and the dielectric constant of PMN (Figure 6.26) are presented for different bias fields and frequencies.

Electric-field-induced critical end points in PMN–PT relaxors and giant electrostriction 177

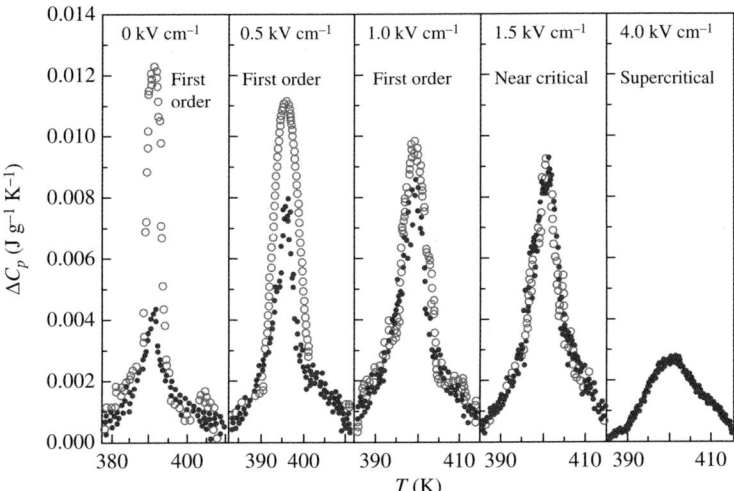

Fig. 6.24 Temperature dependence of the excess heat capacity data obtained in PMN–PT $x = 0.295$ at various bias electric fields, E, shown at the top of each panel. Filled (blue) and open (red) circles represent data obtained in the ac and relaxation mode, respectively. The relaxation runs measure the total enthalpy changes including the latent heat, whereas the ac runs measure the continuous variations of the enthalpy. The nature of the paraelectric to tetragonal phase transition is denoted by 'first order', 'near critical' and 'supercritical' labels. See Plate 2 in the colour plate section.

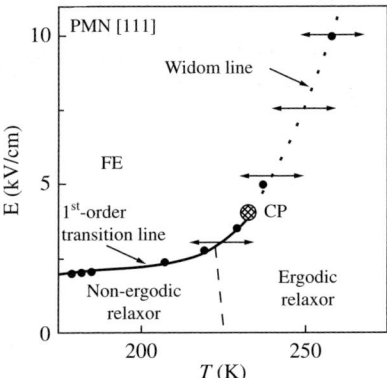

Fig. 6.25 The electric field–temperature (E–T) phase diagram of a PMN single crystal poled along [111]. The solid line represents a first-order transition line. CP stands for the critical end point, which terminates a line of first-order ferroelectric transitions. The dotted line represents the Widom line, a locus of supercritical anomalies emanating from the critical point typically observed in the supercritical region. Double arrows $\leftrightarrow$ show the electric bias field values at which the piezoelectric coefficient d_{31} shown in Fig. 6.27 was measured across the ferroelectric transition and Widom lines.

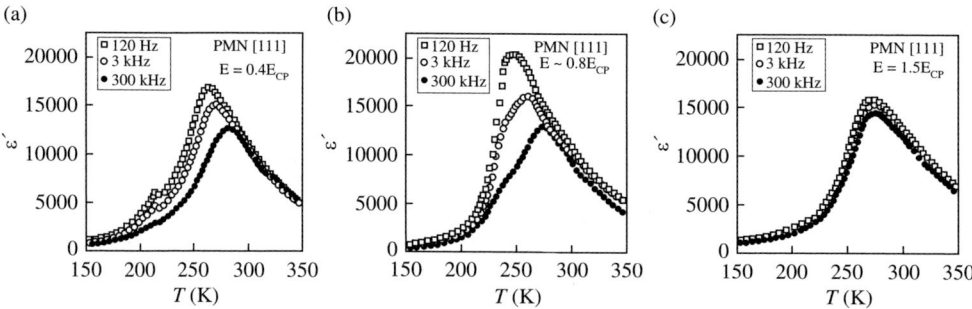

Fig. 6.26 The temperature dependence of the dielectric constant at different frequencies obtained in a PMN single crystal poled along [111] at three different values of the bias electric field: (a) 0.4 $E_{\rm CP}$, (b) 0.8 $E_{\rm CP}$, and (c) 1.5 $E_{\rm CP}$. Here, $E_{\rm CP} = 4.0\,{\rm kV/cm}$.

Fig. 6.27 shows the temperature dependence of the piezoelectric coefficient d_{31} for different regions on the Widom line. In all cases, the piezoelectric coefficient shows a non-critical maximum at the evolution line (including the Widom line). The data can be described by

$$d_{im} = \sum_{jk} 2\varepsilon_{ij} Q_{mjk} P_k \qquad (67)$$

where Q_{mjk} are the elements of the electrostrictive tensor, P_k are the components of the spontaneous polarization, and ε_{ij} are the elements of the dielectric tensor. The

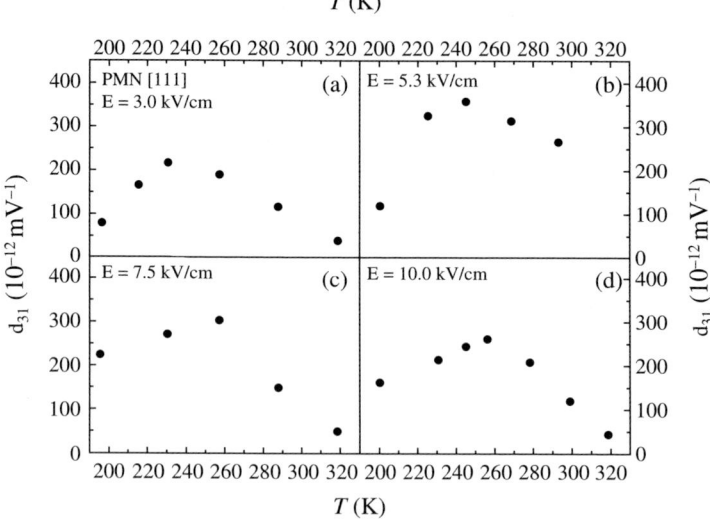

Fig. 6.27 The temperature dependence of the piezoelectric coefficient d_{31} obtained in a PMN single crystal poled along [111] at four different values of the bias electric field below and above $E_{\rm CP} = 4\,{\rm kV/cm}$.

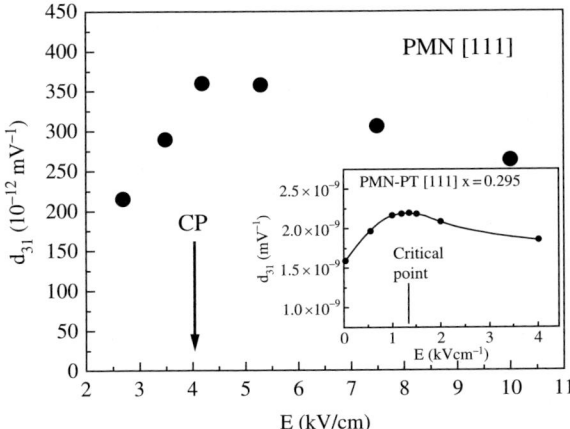

Fig. 6.28 The electric-field dependence of the maximum values of the piezoelectric coefficient d_{31} along the evolution line of a PMN [111] crystal. The arrow denotes the critical field value E_{CP}. The inset shows the electric-field dependence of the piezoelectric coefficient d_{31} along the C–T phase transformation line of a PMN–PT [111] crystal with $x = 0.295$.

electric-field dependence of the thus-obtained maximum values of d_{31} shows itself a clear maximum at the critical point $E = E_{\text{CP}}$ (Fig. 6.28). This is mainly a consequence of the fact that response functions such as the dielectric constant ε are at a maximum at the critical point.

Let us now turn to the PMN–PT system. The points on the temperature–composition (T–x) phase diagram around which the measurements were performed are shown in Fig. 6.22(b). For $\text{PMN}_{0.75}\text{PT}_{0.25}$ and lower concentrations, the critical behaviour is similar to that of pure PMN. The discontinuous jump in P at the ferroelectric–paraelectric transition for $E < E_{\text{CL}}$ disappears for $E \geq E_{\text{CL}}$ and is replaced by a continuous decrease in P with increasing T. The latent heat as well disappears for $E \geq E_{\text{CL}}$. The values of E_{CL} decrease with decreasing disorder.

In the E–T phase diagram, the paraelectric–ferroelectric transition line ends in an isolated critical point above which the difference between the different phases disappears and a continuous supercritical evolution of the response function is observed. The situation gets more complicated with the increase of the PT composition toward the morphotropic boundary ($x \approx 0.35$) (Fig. 6.22(b)). Already for the composition ($x \approx 0.295$) just above the triple point ($x \approx 0.29$), at least four different anomalies have been observed in the temperature dependences of ε, P, and the heat capacity even in the zero field.

The temperature dependence of the dielectric constant ε for a crystal with $x = 0.295$ poled along [110] and measured at a bias field of $E = 0.03\,\text{kV/cm}$ shows (Fig. 6.29) at least three additional phases between the R and T ferroelectric states.

The different phases have been identified by polarized-light microscopy as a monoclinic phase (M_B), an orthorhombic (O) phase, and another monoclinic phase (M_C). The phase sequence R $\to$ M_B $\to$ O $\to$ M_C $\to$ T is similar to what was found in [111]

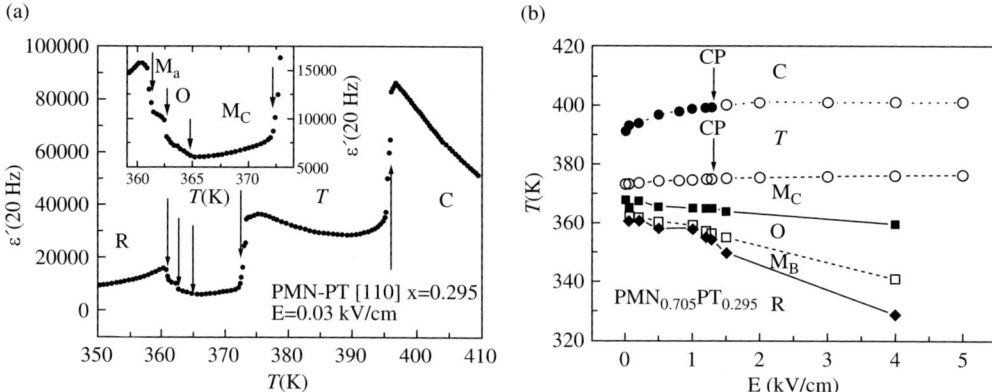

Fig. 6.29 (a) The temperature dependence of the dielectric constant ε of a PMN–PT crystal poled along the [110] direction with $x = 0.295$ and measured at a bias field of $E = 0.03\,\text{kV/cm}$. Note minute steps in the dielectric constant at intermediate monoclinic and orthorhombic phase transitions shown by arrows. (b) The electric-field–temperature phase diagram of a PMN–PT crystal with $x = 0.295$ poled along the [110] direction and obtained in repeated temperature scans at different constant-bias electric fields.

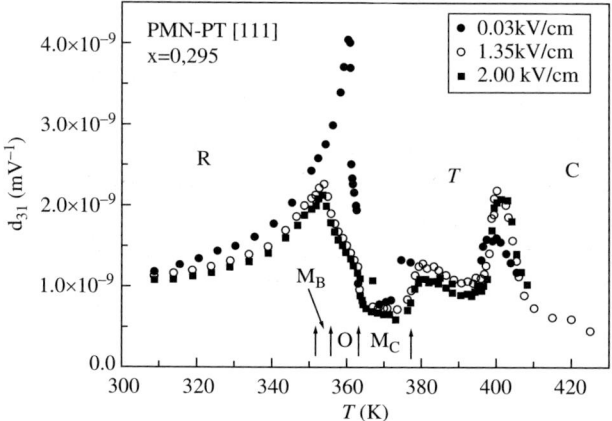

Fig. 6.30 The temperature dependence of the piezoelectric coefficient d_{31} obtained in a PMN–PT single crystal poled along [110] at three different values of the bias electric field. The phase-transition temperatures of intermediate monoclinic and orthorhombic phases obtained at higher bias field values are indicated by arrows.

crystals. The temperature dependence of the piezoelectric coefficient d_{31} for $E \parallel [110]$ is shown in Fig. 6.30.

The electric-field dependence of the transition temperature between the C, T, M_C, O, M_B and R phases for $E \parallel [110]$ is shown in Fig. 6.29(b). The T–M_C transition around 373 K, in particular, is also critical. As shown by the C_p data (Fig. 6.31), it is superimposed on the C–T critical transition at 390 K.

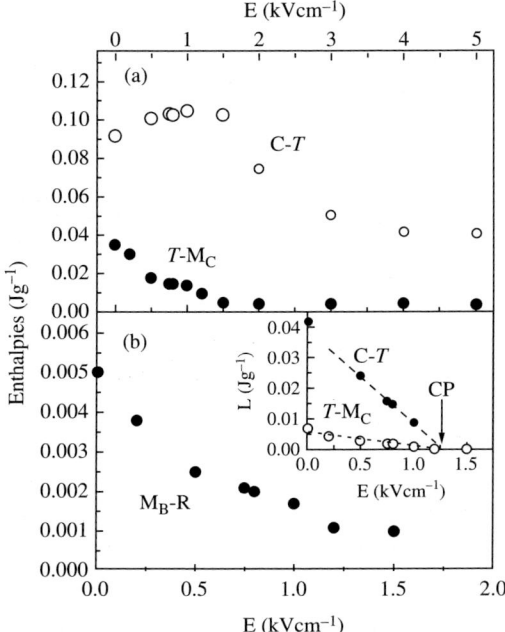

Fig. 6.31 (a) The total enthalpy ΔH for the C–T transition (open circles) and for the T–M$_C$ transition (solid circles) as a function of the electric bias field applied along the [111] direction. (b) The total enthalpy ΔH for M$_B$–R transition as a function of the electric bias field. The arrow indicates the critical value of the electric field. The inset shows the vanishing of the C–T and T–M$_C$ latent heats at the critical field value of $E_{\mathrm{CP}} = 1.3\,\mathrm{kV/cm}$.

The transitions among those phases show up also in the T dependence of the piezoelectric coefficient d_{31}. The corresponding results are shown in Fig. 6.30 for bias fields $E = 0.03\,\mathrm{kV/cm}$, $E = 1.35\,\mathrm{kV/cm}$, and $E = 2\,\mathrm{kV/cm}$. Note that our d_{31} data agree well with the data obtained by Sulc and Pokorny [6.62].

The polarization data for $E \parallel [111]$ again show that the ferroelectric to paraelectric transition line ends in an isolated critical point $E_{\mathrm{CL}} = 1.3\,\mathrm{kV/cm}$ above which supercritical behaviour is observed and the difference between the electric-field-distorted tetragonal and the electric-field-distorted cubic phases disappears.

According to polarized-light microscopy data, the polarization rotates under the applied [111] field at fixed temperature from the T [001] direction into the monoclinic [101] (M$_C$) plane and then into the orthorhombic [101] direction (Fig. 6.22(a)). From there, it goes to the rhombohedral [111] direction via the (101) monoclinic (M$_B$) plane. The common phase for $E \gg E_{\mathrm{CL}}$ should thus be pseudorhombohedral for $E \parallel [111]$.

The above model is strongly supported by the heat-capacity data [6.36, 6.37]. There are two critical points: one exists for the C–T and the other for the T–M$_C$ transition. For the T to C transition at $x = 0.295$, the critical point is at $E_{\mathrm{CP}} = 1.3\,\mathrm{kV/cm}$. The total enthalpy ΔH exhibits a peak at the electric field corresponding to the critical point (Fig. 6.31(a)). The electric field dependence of the enthalpy at the T–M_C transitions is presented in Fig. 6.31(a) for $E \parallel [111]$. The inset of Fig. 6.31(b) shows the electric-field dependence of the latent heat L for the T–M_C transition. Again, the latent heat vanishes at $E_{CP} \approx 1.3\,\mathrm{kV/cm}$. The electric-field dependence of the enthalpy for the M_B–R (for $E = 0$) transition is shown in Fig. 6.31(b). The enthalpy changes for the M_B–O and O–M_C transitions are always below $1\,\mathrm{mJ/g}$. As $P \approx 0.3\,\mathrm{C\,m^{-2}}$ and $\Delta H \approx E \cdot P$, we find that the necessary electric field for the rotation of the polarization decreases close to the critical point by nearly a factor of 10:

$$M_C\text{–}T : E_i = 14\,\mathrm{kV/cm} \to 1.5\,\mathrm{kV/cm}.$$

At the R–M_B transition, the corresponding electric field necessary to induce the polarization rotation is at $E \to E_{CL}$ reduced to less than $0.4\,\mathrm{kV/cm}$:

$$R - M_B : E_i = 2\,\mathrm{kV/cm} \to 0.4\,\mathrm{kV/cm}.$$

The electric-field dependences of the piezoelectric coefficient d_{31} along the C-T phase transformation line are presented in the inset of Fig. 6.28, for [111]. The maximum value of d_{31} is again at the critical point value E_{CP}. For $E \parallel [110]$, the situation is more complex, as shown in Fig. 6.30. Here, it appears as if d_{31} exhibits a maximum value already at the M_B–R phase transition. Fits to the simple power-law ansatz

$$d_{31} = At^{-z} + B \tag{68}$$

with $t = (T - T_C)/T_C$ reveal $z \approx 1.09$, $T_C \sim 376\text{–}382\,\mathrm{K}$ for $B \sim (0.27\text{–}0.57) \times 10^{-9}\,\mathrm{mV^{-1}}$ (Fig. 6.32(a)), i.e. the critical dependence of d_{31} observed below the M_B–R phase transition is actually related to the higher temperature T–M_C transition. The critical dependence of d_{31} is just cut off prior to reaching the critical temperature.

This is also confirmed by data sets obtained at different bias electric fields. As shown in Fig. 6.30, the data sets at different bias fields get slightly suppressed and the cut-off temperature is shifted toward lower temperatures with increasing bias field. If the critical behaviour of d_{31} would be related to the M_B–R phase transition, the opposite effect, i.e. enhancement of the data values at given temperature, should be observed. This is due to the increasingly lowered M_B–R phase-transition temperature with increasing bias field. The decrease of the data values is, on the other hand, in agreement with the fact that the T–M_C transition temperature is increasing with increasing bias field, thus causing downward rescaling of the d_{31} data at a given fixed temperature.

The existence of the critical points in the E dimension of the T–x–E phase diagram of the PMN–PT system results in a significant enhancement of the piezoelectric coefficients as well as a significant decrease of energy costs and electric fields necessary to induce the R–M_B–O–M_C–T polarization rotations (Figs. 6.22 and 6.30) producing the giant lattice strains. The system behaves as being effectively semisoft when $E \to E_{CL}$. The theoretically evaluated E–T critical line agrees qualitatively with the experimental data. The maximum of the piezoelectric response is not achieved at

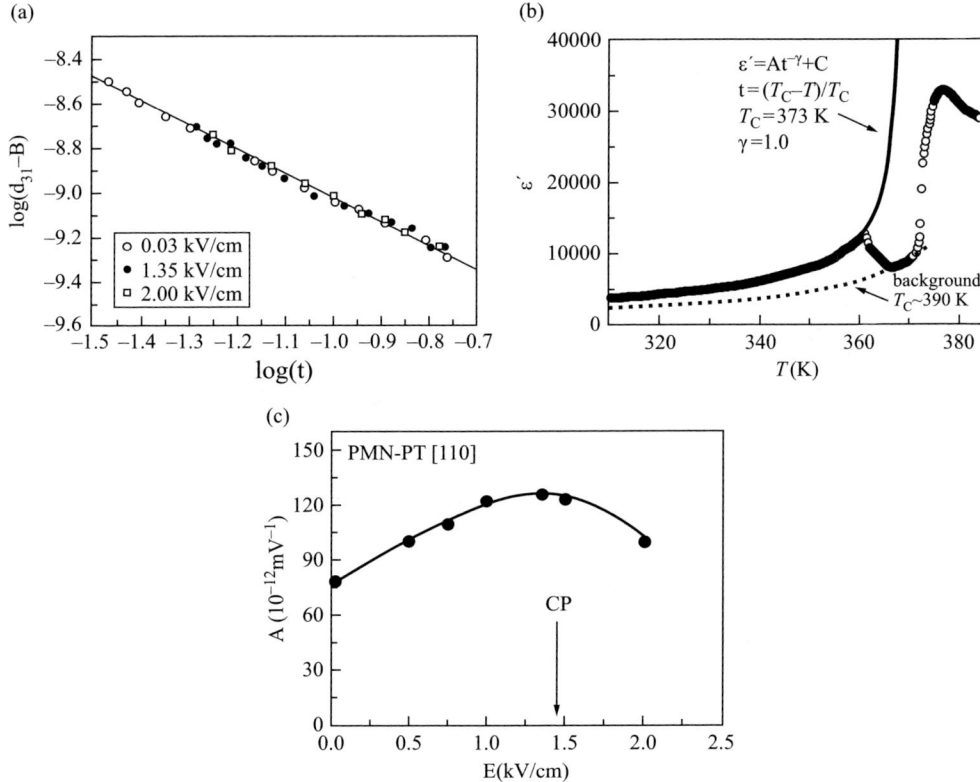

Fig. 6.32 (a) The temperature behaviour of the adjusted piezoelectric coefficient d_{31} obtained on a PMN–PT $x = 0.295$ [110] sample. The straight line indicates that the temperature dependence follows a power-law behaviour. (b) The temperature dependence of the dielectric constant obtained on the same sample at $E = 0.03\,\text{kV/cm}$. The solid line represents a fit to the power law (see the text) after subtraction of the background represented by a dashed line. (c) The electric-field dependence of the amplitude A in the d_{31} power ansatz (Eq. (68)), which exhibits a maximum at the critical field value of E_{CP}.

$E = 0$, but at $E = E_{\text{CL}}$. The electric-field-induced critical points thus provide a new driving force for the giant electromechanical response. In view of the large electric-field dependence of the dielectric constant near the critical point, these systems may also play an important role in electric-field-tunable optoelectric elements.

6.12 Critical end points up to 8th-order terms

In view of the existence of a monoclinic phase in the phase diagram that cannot be described by the Landau expansion up to the sixth order, it seems meaningful to discuss the critical end point in detail based on a simple Landau free energy expanded up to the eighth order [6.38].

We write the free energy as

$$f = \frac{\alpha}{2}p^2 + \frac{\beta}{4}p^4 + \frac{\gamma}{6}p^6 + \frac{\delta}{8}p^8 - pE \tag{69}$$

where α is temperature dependent as

$$\alpha = a(T - T_0), \quad (a > 0) \tag{70}$$

and a, γ, and δ are constants. The scaled polarization p is assumed to be a scalar quantity. For the critical end point to appear in the field-temperature phase diagram, the transition from the paraelectric phase ($p = 0$) to the ferroelectric phase ($p \neq 0$) should be of the first order. With the free-energy function (69), there are two cases conceivable to reproduce the first-order transition, that is, the case that $\gamma > 0$ and $\beta < 0$, and the case that $\gamma < \gamma_c < 0$. For convenience, we summarize the ordinary first-order transition case, where the free energy expansion is truncated at the sixth order, i.e. $\delta = 0$, and then proceed to the special case, where it is truncated at the eighth order.

1) The ordinary case

In this case, in Eq. (69) $\beta < 0, \gamma > 0$ and $\delta = 0$. The transition temperature under zero applied field ($E = 0$), i.e. α_c, and the scaled polarization at α_c, p_c, are obtained from

$$f = 0, \frac{df}{dp} = 0 \tag{71}$$

as

$$\alpha_c = \frac{3\beta^2}{16\gamma}, p_c^2 = -\frac{3\beta}{4\gamma} \tag{72}$$

On the other hand, the critical end point (α_{CE}, E_{CE}) in the temperature–applied field phase diagram, as well as the polarization there, are obtained from

$$\frac{df}{dp} = 0, \frac{d^2f}{dp^2} = 0, \frac{d^3f}{dp^3} = 0 \tag{73}$$

with $E \neq 0$ as

$$\alpha_{CE} = \frac{9\beta^2}{20\gamma} = \frac{12}{5}\alpha_c \tag{74}$$

$$p_{CE}^2 = \frac{-3\beta}{10\gamma} = \frac{2}{5}p_c^2 \tag{75}$$

$$E_{CE} = \frac{6\beta^2}{25\gamma}\sqrt{\frac{-3\beta}{10\gamma}} = \frac{8}{15}|p_{CE}|\alpha_{EC} \tag{76}$$

From the above it is seen that there are certain numerical relations, independent of β and γ, between α_{CE} and α_c, between p_{CE} and p_c, and among α_{CE}, p_{CE}, and E_{CE}. Now, let us proceed to the special case of a first-order transition, where the eighth-order term should be included.

2) The case where $\gamma > 0$ and $\delta > 0$.

In this case, the transition is of first order when $\beta < 0$ and the transition temperature in the absence of the applied field is given as

$$\alpha_c = \frac{\beta^2}{4\gamma} \frac{1}{b^2} [3b - 2 + 2(1-b)\sqrt{1-b}] \tag{77}$$

and the polarization at α_c, p_c, is given as

$$p_c^2 = -\frac{81\alpha_c\delta - 6\beta\gamma}{27\beta\delta - 8\gamma^2} = -\frac{3\beta}{4\gamma} \frac{1 - bc\alpha_c}{1 - b} \tag{78}$$

where

$$b = \frac{27\beta\delta}{8\gamma^2}, \quad c = \frac{4\gamma}{\beta^2} \tag{79}$$

The critical end point (α_{CE}, E_{CE}) as well as p_{CE}^2 can be obtained analytically from Eq. (73). We do not write down these formulas, because the expressions are too complicated for their physical significance. What is important here is that the relations between α_c and α_{CE} and between p_c^2 and p_{CE}^2 are not much different from those given in Eqs. (74) and (75) as it can be proved that

$$\frac{\alpha_{CE}}{\alpha_c} = \frac{12}{5} + O(b) \tag{80a}$$

and

$$\frac{p_{CE}^2}{p_c^2} = \frac{2}{5} + O(b) \tag{80b}$$

where $O(b)$ indicates the quantity of the order of b.

3) The case where $\gamma < \gamma_c < 0$ and $\delta > 0$:

The transition is of first order if $\beta < 0$. It is of first order even if $\beta > 0$ when $\gamma < \gamma_c < 0$. The transition temperature for $E = 0$ is in this case given by:

$$\alpha_c = \frac{16\gamma^3}{3^6 \cdot \delta^2} \left[3b - 2 - 2(1-b)\sqrt{1-b} \right] \tag{81}$$

where $\gamma < \gamma_c = -\sqrt{9\beta\delta/2}$ with $\beta > 0$. p_c^2 is given by expression (78).

In the special case $\beta = 0$, α_c, p_c, and α_{CE}, p_{CE} and E_{CE} are obtained as

$$\alpha_c = -\left(\frac{2}{3}\right)^6 \frac{\gamma^3}{\delta^2}, \quad p_c^2 = -\frac{8\gamma}{9\delta} \tag{82}$$

$$\alpha_{CE} = -\left(\frac{2}{7}\right)^2 \left(\frac{5}{3}\right)^3 \frac{\gamma^3}{\delta^2} \tag{83}$$

$$p_{CE}^2 = -\frac{10\gamma}{21\delta} \tag{84}$$

$$E_{CE} = -\frac{2^5 \times 5^2}{3^3 \times 7^2} \frac{\gamma^3}{\delta^2} \tag{85}$$

The ratio of α_c and α_{CE} is about 4.3, which is very different from 2.4 in Eqs. (74), implying the first-order transition here is qualitatively quite different from the ordinary ones. The transition from the paraelectric phase to the relaxor-type phase in perovskite-type ferroelectrics seems to be of first order, as can be concluded from the existence of the critical end point in the temperature–applied field phase diagram where the monoclinic phase appears in the phase sequence.

In this connection it is important to note that a maximum of the piezoelectric tensor occurs at the critical end point in relaxors in spite of the presence of inherent disorder [6.63]. In the clean limit a divergence of the piezoelectric tensor occurs at the critical point. A small amount of disorder translates the critical point in the temperature–electric-field phase diagram and makes the phase transition from the paraelectric to the ferroelectric phase more diffuse [6.63].

7
Ferroelectric polymers

Electroactive polymers with a high strain response to an applied electric field are attractive for a broad range of applications, such as ultrasonic transducers for medical diagnostics, sonars, robots, and artificial muscles. Typical representatives are the electron-irradiated ferroelectric copolymer poly(vinylidene fluoride-trifluoroethylene) or P(VDF-TrFE) and its terpolymer with chlorofluoroethylene P(VDF-TrFE-CFE). High-energy electron irradiation is needed in order to transform the normal ferroelectric structure into an amorphous one, thus strongly enhancing their electromechanical properties. Recently, it has been demonstrated that these polymers exhibit physical properties, which are analogous to inorganic relaxor ferroelectrics such as PMN, PLZT, etc. [7.1]. Similar to their inorganic counterparts, relaxor-like polymer systems are characterized by slow relaxation and a strong frequency dispersion of the dielectric permittivity. In the low-frequency limit, a broad temperature maximum of the dielectric response is observed, and the longest relaxation time diverges at the freezing temperature T_f according to the Vogel–Fulcher law: $\tau^{-1} = \tau_0^{-1} \cdot \exp[\frac{-E}{T_{max}-T_0}]$.

Following the physical analogy between inorganic relaxor ferroelectrics and relaxor polymers, a theoretical model has been proposed based on the assumption that electron irradiation creates a random set of polar nanoregions (PNRs) embedded in an amorphous matrix [7.2]. Meanwhile, in terpolymers disorder is introduced by the fluctuations of chemical composition alone, and PNRs appear without the need for irradiation.

Ferroelectricity is a property of well-ordered solids that belong to one of the ten pyroelectric crystal classes. It is therefore surprising that ferroelectric polymers (Fig. 7.1) exist too [7.3–7.5]. The best known ferroelectric (FE) polymer is polyvinylidene fluoride (abbreviated as PVDF) [7.3–7.6].

The characteristic properties of this system are:

i) PVDF is a semicrystalline polymer with $\approx$ 50% crystallinity. The monomer is strongly polar [-CH$_2$-CF$_2$-] and contains in addition to the linear carbon backbone two dipolar units, one associated with the -CH$_2$ and the other with the -CF$_2$ unit (Fig. 7.2).

ii) There are many modifications of PVDF. The most important are the anti-polar α-phase (C_{2h}) and the polar β-hase (C_{2v}) both dispersed in the amorphous phase (Fig. 7.2). By stretching, the non-polar α-films (C_{2h}) can be transformed into the polar β-phase (orthorhombic C_{2v}) with two parallel chains per unit cell.

iii) Still another important point is that there must be enough space available to overcome the van der Waals forces and to induce the molecular motion necessary

188 Ferroelectric polymers

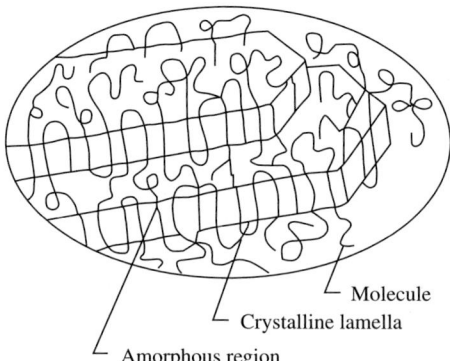

Fig. 7.1 Semicrystalline polymers [7.3].

for a ferroelectric to paraelectric (FE to PE) phase transition. In many cases the polymer melts before this condition is reached. This situation is changed in copolymers VDF/TrFE that exhibit the ferroelectric transition below the melting point T_M. Here, TrFE [-CF$_2$-CFH-] stands for trifluoroethylene and TeFE for tetrafluoroethylene [-CF$_2$-CFH-]. The TeFE and TrFE molecules are randomly distributed in the polymer and can be viewed also as defects. These copolymers

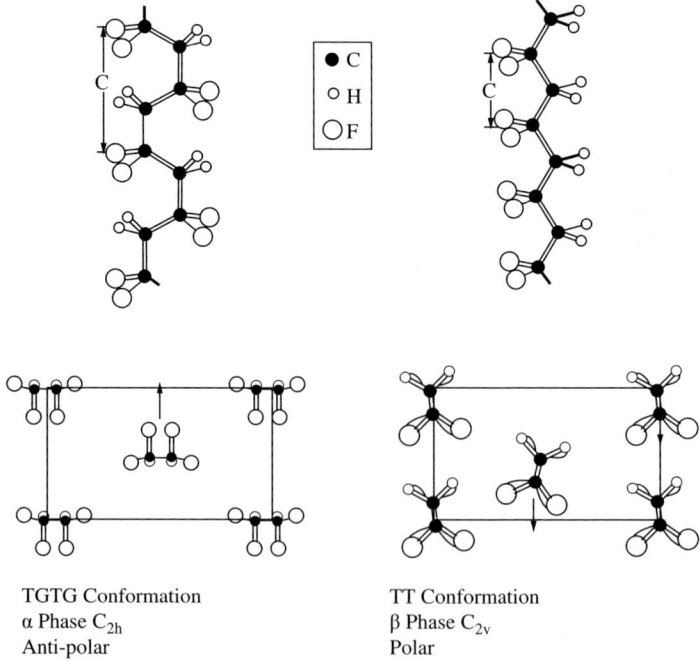

Fig. 7.2 Carbon backbone and the structure of the α- and β-phases of PVDF [7.6].

are important technologically, e.g., as active piezoelectric films. Films can be produced directly from the melt and no stretching is necessary. The degree of crystallinity is higher than in pure PVDF and so are the piezo- and pyroelectric coefficients.

iv) The polymers are highly compressible and hydrostatic pressure is expected to produce large changes in their properties [7.7]. As can be seen from Fig. 7.3, pressure shifts the dielectric peaks to higher temperatures. This can be understood in terms of a closer packing of the polymer chains that hinders the motion necessary for the FE to PE transition. Therefore, higher temperatures are necessary to induce this motion. At still higher pressures there is a large decrease in the area under the ε'' peak, demonstrating a gradual disappearance of the PE to FE motion.

v) Recent dielectric experiments have revealed a striking analogy between inorganic relaxor systems and disordered organic polymers. We show that a statistical ensemble of randomly oriented polar nanoregions embedded in an amorphous matrix can be well described by the SRBRF model, which is exactly soluble [7.2]. The main difference between the SRBRF model for inorganic relaxors and the model for polymer ferroelectrics is in the physical characteristics of polar nanoregions and the nature of interaction among them. The calculated third-order non-linear susceptibility indeed predicts a cross-over between paraelectric behaviour at high temperature and a relaxor-like behaviour near the freezing temperatures.

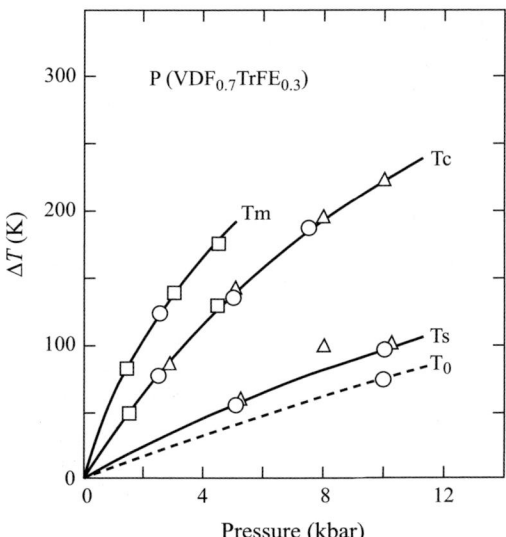

Fig. 7.3 Shifts of the various transition temperatures of the copolymer $P[(VDF)_{0.7}(TrFE)_{0.3}]$ with pressure [7.7]. T_s represents the peak value of ε'', i.e. $T_s = T_{max}$, at 100 kHz. T_0 is the corresponding Vogel–Fulcher 'freezing' temperature $\tau^{-1} = \tau_0^{-1} \cdot \exp[\frac{-E}{T_{max}-T_0}]$. T_c is the PE-FE transition temperature and T_m is the melting temperature.

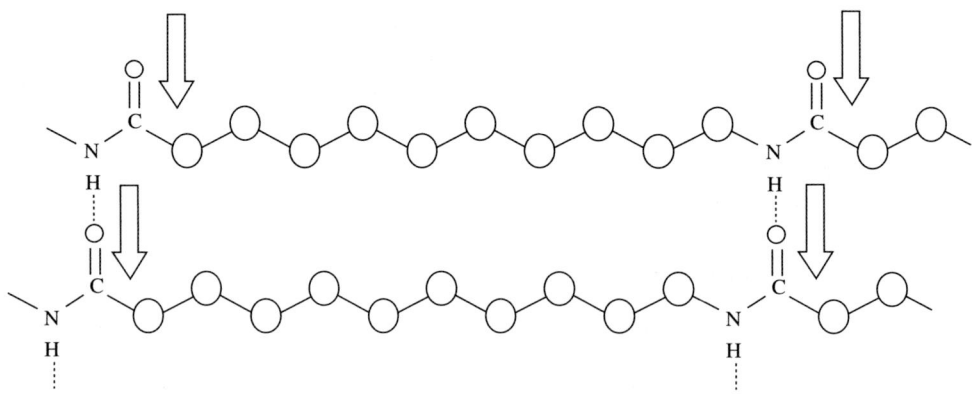

Fig. 7.4 Schematic structure of nylon 11 [7.8].

These results are typical for other ferroelectric polymers too. This is particularly true for the odd [$n = 7, 11$, etc.] nylons that also exhibit ferroelectric transitions (Fig. 7.4).

Odd polyamides – or nylons – are semicrystalline polymers with a hydrogen-bonded sheet structure. The dipole moments are perpendicular to the chains and point in the same direction in all the sheets. The T dependence of the D vs. E hysteresis loop of nylon 11 is shown in Fig. 7.5.

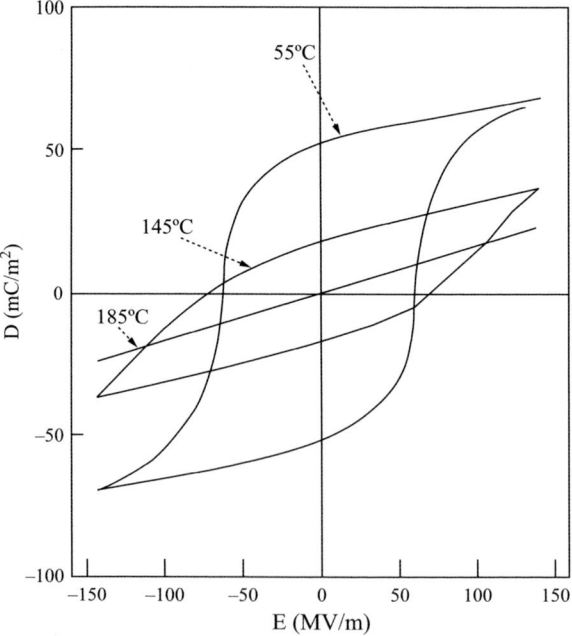

Fig. 7.5 T dependence of the D vs. E hysteresis loop of nylon 11 [7.9].

The theory of ferroelectric polymers can be described by the random-bond–random-field model. It should be mentioned that high-energy radiation, i.e. both electrons and gamma rays, significantly changes the properties of PVDF and TrFE copolymers. The main effect is a large reduction in the degree of crystallinity and a replacement of ferroelectric macrodomains by polar nanodomains. The presence of polar nanodomains makes these systems close to relaxor ferroelectrics [7.2, 7.10, 7.11].

7.1 2D ferroelectricity

Another important development in the field of FE polymers has been the discovery of ferroelectricity in Langmuir–Blodgett (L–B) VDF-TrFE polymer films [7.12]. These films are relatively easy to make and have a thickness of 2 to 500 monolayers. The films have excellent order and are much better suited than semicrystalline thick films to study the FE properties.

In a crystalline Langmuir-Blodgett deposited random copolymer of vinylidene fluoride with trifluoroethylene P(VDF – TrFE 70:30) on graphite, switchable ferroelectric films can be made to a 0.9 nm thickness. The minimum thickness is just two monolayers. Finite size effects thus seem here to impose no practical limitations on thin-film memory capacitors though for some designs tunneling currents may become too large.

In P[VDF – TrFE] 70:30 crystalline films the first-order phase-transition temperature T_c at 108°C on heating and 77°C on cooling was found to be nearly equal to the bulk value, even in films as thin as 10 Å [7.12]. This has been attributed to two-dimensional ferroelectricity where the ferroelectric state is generated only within the plane of the film. The absence of finite-size effects when the film thickness is decreased from 150 Å to 10 Å seems to support this conclusion.

The interphase coupling here should be rather weak and the system can be described by the anisotropic Ising model. Fluctuations in two dimensions should not destroy ferroelectricity – as in the case of an isotropic Heisenberg ferromagnet – because of the anisotropy of the coupling.

The low-temperature second-order phase transition at $T_c \approx 20°C$ in P(VDF – TrFE) films of 30 monolayers or less can be described as a surface layer transition controlled by the interaction with the substrate or the top electrode.

The structure of a L–B P(VDF$_{0.7}$TrFE$_{0.3}$) film is shown in Fig. 7.6.

7.2 Spherical model of relaxor polymers

Let us now treat the SRBRF model of relaxor polymers in greater detail.

7.2.1 Polar nanoregions

We consider the case of a disordered ferroelectric polymer, characterized by a lamellar crystalline structure surrounded by an amorphous configuration of polymer chains [7.3]. The chains are assumed to be twisted and broken into nanosized segments of variable lengths by either irradiation as in P(VDF-TrFE) or by compositional disorder

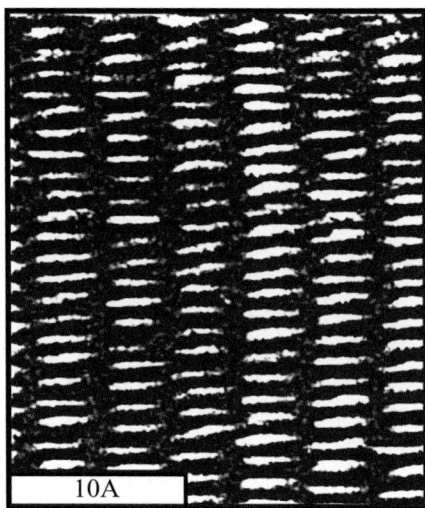

Fig. 7.6 Structure of P(VDF$_{0.7}$TrFE$_{0.3}$) L–B film [7.13].

as in the case of terpolymer P(VDF-TrFE-CFE). For example, it has been shown that electron irradiation breaks up the all-trans $T_{M>3}$ chains in the normal ferroelectric P(VDF-TrFE) polymer, resulting in an irregular structure where the remaining all-*trans* segments are interrupted by T_3G and TG *trans–gauche* bonds [7.14]. Similarly, in terpolymers, due to a much larger van der Waals radius of chlorine (1.8 Å) as compared to hydrogen (1.2 Å) and fluorine (1.35 Å) atoms, the addition of CFE or CFTE monomers favours the formation of non-polar $T_3GT_3\bar{G}$ bonds against polar all-*trans* bonds between the monomers [7.15]. Thus, in each case the breaking up of the ferroelectric layered structure results in a random network of polar nanoregions embedded in an amorphous matrix, which are responsible for the observed relaxor behaviour.

In a simplified description, we will think of a polarized nanoregion as a compact lamellar array of polar nanosegments, with a net dipole moment perpendicular to the segments [7.3]. The magnitude of the dipole moment depends on the volume of the region and on the degree of ordering of the dipoles, and can thus vary over a wide range. Since the orientations of the dipole moments will also change across the sample, we are dealing with a statistical ensemble of randomly oriented dipoles embedded in a continuous amorphous medium. With the ith nanoregion, $I = 1, 2, \ldots, N$, we therefore associate a vector $\vec{S}_i = S_i \vec{\sigma}_i$, where the length of the dipole moment is represented by the dimensionless scalar variable S_i, and its orientation is described by an m-dimensional unit vector $\vec{\sigma}_i$, where m is the effective dimensionality of the embedding medium. For $N \gg 1$, we can assume a continuous probability distribution of S on the interval $(0, \infty)$. For convenience, we choose the generalized gamma distribution having the form

$$\rho(S) = A_m S^{m-1} \exp(-S^2/2a^2) \tag{1}$$

where $A_m = 2(2)^{-m/2}/\Gamma(m/2)$ and the parameter a determines the distribution width. We can readily evaluate the moments $\langle S^p \rangle$, where p is some positive integer,

$$\langle S^p \rangle = 2^{p/2} a^p \, \Gamma\left(\frac{m+p}{2}\right) / \Gamma\left(\frac{m}{2}\right) \quad (2)$$

The first two moments are $\langle S \rangle = 2^{1/2} a \Gamma((m+1)/2)/\Gamma(m/2)$ and $\langle S^2 \rangle = ma^2$. Returning to the discrete case we find $\sum_{i=1}^{N} (\vec{S}_i)^2 = ma^2 N$. We can now rescale the fields $\vec{S}_i \to \vec{S}_i/a$ and obtain

$$\sum_{i=1}^{N} (\vec{S}_i)^2 = mN \quad (3)$$

This is equivalent to a generalized spherical condition for the m-component order parameter field $\vec{S}_i$ with the width of the corresponding distribution $\rho(S)$ equal to unity. We will, therefore, describe the relaxor polymer by the m-component SRBRF model in which the order-parameter field $\vec{S}_i$ satisfies the spherical condition (3).

7.2.2 Free energy

The free energy of a quenched disordered system described by the spherical model can be calculated exactly either by using the eigenvalue spectrum of the random interaction matrix or by applying the replica method [7.16].

The Hamiltonian of a relaxor polymer is now written as

$$\mathcal{H} = -\frac{1}{2} \sum_{i,j=1}^{N} \sum_{\mu=1}^{m} J_{ij} S_{i\mu} S_{j\mu} - \sum_{i=1}^{N} \sum_{\mu=1}^{m} (h_{i\mu} + gE_\mu) S_{i\mu} \quad (4)$$

As usual, J_{ij} are quenched random interactions between the polar nanoregions i and j, which are infinitely ranged and have a Gaussian probability distribution such that

$$\lfloor J_{ij} \rfloor_{av} = J_0/N, \; [J_{ij}^2]_{av}^c = J^2/N \quad (5)$$

In the last term of Eq. (4), $h_{i\mu}$ are components of local random fields with zero mean and second cumulant

$$[h_{i\mu} h_{j\nu}]_{av}^c = \delta_{ij} \delta_{\mu\nu} \Delta \quad (6)$$

E_μ is the applied field, and g an effective dipole moment. When $E = 0$, the model is characterized by three parameters, namely, J_0, J, and Δ.

In inorganic relaxors, compositional fluctuations give rise to so-called chemical clusters, i.e. chemically ordered regions with a net non-zero electric charge, which act as sources of random electric fields. This is to be contrasted with relaxor polymers, where such charged regions are less likely to be created by the random disorder. Therefore, we expect that random electric fields will be weaker in polymers. On the other hand, the breaking up of the layered ferroelectric structure results in large strain fields, which have a different symmetry from the electric fields, but may also affect

194 Ferroelectric polymers

the polar nanoregions through piezoelectric and electrostrictive coupling. Thus, it is not surprising that strong electrostriction is observed in some polymer composites.

For a general direction of the electric field $\vec{E} = (E_1, E_2, ..., E_m)$, where we choose $E_1 = E$ and $E_2 = ... = E_m = 0$, we obtain three types of thermodynamic averages, namely,

$$\langle S_{\mu\alpha} \rangle = P_{\mu\alpha} \tag{7}$$

$$q_{\mu\nu\alpha\beta} = \langle S_{\mu\alpha} S_{\nu\beta} \rangle \equiv \delta_{\mu\nu} q_{\mu\alpha\beta}, \, q_{\mu\alpha\beta} = \langle S_{\mu\alpha} S_{\mu\beta} \rangle \tag{8}$$

$$\langle S_{\mu\alpha} S_{\nu\alpha} \rangle = r_{\mu\nu\alpha} \equiv \delta_{\mu\nu} r_{\mu\alpha}, \, r_{\mu\alpha} \equiv \langle S_{\mu\alpha}^2 \rangle \tag{9}$$

The $\mu \neq \nu$ averages are zero by symmetry.

It is well known that for a random spherical model the replica symmetric solution is exact. Thus we may choose the longitudinal (L) and transverse (T) components of the order parameters (8)–(10) as follows [7.17, 7.18]:

$$P_{\alpha 1} = P_L \equiv P, \quad P_{\alpha 2} = ... P_{\alpha m} = P_T = 0 \tag{10}$$

$$q_{1\alpha\beta} = q_L(1 - \delta_{\alpha\beta}), \quad q_{2\alpha\beta} = ... = q_{m\alpha\beta} = q_T \tag{11}$$

$$r_{1\alpha} = r_L, \quad r_{2\alpha} = ... = r_{m\alpha} = r_T \tag{12}$$

7.2.3 Order parameters

The equilibrium values of the order parameters $q_{L,T}, r_{L,T}$, and P are determined by the saddle point conditions $\partial f/\partial q_{L,T} = \partial f/\partial r_{L,T} = \partial f/\partial P = 0$.

The saddle-point conditions now lead to the following relations:

$$q_L = \beta^2 \lfloor J^2 \tilde{q}_L + (J_0 P + gE)^2 \rfloor B_L^2 \tag{13a}$$

$$q_T = \beta^2 J^2 \tilde{q}_T B_T^2 \tag{13b}$$

$$r_L - q_L = B_L \tag{13c}$$

$$r_T - q_T = B_T \tag{13d}$$

and

$$q_L + B_L + (m-1)(q_T + B_T) = m \tag{14}$$

It can be shown that Eqs. (13c) and (13d) imply

$$r_L - q_L = r_T - q_T \tag{15}$$

and thus $B_L = B_T \equiv B$, with B to be determined below.

From Eqs. (13c), (13d), and (14) we find

$$r_L + (m-1)r_T = m \tag{16}$$

The field-cooled polarization is given by the equation

$$P = \beta(J_0 P + gE)B, \tag{17}$$

so that Eqs. (13a) and (13b) can be rewritten as

$$q_\text{L} = \beta^2 J^2 \tilde{q}_\text{L} B^2 + P^2 \tag{18a}$$

$$q_\text{T} = \beta^2 J^2 \tilde{q}_\text{T} B^2 \tag{18b}$$

The order parameters $q_\text{L,T}$ and $r_\text{L,T}$ are in general the diagonal elements of $m \times m$ matrices $\mathbf{q}$ and $\mathbf{r}$, respectively. It is convenient to consider the corresponding invariants

$$q \equiv \frac{1}{m} \operatorname{Tr} \mathbf{q} = \frac{1}{m}[q_\text{L} + (m-1)q_\text{T}] \tag{19}$$

$$r \equiv \frac{1}{m} \operatorname{Tr} \mathbf{r} = \frac{1}{m}[r_\text{L} + (m-1)r_\text{T}] = 1 \tag{20}$$

where the last relation follows from Eq. (16). The parameter q plays the role of the Edwards–Anderson (E–A) order parameter in dipolar glasses [7.18].

Equations (13)–(16) now yield

$$B = (1-q) \tag{21}$$

Thus,

$$q = \beta^2 J^2 \tilde{q}(1-q)^2 + P^2/m \tag{22}$$

and

$$P = \beta(J_0 P + gE)(1-q). \tag{23}$$

Analogous results have been derived earlier using symmetry arguments for inorganic relaxors ($m=3$) in a field $\vec{E}\|[111]$ and for $\vec{P} = P(1,1,1)$[7.19].

It should be noted that in analogy to spin and dipolar glasses the quantities q and r are not thermodynamic order parameters, and thus cannot be derived by minimizing the free energy, but rather follow from the saddle-point conditions employed in evaluating the free energy.

7.3 Dielectric susceptibility

7.3.1 Longitudinal and transverse susceptibilities

The linear longitudinal static field-cooled susceptibility $\chi_{1,\text{L}}$ of a relaxor polymer is obtained by differentiating the physical polarization $P = (g/v_0)$ with respect to the field E, where we introduce $v_0 = V/N$ as the effective average volume of a polar nanoregion. Equation (23) yields

$$\chi_{1,\text{L}} = \left(\frac{g^2}{v_0}\right) \frac{\beta(1-q)}{1 - \beta J_0 (1-q)} \tag{24}$$

The corresponding transverse susceptibility $\chi_{1,\text{T}}$ follows from the free energy by differentiating with respect to an infinitesimal field along any of the perpendicular directions $\mu \neq 1$, say, E_2,

$$\chi_{1,\text{T}} = \left(\frac{g^2}{v_0}\right) \beta \left[\langle S_2^2 \rangle - \langle S_2 \rangle^2\right]_\text{av} \tag{25}$$

where $[\cdots]_{\mathrm{av}}$ again represents the random average. It is easily seen that the order parameters $q_{L,T}$ and $r_{L,T}$ are given by

$$q_{\mathrm{L}} = \lfloor \langle S_1 \rangle^2 \rfloor_{\mathrm{av}}, \quad q_{\mathrm{T}} = \lfloor \langle S_2 \rangle^2 \rfloor_{\mathrm{av}} \tag{26}$$

$$r_{\mathrm{L}} = \lfloor \langle S_2^2 \rangle \rfloor_{\mathrm{av}}, \quad r_{\mathrm{T}} = \lfloor \langle S_1^2 \rangle \rfloor_{\mathrm{av}} \tag{27}$$

Thus,

$$\chi_{1,\mathrm{T}} = \left(\frac{g^2}{v_0}\right)\beta(1-q) \tag{28}$$

and from Eq. (24) we obtain the relation

$$\chi_{1,\mathrm{L}} = \frac{\chi_{1,\mathrm{T}}}{1 - J_0(v_0/g^2)\chi_{1,\mathrm{T}}} \tag{29}$$

In Fig. 7.7, the calculated temperature dependence of the static field-cooled longitudinal susceptibility $\chi_1 \equiv \chi_{1,\mathrm{L}}$ and the corresponding order parameter q obtained from Eq. (22) are shown for $J_0/J = 0.9$ and three representative values of the random field parameter Δ/J^2. Figure 7.8 shows a comparison between the experimental and theoretical values of the static field-cooled dielectric constant $\varepsilon_s = 1 + \chi_1/\varepsilon_0$ in electron-irradiated copolymer P(VDF-TrFE) [7.20] and in terpolymer P(VDF-TrFE-CFE) [7.21]. The solid lines were calculated from Eq. (24) and fitted to the data using a set of fit parameters listed in the caption, and by adjusting the amplitude of χ_1. The agreement between the experimental and theoretical values is reasonably good. It should also be noted that the fit values $J/k = 276 \pm 5\,\mathrm{K}$ and $266 \pm 16\,\mathrm{K}$ are close

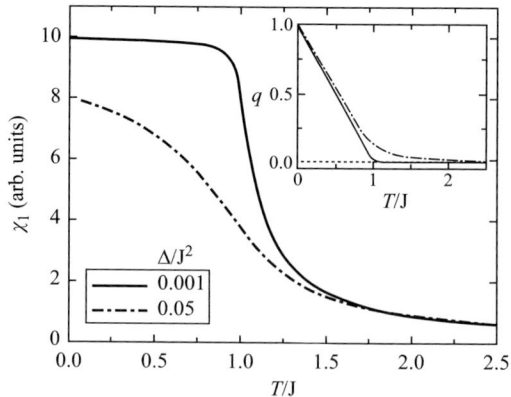

Fig. 7.7 Calculated temperature dependence of the static linear field-cooled longitudinal dielectric susceptibility of a relaxor polymer in the zero-field limit for two values of the random field strength Δ/J^2. Inset: Temperature dependence of the Edwards–Anderson order parameter.

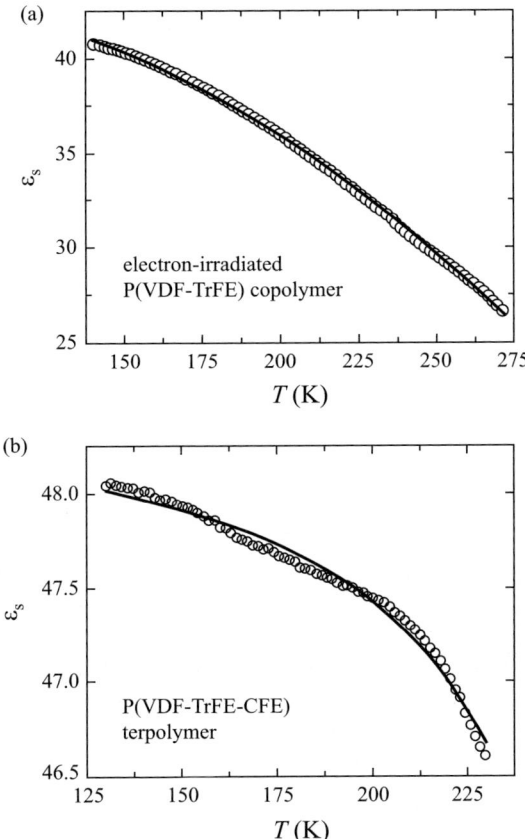

Fig. 7.8 Comparison between calculated (solid lines) and experimental (circles) static field cooled dielectric constants for two relaxor polymer systems, as indicated. (a) Data: [7.20]. Solid line: Fit with $J/k = 276 \pm 5\,\mathrm{K}$, $J_0/k = 234\,\mathrm{K}$, $\Delta/J^2 = 0.059$. (b) Data: [7.21]. Solid line: $J/k = 266 \pm 16\,\mathrm{K}$, $J_0/k = 113\,\mathrm{K}$, $\Delta/J^2 = 0.008$.

to the experimental values for the freezing temperature $T_\mathrm{f} = 277 \pm 2\,\mathrm{K}$ and $269 \pm 2\,\mathrm{K}$, respectively, determined from the temperature-frequency plots [7.20, 7.21]. The random field strength $\Delta/J^2 = 0.059$ in electron-irradiated P(VDF-TrFE) is much larger than $\Delta/J^2 = 0.008$ for P(VDF-TrFE-CFE), suggesting that the degree of local disorder created by electron irradiation may be stronger than the intrinsic compositional disorder occurring in the terpolymer.

In vector spin glasses, the transverse order parameter q_T vanishes above the so-called Gabay–Toulouse (G–T) line in the (E, T)-plane [7.22]. In the present case, one has $q_\mathrm{T} \neq 0$ at all temperatures and fields due to the presence of random fields, and there can be no GT line. However, even without random fields ($\Delta = 0$) it can be shown that there is no GT line in the spherical model.

To illustrate the above point further, let us consider the simpler case $\Delta = 0$ and $J_0 = 0$, but $E \neq 0$. Let us assume that $q_T \neq 0$. According to Eq. (18), in which now $\tilde{q}_{L,T} = q_{L,T}$, we have $\beta^2 J^2 B_T^2 = 1$ or

$$r_T - q_T = \frac{1}{\beta J} \tag{30}$$

Since $r_L - q_L = r_T - q_T$, we see that $r_L - q_L = 1/(\beta J)$, too, and Eq. (18a) cannot be fulfilled unless $P = E = 0$. Thus, for $E \neq 0$ we cannot have $q_T \neq 0$ at any temperature, and there can be no transition from $q_T = 0$ to $q_T \neq 0$, i.e. no GT line.

It should finally be noted that a general feature of classical continuous models – and hence of the SRBRF model – is a negative value of the entropy in the limit $T \to 0$. Thus, the model predictions should be treated with some caution on approaching the zero-temperature limit.

7.3.2 Spontaneous polarization

A second illustrative example is the case $E = 0$ with both $\Delta \neq 0$ and $J_0 \neq 0$. For $J_0^2 < J^2 + \Delta$ there is no spontaneous polarization at any temperature. By symmetry, we have $r_T = r_L = 1$ and $q_T = q_L = q$, where q is the solution of

$$q = \beta^2 J^2 (q + \Delta/J^2)(1 - q)^2 \tag{31}$$

We can show that for $J_0^2 > J^2 + \Delta$ long-range order (LRO) exists and the spontaneous polarization $\vec{P}$ is non-zero at $T < T_c$.

The E–A order parameter at $T < T_c$ is obtained from Eq. (19) as

$$q = 1 - \frac{kT}{J_0} \tag{32}$$

and the spontaneous polarization follows from Eq. (22) as

$$P^2 = m \left[\left(1 - \frac{kT}{J_0}\right) \left(1 - \frac{J^2}{J_0^2}\right) - \frac{\Delta}{J_0^2} \right] \tag{33}$$

The critical temperature T_c is, therefore, given by

$$kT_c = J_0 \left(1 - \frac{\Delta}{J_0^2 - J^2}\right) \tag{34}$$

The spontaneous polarization at $T = 0$ is

$$P(0)^2 = m \left(1 - \frac{J^2 + \Delta}{J_0^2}\right) \tag{35}$$

This shows that LRO exists only for $J_0^2 > J^2 + \Delta$.

7.3.3 Non-linear susceptibility

We introduce the third-order non-linear longitudinal susceptibility χ_3 by the usual relation $P = \chi_1 E - \chi_3 E^3$, where the indices L have been omitted. By performing the derivatives of q and P with respect to E in Eqs. (22) and (23), respectively, we obtain in the limit $E \to 0$

$$\chi_3 = \frac{(kTv_0^3/mg^4)\chi_1^4}{(1-q)^2\left[1-\beta^2 J^2(1-q)(1-3q-2\Delta/J^2)\right]} \tag{36}$$

This expression corresponds to the static field-cooled third-order non-linear susceptibility of a relaxor. It is useful to introduce the scaled third-order non-linear susceptibility $a_3 = \chi_3/\chi_1^4$, [7.23], i.e.

$$a_3 = \frac{kTv_0^3/mg^4}{(1-q)^2\left[1-\beta^2 J^2(1-q)(1-3q-2\Delta/J^2)\right]} \tag{37}$$

By determining a_3 experimentally, we can simply discriminate between the paraelectric, ferroelectric, and relaxor behaviour of a given system. For example, in a mean-field-type ferroelectric χ_3 diverges, but a_3 is finite at T_c, whereas in a relaxor we expect the same type of behaviour to occur in both χ_3 and a_3.

It should be emphasized that the above model of a relaxor must include random bonds – i.e. a non-zero parameter J– in addition to random fields in order to describe the anomaly in the scaled non-linear response.

The temperature dependence of the static third-order non-linear response $a_3 = \chi_3/\chi_1^4$ is shown in Fig. 7.9 for various values of the random field strength Δ/J^2, and for $J_0 = 0$. As is known from the theory of dipolar glasses and inorganic relaxors, for small values of Δ/J^2 the non-linear susceptibility has a nearly divergent behaviour at the freezing temperature $T_f \cong J/k$, whereas in spin glasses $(\Delta = 0)\chi_3$ actually diverges as $\approx |T - T_c|^{-1}$ at the glass-transition temperature $T_g = J/k$. According to the Landau theory for homogeneous ferroelectric systems, χ_3 should diverge at the critical temperature T_c as $\chi_3 \approx |T - T_c|^{-4}$. This type of behaviour is also found in the present model for $J_0^2 > J^2 + \Delta$. For a trivial paraelectric system with $J^2 = J_0^2 = \Delta = 0$, we see that $\chi_3 \approx T$. On the other hand, in the general case of a relaxor polymer with all parameters J, J_0, and Δ non-zero and $J_0^2 < J^2 + \Delta$, we find for $T \gg J/k$ that $q \ll 1$, and Eq. (36) predicts a linear temperature dependence at high temperatures, similar to the paraelectric case.

The low-frequency third-order non-linear dielectric response χ_3 and the corresponding scaled response a_3 have been measured in the electron-irradiated relaxor copolymer P(VDF-TrFE) [7.20] and in the terpolymer P(VDF-TrFE-CFE) [7.21]. Specifically, in P(VDF-TrFE) the non-linear susceptibility χ_3 shows a peak near $\sim 300\,\text{K}$, whereas the scaled non-linear response a_3 shows a cross-over between the paraelectric behaviour $a_3 \sim T$ at high temperatures and a relaxor-type behaviour $a_3 \sim (T - J/k)^{-1}$ near $\sim 300\,\text{K}$ [7.20]. Figure 7.10 shows the experimental data for a_3 in electron-irradiated P(VDF-TrFE) [7.20], which is in qualitative agreement with the theoretical predictions.

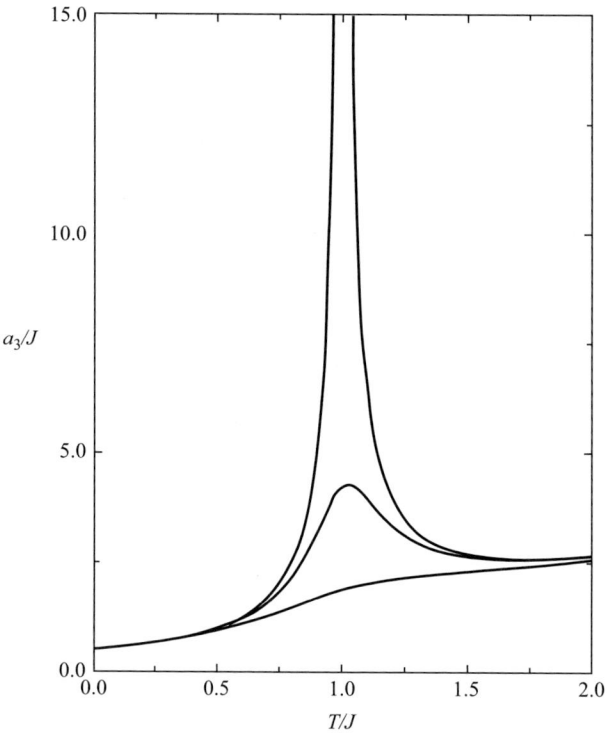

Fig. 7.9 Scaled third-order non-linear response $a_3 = \chi_3/\chi_1^4$ in the SG phase ($J_0 = 0$) as a function of temperature plotted for various values of the random-field strength Δ. Top to bottom: $\Delta/J^2 = 0, 0.001, 0.01, 0.1$.

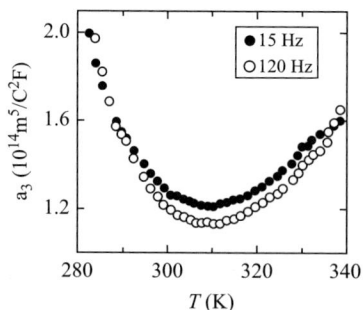

Fig. 7.10 Experimental data for a_3 in electron irradiated P(VDF-TrFE) [7.20] obtained by the 3 ω response.

7.4 Electrostriction

It has recently been shown that relaxor polymers and their all-organic composites exhibit a high value of electrostriction, namely, large mechanical strains can be generated by application of an electric field. Typically, the copper phtalocyanine (CuPc) oligomers have been dispersed in the electrostrictive copolymer P(VDF-TrFE) matrix, leading to a high net dielectric constant, while retaining the elastic modulus of the copolymer as well as its flexibility [7.24]. It has been suggested that the ultrahigh strain response in relaxor polymers is due to the expansion of polar nanoregions under an electric field, coupled with a large difference in the lattice strain between the polar and unpolar phases [7.25]. Alternatively, an exchange coupling between CuPc and the copolymer matrix could lead to a dramatic enhancement of the electrostriction and dielectric constant in the CuPc-P(VDF-TrFE) composite [7.26, 7.27].

Here, we consider a three-dimensional system with Cartesian indices $\mu, \nu, \ldots,$ $= 1, 2, 3$. The electrostrictive coefficients $Q_{\mu\nu\kappa\lambda}$ are defined by the relation

$$u_{\mu\nu} = \sum_{\kappa\lambda} Q_{\mu\nu\kappa\lambda} P_\kappa P_\lambda \tag{38}$$

where $u_{\mu\nu}$ is the strain tensor and P_κ, etc., are components of the dielectric polarization induced by the applied field, i.e.

$$P_\mu = \sum_\nu \chi_{\mu\nu} E_\nu \tag{39}$$

with $\chi_{\mu\nu}$ representing the longitudinal field-cooled dielectric susceptibility tensor.

The electrostrictive coefficients can be calculated from the thermodynamic Maxwell relation

$$Q_{\mu\nu\kappa\lambda} = -\frac{1}{2}\left(\frac{\partial \chi_{\mu\nu}^{-1}}{\partial p_{\kappa\lambda}}\right)_{P,T} \tag{40}$$

where $\chi_{\mu\nu}^{-1}$ are components of the inverse susceptibility tensor and $p_{\kappa\lambda}$ of the stress tensor. From Eq. (24) we have

$$\chi_{\mu\nu}^{-1} = \delta_{\mu\nu}\left(\frac{v_0}{g^2}\right)\frac{1 - \beta J_0(1-q)}{\beta(1-q)} \tag{41}$$

At constant polarization P, the E–A order parameter q is independent of J_0 according to Eq. (22). The stress dependence of $\chi_{\mu\nu}$ is expected to arise from the parameters J_0 and g^2/v_0. In the following we will limit ourselves to the case of hydrostatic pressure, where $p_{\mu\nu} = -p\delta_{\mu\nu}$. At high temperatures, the static dielectric constant ε_s behaves asymptotically as $\varepsilon_s \sim C/(T - T_0)$, where $T_0 = J_0/k$ plays the role of an effective Curie–Weiss temperature and $C = g^2/(k\varepsilon_0 v_0)$ is the corresponding Curie constant. Thus, by measuring $d\varepsilon_s/dp$ at high temperatures an estimate for dT_0/dp and hence for dJ_0/dp can be obtained.

From Eq. (41) we derive an expression for the hydrostatic electrostriction constant $Q_\mathrm{h} = Q_{33} + 2Q_{13}$ [7.28]

$$Q_\mathrm{h} = -\frac{1}{2}\left(\frac{kv_0}{g^2}\right)\frac{dT_0}{dp} + \frac{1}{2\varepsilon_\mathrm{s}\varepsilon_0}\left(\frac{1}{v_0}\frac{dv_0}{dp} - \frac{2}{g}\frac{dg}{dp}\right) \qquad (42)$$

as before.

The first term is expected to be dominant in inorganic relaxors of PMN type [7.29] where $dT_0/dp < 0$ and $Q_\mathrm{h} > 0$ [7.30]. In electroactive polymers, such as P(VDF-TrFE), however, we have $Q_\mathrm{h} < 0$ [7.28], and the pressure dependence of the parameters v_0 and g must obviously be considered. We assume that $dv_0/dp < 0$, i.e. $v_0^{-1}dv_0/dp \approx -1/B$, where B is the bulk modulus. Adopting the value [7.31] $B \cong 0.8 \times 10^9 \,\mathrm{N/m^2}$ as well as [7.32] $\varepsilon_\mathrm{s} \cong 25$ for 60 Mrad irradiated conventional electrostrictive polymer P(VDF-TrFE) at room temperature, we can estimate the contribution of the second term in Eq. (42) to be of the order $Q_\mathrm{h} \sim -2.8 \mathrm{m^4/C^2}$. A comparison with the experimental value [7.28] $Q_\mathrm{h} = -6 \mathrm{m^4/C^2}$ suggests that the contribution of the last term in Eq. (42) should also be included; however, there is no estimate of the dg/dp available at this time. The above value of Q_h is roughly two orders of magnitude larger than in inorganic relaxors [7.30], and Q_h is hence referred to as 'giant electrostriction.'

The main differences between the present model and the SRBRF model of inorganic relaxors are in the physical character of polar nanoregions, the nature of the interactions between them, and the origin of the random fields. Formally, the two models are equivalent; however, the values of the physical parameters characterizing the model can vary to a significant degree. For example, relaxor polymers are usually softer than the inorganic systems and thus have a greater value of compressibility, which strongly affects the hydrostatic electrostriction constant.

8
Electrocaloric effect in ferroelectrics and ferroelectric thin films

Alternative cooling technologies have generated a lot of interest in recent decades. The electrocaloric (EC) effect is the change in temperature [8.1–8.3] of a system upon the application or withdrawal of an electric field under adiabatic conditions and the corresponding change in entropy under isothermal conditions. The EC effect is of interest for applications in solid-state devices such as chip cooling, temperature regulation for electronic devices and possibly domestic as well as industrial refrigeration [8.4]. It is environmentally friendly and may provide an alternative to the presently used vapour-compression (Fig. 8.1) approach. In addition there are no moving parts and the EC effect is more energy efficient. So far, it has not been commercially exploited as the observed EC effects were too small and limited by the breakdown strength (Fig. 8.2) of the material. The required temperature changes are $\Delta T \approx 30$–$50\,\mathrm{K}$, whereas until recently only ΔT of a few K were achieved. This, however, may no longer be true [8.4, 8.5]. It has been reported that large EC responses could be obtained [8.4–8.6] in thin films with controlled misfit strains as well as in the vicinity of critical points in relaxors accompanying the giant electromechanical response [8.7].

A simple explanation is as follows: The application of an external electric field changes the disordered dipolar state into an ordered one resulting in a reduction of the entropy of the dipolar subsystem (Fig. 8.1). If this is done under adiabatic conditions the change in the dipolar entropy ΔS is compensated by an increase in lattice vibrations, i.e. by an increase of the temperature ΔT of the material. If the sample is then allowed to cool down and the electric field is subsequently removed under adiabatic conditions we again get a disordered dipolar state with an increased value of entropy. This results in cooling the sample, i.e. a decrease of sample's heat $Q_\mathrm{C} = T \cdot \Delta S$.

In Fig. 8.1. an EC heating/cooling cycle is compared with a corresponding heating/cooling cycle of a vapour compressor where the pressure of the gas is changing.

In the refrigeration cycle we have first an isothermal entropy change ΔS. The working material must absorb entropy from the load to be cooled while in thermal contact with the load. The material is then isolated from the load while the temperature is increased because of the applied electric field, i.e. we have an additional temperature change ΔT. The material is then placed in thermal contact with the heat sink, and the entropy that was absorbed from the cooling load is transferred to the heat sink. The temperature of the sample is reduced back to the temperature of the cooling load and the process is repeated.

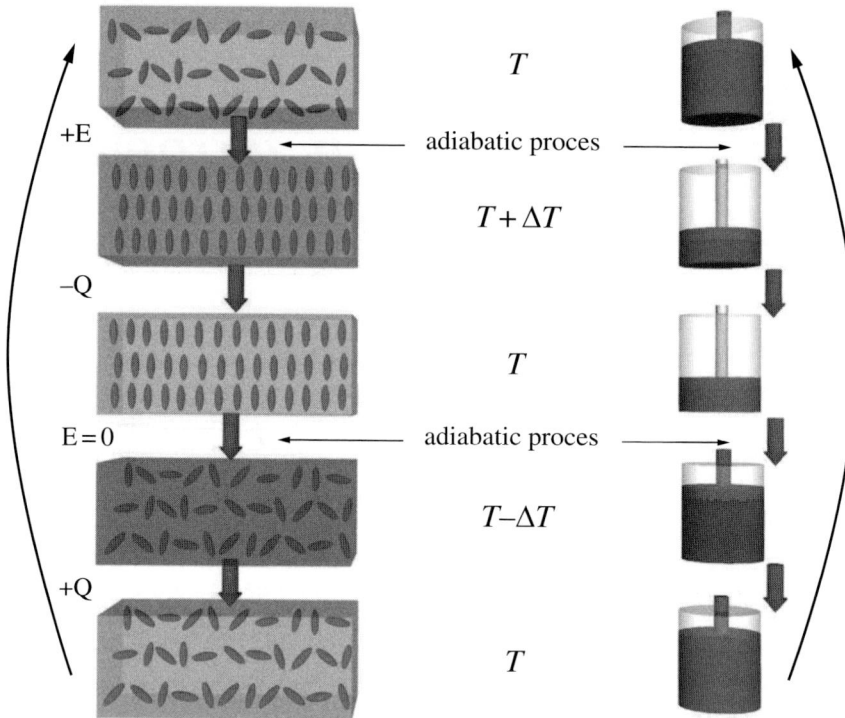

Fig. 8.1 Analogy between EC cooling and a vapour-compression cooling cycle in a heat engine. Here, $-Q$ is the heat ejected from the EC module and $+Q$ is the heat absorbed (courtesy of Z. Kutnjak and B. Rožič). See Plate 3 in the colour plate section.

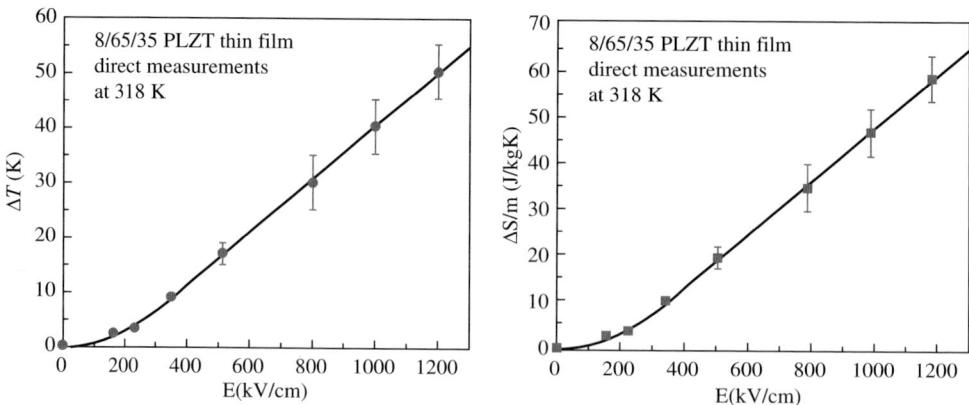

Fig. 8.2 E-field dependence of the EC effect in thin films of 8/65/35 PLZT ceramics near room temperature.

The largest electrocaloric temperature change ΔT ever reported in a solid material has been found in a 450-nm thin film of 8/65/35 PLZT [8.8]. Here ΔT exceeds 40 K at 1.25 MV/cm and the effective heat change $Q_C = C_p \cdot \Delta T$, which can be related to the isothermal entropy change ΔS via $Q_C = T \cdot \Delta S$, is 17 kJ/kg (Fig. 8.2).

The reversible change in the internal energy U of an elastic dielectric solid is given by

$$dU = TdS + EdD + \sigma du \tag{1}$$

where dS, dD and du are the changes in entropy, dielectric displacement and strain. T, E and σ are temperature, electric field and stress. The total free energy is

$$G = U - TS - ED - \sigma u \tag{2}$$

The differential of G is

$$dG = -SdT - DdE - ud\sigma \tag{3}$$

yielding

$$dG = \left(\frac{\partial G}{\partial T}\right)_{\sigma, E} dT - \left(\frac{\partial G}{\partial E}\right)_{T,\sigma} dE - \left(\frac{\partial G}{\partial \sigma}\right)_{T,E} d\sigma \tag{4}$$

where

$$\left(\frac{\partial G}{\partial T}\right)_{\sigma, E} = -S, \quad \left(\frac{\partial G}{\partial E}\right)_{T,\sigma} = -D, \quad \left(\frac{\partial G}{\partial \sigma}\right)_{T,E} = -u \tag{5}$$

Differentiating Eq. (5) we obtain the following identities relating the material compliances of the system:

$$-\left(\frac{\partial^2 G}{\partial T \partial u}\right)_E = \left(\frac{\partial S}{\partial u}\right)_{E,T} = \left(\frac{\partial u}{\partial T}\right)_{\sigma,T} \tag{6a}$$

$$-\left(\frac{\partial^2 G}{\partial \sigma \partial E}\right)_T = \left(\frac{\partial u}{\partial E}\right)_{\sigma,T} = \left(\frac{\partial D}{\partial \sigma}\right)_{E,T} \tag{6b}$$

$$-\left(\frac{\partial^2 G}{\partial E \partial T}\right)_\sigma = \left(\frac{\partial D}{\partial T}\right)_{\sigma,E} = \left(\frac{\partial S}{\partial E}\right)_{\sigma,T} \tag{6c}$$

Expression (6c) provides the most general definition of the EC effect in dielectrics. While the entropy is continuous if there is no phase transition, there is a deflection point in the $S = S(T)$ relation for a second-order phase transition and a jump for a first-order phase transition. The largest EC effect should be thus observed near a phase transition [8.9].

For a linear dielectric

$$D = \varepsilon_0 E + P = \varepsilon \varepsilon_0 E, \text{ so that } P = (\varepsilon - 1)\varepsilon_0 E \tag{7a}$$

and the polarization dependent part of the free energy is

$$G_P = \frac{1}{2\varepsilon\varepsilon_0} P^2 \tag{7b}$$

where ε is the dielectric permittivity of the material and ε_0 the permittivity of the vacuum. For a ferroelectric the displacement includes both the induced and the spontaneous polarization

$$D = \varepsilon_0 E + P, \quad P = P_{\text{ind}} + P_{\text{s}} \tag{7c}$$

In a ferroelectric the Maxwell relation (6c) is usually expressed as

$$\left(\frac{\partial P}{\partial T}\right)_E = \left(\frac{\partial S}{\partial E}\right)_T \tag{8}$$

The polarization dependent part of the free energy density for a bulk unclamped, stress free sample is [8.9]:

$$G_P = G_0 + a_1 P^2 + a_{11} P^4 + a_{111} P^6 - EP \tag{9}$$

Here, $a_1 = (T-T_c)/(2C\varepsilon_0)$ where C is the Curie constant and T_c the Curie temperature. a_{11} and a_{111} are higher-order stiffness coefficients.

In the presence of an in-plane stress ($\sigma_1 = \sigma_2 \neq 0, \sigma_3 = \sigma_4 = \sigma_5 = \sigma_6 = 0$) expression (9) becomes

$$G_{\text{film}} = G_0 + a_1 P^2 + a_{11} P^4 + a_{111} P^6 - EP + G_{\text{elastic}} \tag{10}$$

Expression (10) is appropriate for a thin film. Here

$$G_{\text{elastic}} = \tilde{C}(u_m - Q_{12} P^2)^2 \tag{11a}$$

where u_m is the in-plane polarization free misfit strain

$$u_m = \frac{a_{\text{substrate}} - a_{\text{film}}}{a_{\text{substrate}}} \tag{11b}$$

Q_{ij} is the electrostrictive coefficient and $\tilde{C}$ an effective elastic modulus. The modified free-energy density is now

$$G_{\text{film}} = G_0 + \tilde{a}_1 P^2 + \tilde{a}_{11} P^4 + a_{111} P^6 - EP + \tilde{G}_{\text{elastic}} \tag{12}$$

where the modified dielectric stiffness coefficients are

$$\tilde{a}_1 = a_1 - 2u_m Q_{12} \tilde{C} \tag{13a}$$
$$\tilde{a}_{11} = a_{11} + Q_{12}^2 \tilde{C} \tag{13b}$$

and

$$\tilde{G}_{\text{elastic}} = u_m^2 \tilde{C} \tag{13c}$$

The internal stresses and the clamping effect of the substrate thus change T_c and the T dependence of the polarization. This can be obtained for $E = 0$ from the equilibrium condition

$$\frac{\partial G_{\text{film}}}{\partial P} = 0 \tag{14a}$$

as

$$\tilde{a}_1 + 2\tilde{a}_{11}P^2 + 3a_{111}P^4 = 0 \tag{14b}$$

For a constant applied field and stress, the excess entropy S^{xs} and the excess specific heat ΔC are obtained from:

$$S_{E,\sigma}^{xs} = -T\left(\frac{\partial G}{\partial T}\right)_{E,\sigma} \tag{15a}$$

and

$$\Delta C_{E,\sigma}(T, E, u_m) = -T\left(\frac{\partial^2 G}{\partial T^2}\right)_{E,\sigma} \tag{15b}$$

Using expression (6c) we can determine the reversible adiabatic temperature change ΔT in the ferroelectric due to the change in the electric field ΔE:

$$\Delta T = -T\int_{E_1}^{E_2} \frac{1}{C_{E,\sigma}}\left(\frac{\partial P}{\partial T}\right)_{E,\sigma} dE \tag{16a}$$

Here, $\Delta E = E_2 - E_1$ is the difference in the applied electric field. It is clear that the heat capacity $C_{E,\sigma}$ (i.e. volume specific heat at constant field $C_E = \rho \cdot c_E$) and the polarization P are functions of T, E and u_m so that $\Delta T = \Delta T(T, E, u_m)$. Both ΔT and ΔS are key parameters for the EC effect. They are related to the effective heat change $Q_C = C_p \cdot \Delta T$ and the isothermal entropy change $Q_C = T \cdot \Delta S$. They are also related to the 'refrigerant capacity (RC)' used to compare the performance of different refrigerants. Here $RC = \Delta T_{hC} \cdot \Delta S_E$ where ΔS_E is the entropy change between the cold (T_C) and hot (T_h) reservoirs and $\Delta T = T_h - T_C$.

Changing the integration variables in Eq.(16a) from dE to $dP(E)$ via $\left(\frac{\partial P}{\partial T}\right)_E = -\left(\frac{\partial E}{\partial T}\right)_P \left(\frac{\partial P}{\partial E}\right)_T$ we obtain

$$\Delta T = T\int_{P_0}^{P} \frac{1}{C_E}\left(\frac{\partial E}{\partial T}\right)_P dP \tag{16b}$$

where $P_0 = P(0, T)$ and $P = P(E, T)$. From $F = F_0 + \frac{1}{2}aP^2 + \frac{1}{4}bP^4 - EP$ and the equilibrium condition $\left(\frac{\partial F}{\partial P}\right)_T = 0$ we obtain $E = aP + bP^3 + ...$ so that Eq. (16b) can be integrated to yield

$$\Delta T = \frac{T}{C_E}\left[\frac{1}{2}a_1\left(P^2 - P_0^2\right) + \frac{1}{4}b_1\left(P^4 - P_0^4\right) + ...\right] \tag{16c}$$

Here, $a_1 = da/dT$ and $b_1 = db/dT$. We also find

$$\Delta T = -\frac{T}{C_E}\Delta S_P \qquad (16d)$$

with $\Delta S_P = S_1(P) - S_1(P_0)$ and $S_1(P) = -\left(\frac{1}{2}a_1 P^2 + \frac{1}{4}b_1 P^4 \ldots\right)$.

Figure 8.3 shows that the EC efficiency $\Delta T/\Delta E$ as a function of E is best near the critical end point in relaxors. This is analogous to the electromechanical response d_{31} in PMN [111] which is also the largest (Fig. 8.5) near the critical end point [8.7].

In ferroelectrics the EC effect is largest at the phase transition temperature, whereas in relaxors it is at the critical end point $T = T_{CP}$ and $E = E_{CP}$. For practical purposes it is, however, important to have a large EC effect in a wide temperature and electric field range. Such a possibility is offered by relaxors and relaxor thin films.

It should be noticed that in relaxors and relaxor terpolymers the directly measured EC effect is much higher than deduced from the Maxwell relation, Eq. (8). This is likely caused by the non-ergodic behaviour of relaxor ferroelectric polymers and relaxors, while the Maxwell relations are strictly valid for ergodic equilibrium systems [8.8]. Another fact to be stressed is that according to the conventional theory the maximum

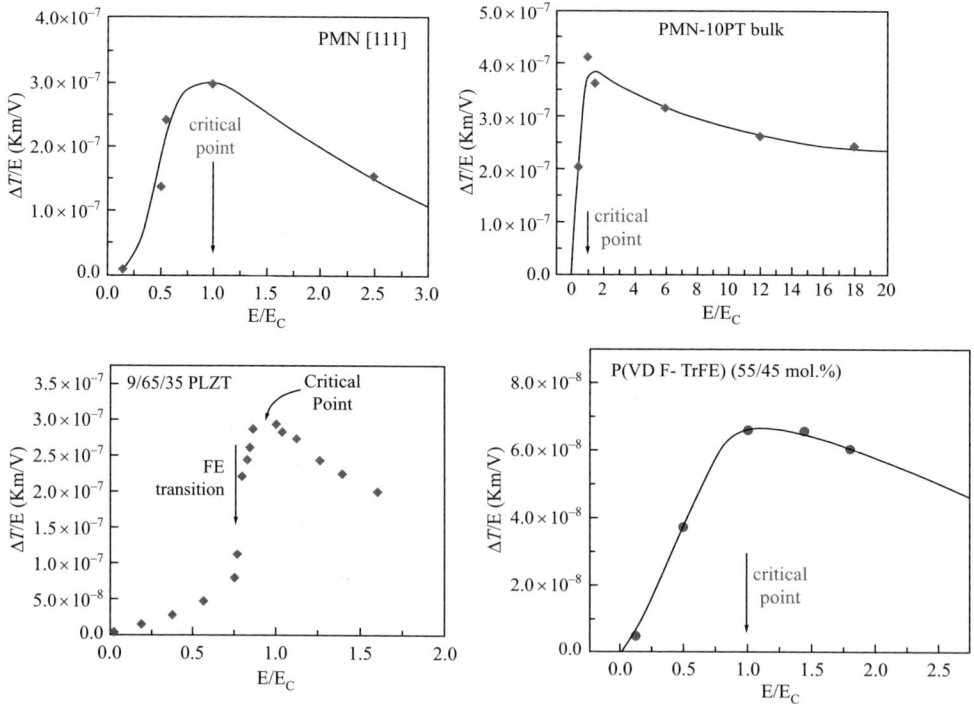

Fig. 8.3 EC effect in bulk PMN with $E_{\|[111]}$, bulk PMN-30PZT, PLZT ceramics and P[VDF-TrFE] polymer film [8.8]. The solid lines are a fit to SRBRF model [8.10].

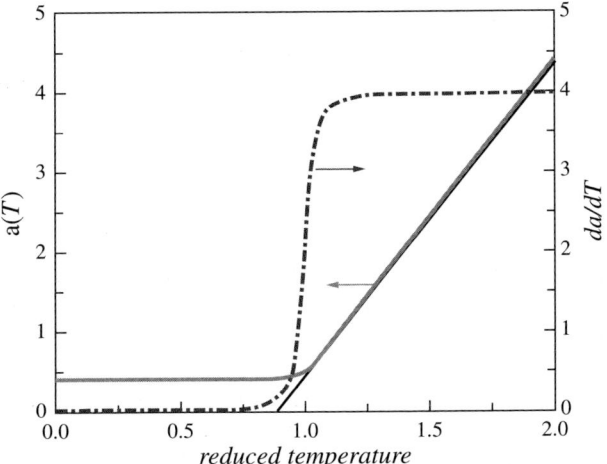

Fig. 8.4 Temperature dependence of the $a(T)$ and da/dT according to the SRBRF model [8.10, 8.12]. The solid black line denotes $a(T)$ for normal ferroelectrics.

of $\Delta T/E$ should occur well below the field $E_{\rm CP}$ corresponding to the critical end point (CP). It should occur at the point where $(dP/dT)_E$ exhibits a discontinuous change. As seen from Fig. 8.3 this is not the case. The glassy nature of relaxors, as described by the SRBRF model [8.10, 8.11], shifts the maximum of $\Delta T/E$ close to $E_{\rm CP}$. According to the SRBRF model [8.10, 8.11] the EC effect in relaxors can be described as follows:

The free energy of relaxor ferroelectric can be written in Landau form as

$$F = \frac{1}{2}aP^2 + \frac{1}{4}bP^4 + \frac{1}{6}cP^6 + \cdots - EP \tag{17}$$

where, however, $a(T)$ has to be positive (Fig. 8.4) at all temperatures [8.12].

For simplicity we consider a scalar order parameter P. We assume that $a = a(T)$, whereas $b < 0$ and $c > 0$ are temperature independent. In normal ferroelectrics $a \propto (T - T_c)$, and for $b > 0$, $E = 0$ the phase transition at T_c is of second order. Similarly, $b < 0$ leads to a first-order transition at $a_c^I = \frac{3}{16}b^2/c$. For $E \neq 0$ a critical point (CP) is located [8.11] at $a(T_{\rm CP}) = \frac{9}{20}b^2/c$ and $E_{\rm CP} = \frac{6}{25}\frac{b^2}{c}\sqrt{\frac{3|b|}{10c}}$.

In a relaxor ferroelectric, however, no spontaneous long-range order exists, since the coefficient $a(T)$ never drops to zero, as illustrated in Fig. 8.4. Its explicit form is given by the SRBRF model of relaxor ferroelectrics [8.11],

$$a(T) = \chi(T)^{-1} = \varepsilon_0^{-1}\theta^{-1}\left(\frac{T}{1-q} - T_0\right) \tag{18}$$

where $\chi = (\partial P/\partial E)_T$ is the quasi-static field-cooled susceptibility, and the glass order parameter $0 < q < 1$ is a measure of glassy disorder; θ is an effective Curie constant

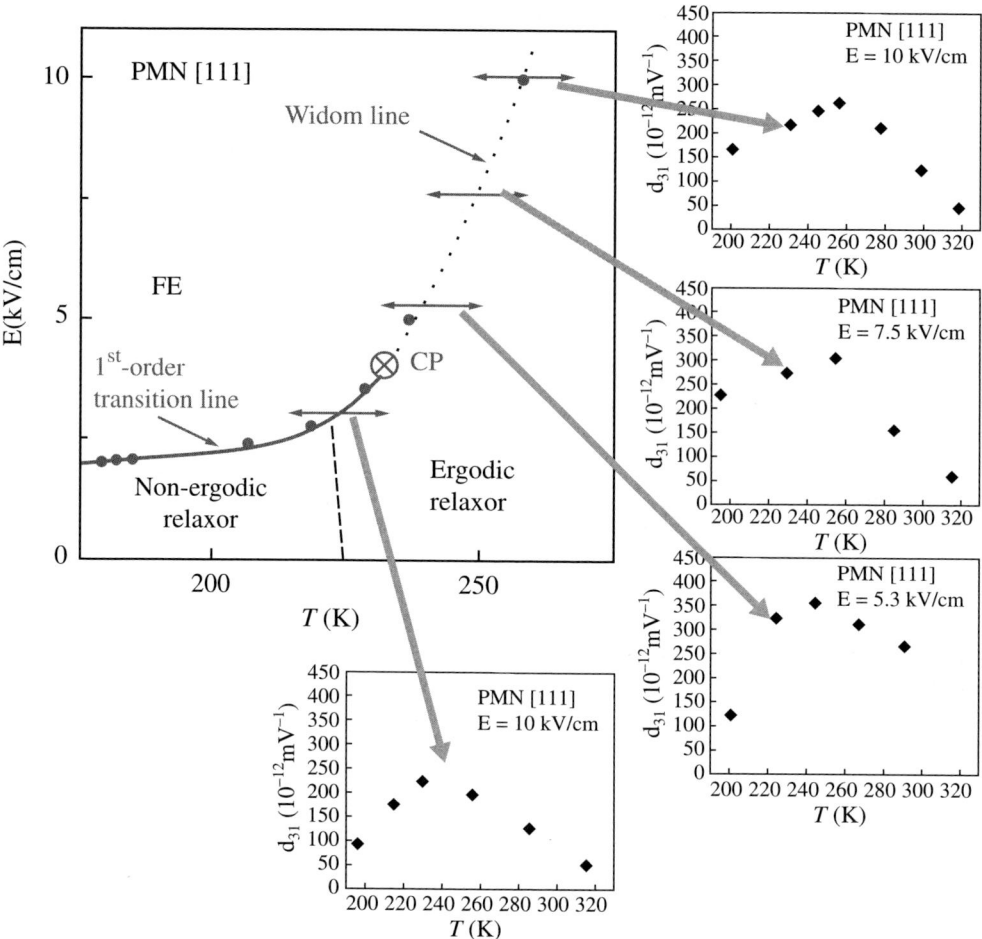

Fig. 8.5 T dependence of the electromechanical response d_{31} in PMN[111] along the first-order transition line and the Widom line [8.7].

and $T_0 = J_0/k$ is related to the average interaction between polar nanoregions J_0. $\theta = \frac{g^2}{vk\varepsilon_0}$ is expressed in terms of the average squared dipole moment of the polar nanoregions $g^2 = \sum_{i=1}^{N} g_i^2/N$ and the average volume $v = V/N$ of a nanocluster. The susceptibility χ_1 is derived from the SRBRF model as $\chi_1 = \frac{\theta(1-q)}{T-T_0(1-q)}$. For $a(0) > a_c$ the first-order transition is likewise suppressed. When $E \neq 0, P(T)$ is obtained by direct minimization of F. It first increases monotonically with E and then makes a jump at E_c, implying a first-order transition. At $E = E_{CP}$ the critical end point is reached, whereas for $E > E_{CP}$ the system is in the supercritical regime with a smooth field and temperature dependence of $P(E,T)$. It should be noted that in relaxors the crucial coefficient is $a_1(T)$; i.e. $\frac{\partial}{\partial T}\left(\frac{T}{1-q}\right)$.

The fact that the EC effect and in particular the entropy change is larger in relaxors than in normal ferroelectrics can be most easily explained as follows.

We consider two systems, a normal ferroelectric (FE) and a relaxor. Both have the same volume V, but contain different numbers of dipolar entities that are responsible for their dielectric properties, N_f and N_r, respectively. Below the FE critical temperature T_C, N_f represents the total number of FE domains, whereas N_r refers to the number of polar nanoregions (PNRs) in the relaxor. We expect that $N_r > N_f$. Let the average dipole moment associated with a domain be g_f, and similarly g_r the average dipole moment of the PNRs. The maximum polarization of the FE and relaxor are thus

$$P_f^{\max} = \frac{N_f g_f}{V} = \frac{g_f}{v_f} \text{ and } P_r^{\max} = \frac{N_r g_r}{V} = \frac{g_r}{v_r} \quad (19)$$

respectively, where we have introduced $v_f \equiv N_f/V$ and $v_r \equiv N_r/V$ as the average volume belonging to the corresponding dipolar entities. Clearly, the maximum polarization implies a state where all domains or all PNRs are oriented in the same direction. In the following, we will assume that $P_f^{\max} = P_r^{\max}$, thus

$$\frac{g_f}{v_f} = \frac{g_r}{v_r} \quad (20)$$

We now consider the entropy density of the system assuming that there are N_+ entities oriented in the 'up' and N_- in the 'down' orientation, where we skip the indices f or r. Of course, $N_+ + N_- = N$. In terms of concentrations $c_+ = N_+/N$ and $c_- = N_-/N$, the entropy is given by the standard result

$$S = -N k_B (c_+ \log c_+ + c_- \log c_-) \quad (21)$$

We now introduce the dimensionless polarization $p \equiv c_+ - c_-$ and using the relation $c_+ + c_- = 1$ we have $c_\pm = (1 \pm p)/2$. Thus, Eq. (21) can be rewritten as

$$S_f = N_f k_B \left[\log 2 - \frac{1}{2}(1-p) \log(1-p) - \frac{1}{2}(1+p) \log(1+p) \right] \quad (22)$$

The physical polarization P is related to p via $P = \frac{g}{v} p$, from which we recover Eqs. (19) if we set up $p^{\max} = 1$.

Expanding the r.h.s. of Eq. (22) in powers of $p^2 = P^2(p/g)^2$ and restoring the indices f, we find the entropy of a FE system as

$$S_f = N_f k_B \log 2 - \frac{1}{2} N_f k_B \left(\frac{v_f}{g_f}\right)^2 P^2 + O(P^4) \quad (23)$$

By introducing the 'Curie' constant

$$\theta_f = \frac{g_f^2}{v_f k_B \varepsilon_0} \quad (24)$$

and dividing Eq. (23) by the volume V we obtain the entropy density S of a FE,

$$S_f = \frac{k_B}{v_f} \log 2 - \frac{1}{2}\beta_f P^2 + \cdots \tag{25}$$

where $\beta_f = 1/\varepsilon_0 \theta_f$. The temperature change due to the EC effect is given by

$$\Delta T = -\frac{T}{\rho C_E}\Delta S \tag{26}$$

If the initial field is $E = 0$, the temperature change after applying a finite field E is given by

$$\Delta T = -\frac{T}{\rho C_E}\frac{1}{2}\beta_f \left[P(E,T)^2 - P(0,T)^2\right] + \cdots \tag{27}$$

Equation (27) implies that for a given difference of P^2 the EC effect is proportional to the inverse Curie constant of the system, which using Eq. (24) can be rewritten as

$$\theta_f^{-1} = \frac{1}{v_f} k_B \varepsilon_0 \left(\frac{v_f}{g_f}\right)^2 = \frac{N_f}{V} k_B \varepsilon_0 \left(\frac{v_f}{g_f}\right)^2 \tag{28}$$

We realize that the EC effect is directly proportional to the number of dipolar entities, for example, the number of domains in a FE. A comparison with relaxors is possible if we modify expression (25) for the entropy density accordingly. The theory of relaxors based on the SRBRF model leads to the following result for S_r:

$$S_r = \frac{k_B}{v_r} \log 2 - \frac{1}{2}\beta_r \left[\frac{\partial}{\partial T}\left(\frac{T}{1-q}\right)\right] P^2 + \cdots \tag{29}$$

where $\beta_r = 1/\varepsilon_0 \theta_r$. Thus, β_f in Eq. (27) will be replaced by $\beta_r a'$, suggesting that the EC effect in relaxors will also be proportional to the number of PNRs, however, with an extra factor $a' \equiv \frac{\partial}{\partial T}\left(\frac{T}{1-q}\right)$. At high temperatures, $q \ll 1$ and thus $a' \cong 1$, and we can trivially compare Eqs. (28) and (29). Since by assumption $N_r > N_f$, the EC effect in relaxors will be larger than in ordinary FEs. Below the freezing temperature, however, a' can be very small and the EC effect of relaxors will be strongly reduced, as observed experimentally.

Large EC effects have been achieved in PZT thin films, PMN-30PT bulk ceramics, PMN-35PT bulk as well as thick ceramic films, PMN bulk ceramics and single crystals, P[VDF-TrFE] (poly (vinylidene-trifluoroethylene)) 55/45 polymers, P[VDF-TrFE-CFE] (CFE – chlorofluoroethylene) terpolymers and PLZT thin films (Fig. 8.6). In the relaxor ferroelectric terpolymer P[VDF-TrFE-CFE] a $\Delta T = 12\,\text{K}$ and a Q_C of $15\,\text{kJ/kg}$ can be obtained at room temperature.

It should be noted that the EC effect can be amplified by using a cooling line of ECs in series (Fig. 8.7) and applying asymmetric electrical pulses.

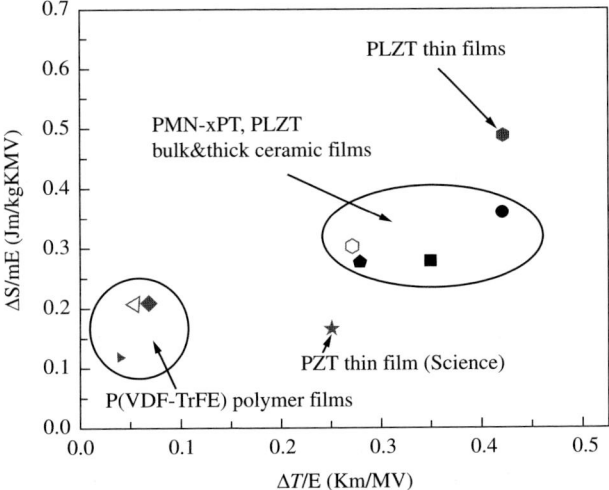

Fig. 8.6 Maximum 'efficiencies' $\Delta T/\Delta E$ and $\Delta S/\Delta E$ for various samples [8.8].

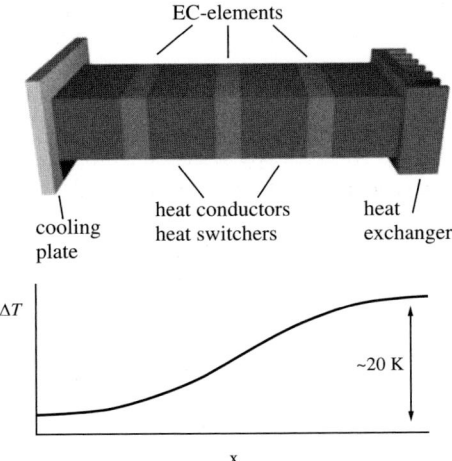

Fig. 8.7 EC cooling line [8.8] composed of several EC and heat switcher elements. By turning on and off the heat switchers heat can be pumped from the cold to the hot end (courtesy of Z. Kutnjak and B. Rožič).

Two things should be pointed out:

- The experimentally observed electrocaloric effect in relaxors is larger than that predicted by the Maxwell relations. This is so as relaxors are non-ergodic systems, whereas the Maxwell relations apply to ergodic systems only.
- Another thing to be remembered is that the entropy change and the EC effect is mainly determined by the magnitude of the entropy of the disordered dipolar

state. In view of the large number of possible equivalent orientations of polar nanoclusters this is quite large in relaxors. PLZT 8/65/35 thin films show a $\Delta T \cdot \Delta S \sim 2000\,\text{J/kg}$, which is two times or more higher than previously reported values for EC effect in ferroelectrics as well as giant magnetocaloric materials. The maximum EC temperature change in the saturation regime is [8.12]

$$\left(\frac{\Delta T}{T}\right)_{\text{max}} = \frac{R \ln \Omega}{C_{\text{m}}} \qquad (30)$$

where $R = N_{\text{Avogadro}} \cdot k$ and $C_{\text{m}} = N_{\text{Avogadro}} \cdot v_0 C_E$ is the molar specific heat. Ω is the configuration number. Expression (30) is valid for the electrocaloric effect in ferroelectrics and relaxors as well as the magnetocaloric effect in the saturation regime, i.e. at high enough electric, respectively, magnetic fields.

9
Ferroelectric thin films

Until 1980, ferroelectric research was centred on the properties of single crystals, on the development of new theoretical models, and on the discovery of new ferroelectric systems. Applications were rather limited. The most successful applications were piezoelectric transducers, pyroelectric detectors, $BaTiO_3$ capacitors and actuators. Because of the high cost of single crystals, devices were usually produced from bulk ceramics.

The situation in the field of applications changed after 1984 when thin-film ferroelectrics were developed and integrated into semiconductor chips [9.1, 9.2]. The miniaturization requirements of nanoengineering for ferroelectric memories with high information density have reopened the question how the ferroelectric properties of a system are changed when the dimensions reach the nanometre region [9.2]. The old question of how many unit cells are necessary such that ferroelectricity, which is a collective phenomenon, does not disappear, therefore became important not only for basic physics but also for technology.

The importance of thin-film technology can be best seen from the following: the polarization reversal field – 'coercive field' for a typical ferroelectric is $\approx 50\,kV/cm$. In a 1-mm bulk device this corresponds to a voltage of $\sim 5\,kV$ which is much too high for, e.g., mobile telephones.

For submicrometre films this voltage is reduced to $< 5\,V$ (Fig. 9.1), allowing for integration into most silicon chips. The introduction of thin-film technology has thus completely changed the situation concerning applications. The crucial question in the field of integrated ferroelectrics is what is the thinnest ferroelectric layer that can still yield a stable out-of-plane remanent polarization [9.3–9.5]. Such a polarization is necessary for memory applications [9.2] and tunnel junctions [9.6–9.9].

Early theoretical work [9.4] at IBM indicated that with semiconducting electrodes, depolarization fields would destroy the polarization of layers thinner than 400 nm. With metallic electrodes, on the other hand, layers as thin as 4 nm should maintain switching. Later theories proposed a limit of 2 nm [9.5]. Experiments, on the other hand, seemed to show functioning layers thinner than 0.9 nm [9.10, 9.11].

Current experimental techniques have allowed the detection of ferroelectricity in perovskite films – placed between two metallic electrodes in a short circuit – down to a thickness of about 6–10 unit cells (2.4–4 nm) [9.12, 9.13]. The depolarizing electrostatic field caused by dipoles at the ferroelectric/metal interfaces is responsible for the disappearance of the ferroelectric state below some critical film thickness [9.9–9.18]. This disappearance amounts to a size-induced ferroelectric–paraelectric phase transition [9.13, 9.19]. It should be mentioned that the generation of a depolarizing field is

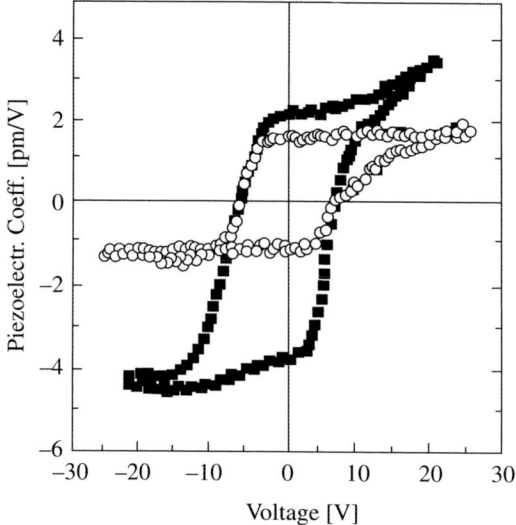

Fig. 9.1 Hysteresis loops of 1 µm (solid squares) and 100 nm (empty circles) large PZT cells measured via atomic force microscopy (AFM) in the piezoelectric mode [9.3].

inevitable because of the finite value of the Thomas–Fermi screening length for the free carriers inside the electrodes that compensate the polarization bound charges at the FE electrode interface.

According to Pertsev and Kohlstedt [9.20] the ferroelectric state is stabilized in heterostructures involving strained epitaxial films (Figs. 9.2 and 9.3).

For practical applications another important quantity is the memory retention time, i.e. the polarization decay time. Ferroelectric random access memories require the retention of a certain value of polarization ΔP for about 10 years [9.1]. This means that the retention loss due to thermodynamic nucleation of reversed domains must be very small. As the barrier U^* for such nucleation decreases with the reduction of the film thickness and the corresponding increase in the depolarizing field, this imposes a new fundamental size limit on ferroelectric devices: $U^* \geq 40\, k_\mathrm{B} T$. For BaTiO$_3$ films the retention time of $\approx$ 10 years requires a film thickness of 40 nm. This is much more than the 'critical thickness' for homogeneous ferroelectricity which is $\approx$ 2 nm [9.13].

It should be noted that the polarization decay follows a power-law behaviour

$$\Delta P(t) \propto t^{-n}$$

As the films become thinner, the value of n increases. In Jo et al. [9.21] a universal scaling relation between n and the barrier height for nucleation of reverse domains $U^*/k_\mathrm{B} T$ is presented (Fig. 9.4). This universal scaling relation seems to demonstrate that in thin films the polarization switching process is indeed governed by thermal fluctuations leading to nucleation. This is not the case in bulk ferroelectrics where Landauer has shown that a thermodynamic nucleation process does not play a role in FE domain switching as $U^* > 10^8\, k_\mathrm{B} T$ at an electric field of 1 kV/cm [9.22].

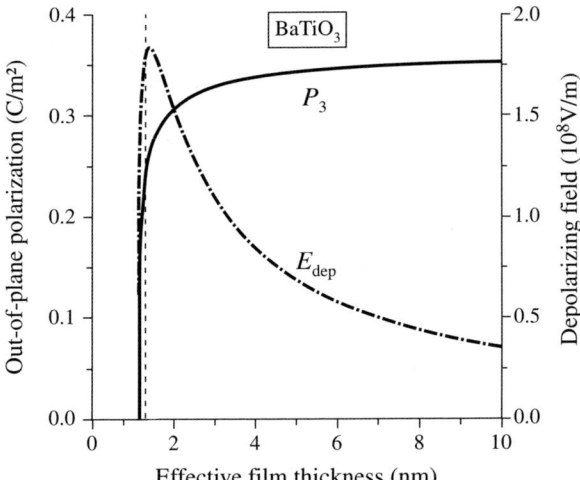

Fig. 9.2 Thickness dependence of the out-of-plane polarization and depolarizing field in ultrathin BaTiO$_3$ (BT) films grown on SrTiO$_3$. The misfit strain is taken to be -26×10^{-3} for BT films; $T = 25\,°C$. The dashed line shows the thickness below which the single domain state becomes unstable [9.9, 9.13–9.20].

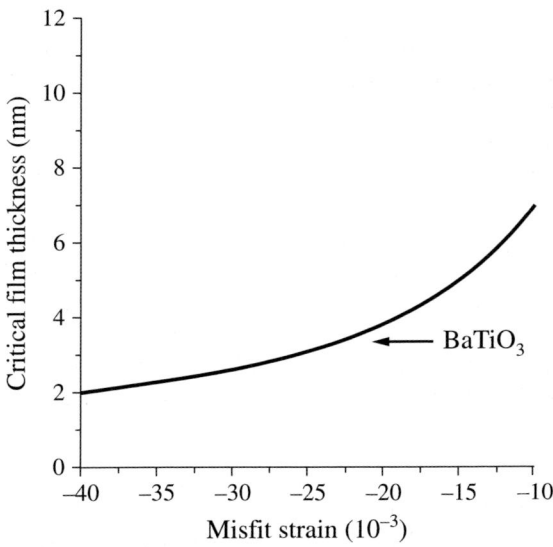

Fig. 9.3 Variation of the critical thickness $t_{c,\max}$ with the misfit strain calculated for epitaxial BT films sandwiched between SrRuO$_3$ electrodes (interfacial capacitance $c_i = 0.444\,\text{F/m}^2$). The temperature equals $25\,°C$, and the strain range is restricted to guarantee the stability of the FE phase [9.20].

218 Ferroelectric thin films

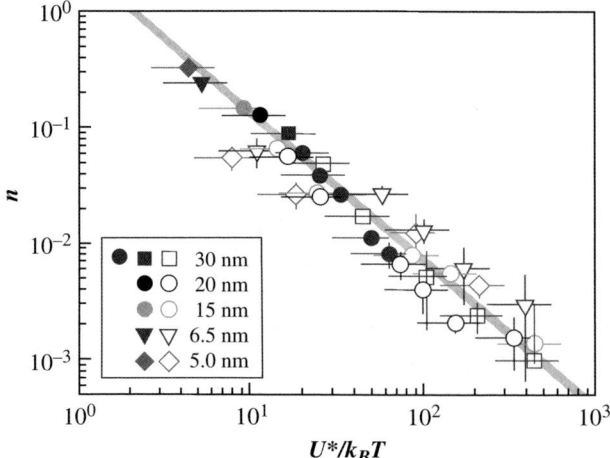

Fig. 9.4 A scaling relation between n and $U^*/k_\mathrm{B}T$. The open and solid symbols come from the experimental $\Delta P(t)$ data with and without E_ext, respectively [9.21].

NEC has reported polarization switching for PZT capacitors of $0.7 \times 0.7 \times 0.2\,\mathrm{nm}^3$ electrodes of bismuth oxide. The first commercially successful integration of a FE film on a semiconductor integrated circuit – $Ba_xSr_{1-x}TiO_3$ on GaAs – was achieved in 1988 [9.2]. This was a low-loss amplifier integrated circuit for 800 MHz–2.3 GHz operation used for mobile digital telephones. These BST/Ga chips were 50 times smaller than their predecessors and reached a production level of 270 million in 2001. In 1993 'fatigue-free' $SrBi_2Ta_2O_9$ (SBT) films were invented and in 1997 used in 'smart' cards. More than one million were produced in the first year.

Another possibility is a FET with a ferroelectric film as a gate. The source–drain current of the FET is quite different for +P and –P states of the FE gate. This cell can be thus read simply by monitoring the current without switching. Only when writing in a new bit does the FE gate have to be switched (Fig. 9.5).

Ferroelectric thin film devices are now used [9.2, 9.4] in many applications such as:

- thermal infrared switches;
- second-harmonic generation;
- infrared sensors;
- high-permittivity storage capacitors in DRAMs (dynamic random access memories) for use in cellular phones;
- piezoelectric microactuators;
- non-volatile ferroelectric random access memories (NVFRAMs) for memory elements.

Ferroelectric random access memories (FERAMs) are permanently keeping the stored information and thus make periodic rejuvenation of the information used in DRAMs unnecessary.

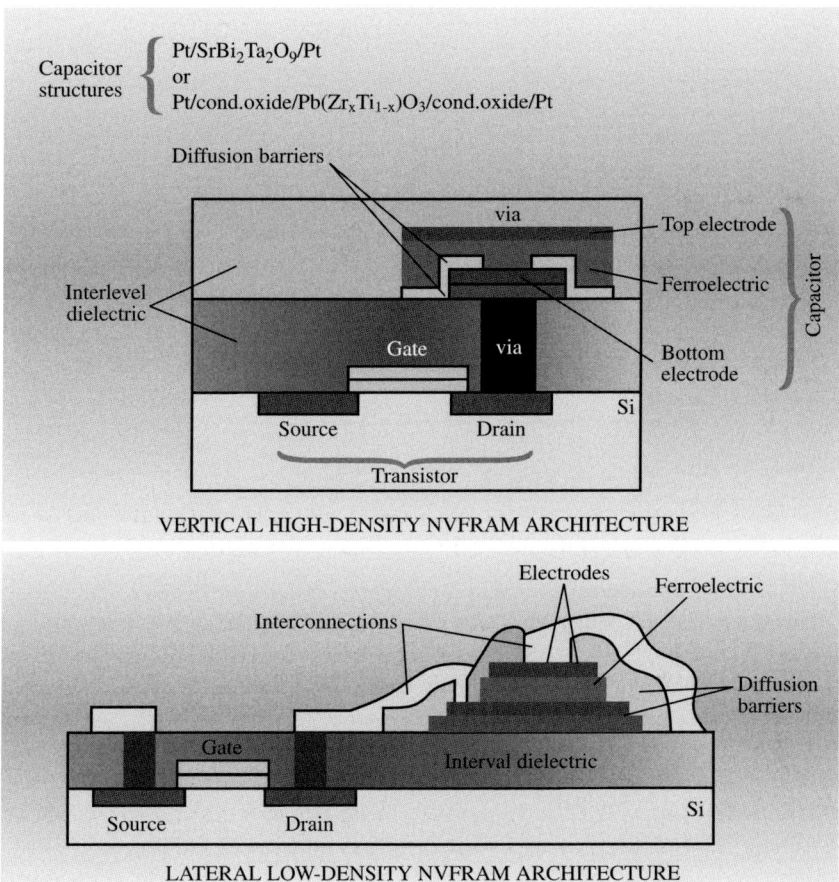

Fig. 9.5 Schematic diagram of two kinds of non-volatile ferroelectric random-access memory (NVFRAM) architecture. The high-density architecture (top) is designed for use as computer memory. The low-density architecture (bottom) is for smart cards and other applications of embedded memories, such as microprocessor controllers. Both architectures are examples of 1T-1C design—that is, each memory cell contains one transistor and one capacitor. This configuration helps to prevent crosstalk between adjacent cells [9.10].

The main thrust is in memory elements where the dependence on thin-film thickness is critical. It is therefore of some interest to consider the simple Tilley–Žekš (T–Z) model of phase transitions in ferroelectric films [9.11, 9.23].

9.1 The Tilley–Žekš model

Let us consider a thin film that extends from $-L/2$ to $L/2$ and makes a second-order phase transition in the absence of surface effects (i.e. in the bulk). We are looking

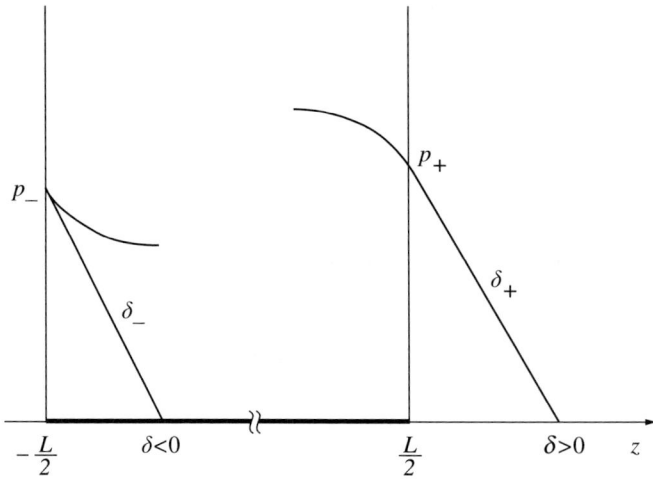

Fig. 9.6 The surface polarizations p_+ and p_- and the extrapolation lengths δ_+ and δ_-.

for the transition temperature and the polarization profile in the thin film (see also reference [9.2]).

The Landau free energy can be in such a case written as:

$$F = \frac{1}{L}\left\{\int_{-L/2}^{L/2}\left[\frac{\alpha}{2}p^2 + \frac{1}{4}\beta p^4 + \frac{1}{2}K\left(\frac{dp}{d\xi}\right)^2\right]d\xi + \frac{K}{2}\left(\frac{p_+^2}{\delta_+} + \frac{p_-^2}{\delta_-}\right)\right\} \quad (1)$$

The K term describes the contribution due to the spatial modulation inside the film, whereas the surface contribution is given by the surface polarizations p_+ and p_- and the extrapolation length δ. It is assumed that

$$\left.\begin{array}{l}\alpha = a(T - T_0) \\ \beta > 0 \\ K > 0\end{array}\right\} \quad (2)$$

The surface polarizations p_+ and p_- and the extrapolation lengths δ_+ and δ_- are shown in Fig. 9.6. They are regarded as material constants like β and K and can be positive or negative. For a positive δ the surface polarization is smaller than in the bulk, while it is larger for a negative δ.

To find the transition temperature α_c and the polarization profile $p(z)$ below T_c, we have to minimize the free energy

$$\frac{\delta F}{\partial p(z)} = 0 \quad (3)$$

for given boundary conditions

$$\frac{\partial p_-}{\partial z} = \frac{p_-}{\delta_-} \text{ at } z = -L/2 \qquad (4a)$$

$$\frac{\partial p_+}{\partial z} = \frac{-p_+}{\delta_+} \text{ at } z = L/2 \qquad (4b)$$

The corresponding Euler–Lagrange equation is

$$\alpha p + \beta p^3 - K\frac{\partial^2 p}{\partial z^2} = 0 \qquad (5)$$

The integration yields

$$\frac{K}{2}\left(\frac{\partial p}{\partial z}\right)^2 = \frac{\alpha}{2}p^2 + \frac{\beta}{4}p^4 + c \qquad (6)$$

where c is an integration constant.

I. Let us first look into the **symmetric case where $\delta_+ = \delta_- = 0$** and the extrapolation length is zero. This corresponds to the case where the surface layer is infinitesimally thin.

Here, the maximum of the polarization p_c is at the centre of the film, where $\frac{dp}{dz} = 0$. We get $p(0) = p_c$ and the integration constant

$$c = \frac{\alpha}{2}p_c^2 + \frac{\beta}{4}p_c^4 \qquad (7)$$

The polarization of the bulk is

$$p_b^2 = -\frac{\alpha}{\beta} \qquad (8)$$

$p(z)$ is finally obtained as

$$p(z) = p_c \, sn\sqrt{\frac{-\alpha}{K(1+k^2)}}\left(z + \frac{L}{2}\right), z < 0 \qquad (9)$$

where sn is the Jacobian elliptic function and k^2 is

$$k^2 = \frac{p_c^2}{2p_b^2 - p_c^2} \qquad (10a)$$

The corresponding polarization profiles are shown in Fig. 9.7. The thickness dependence of the average polarization is shown in Fig. 9.8. The critical thickness L_c below which p_c becomes zero and ferroelectricity is totally lost is

$$L_c = \sqrt{\frac{-K}{\alpha}}\pi \qquad (10b)$$

The thickness transition is of second order (Fig. 9.7). Here, the 'thickness transition' is defined by the disappearance of the average polarization.

The thickness dependence of the average polarization $\bar{p}$ is shown in Fig. 9.8. The thickness L can be expressed as

$$L = 2\sqrt{\frac{-\kappa}{\alpha}}\sqrt{1+k^2}K(k) = 2\xi\sqrt{1+k^2}K(k) \qquad (10c)$$

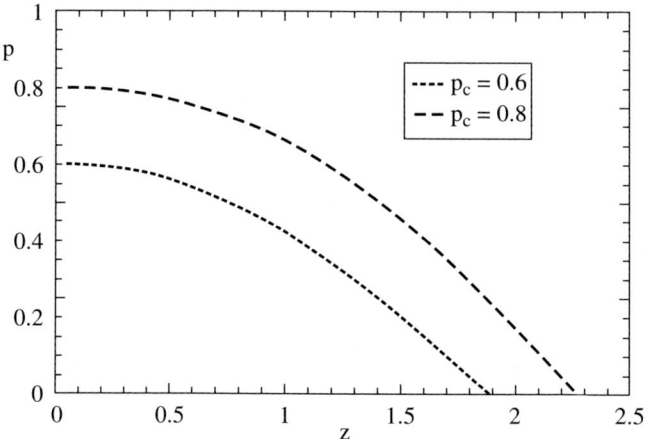

Fig. 9.7 Polarization profiles for $\delta = 0$. The z-coordinate, where p vanishes, corresponds to $z = L/2$ [9.11].

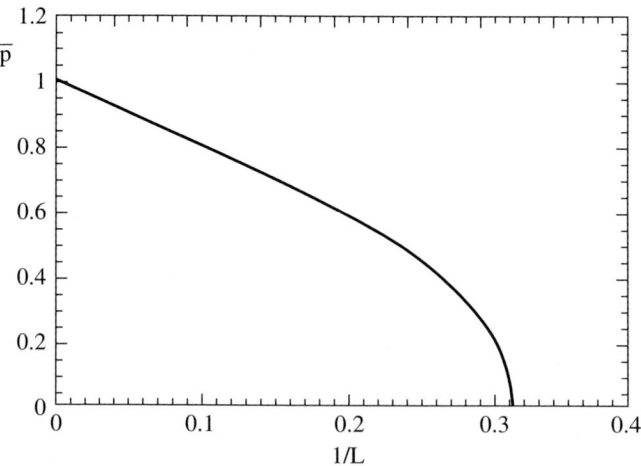

Fig. 9.8 The average polarization as a function of inverse film thickness for $\delta = 0$ [9.11].

where $K(k)$ is the complete elliptic integral of the first kind. This implies that the relationship between the length scales L and $\xi = \sqrt{\frac{\kappa}{|\alpha|}}$ can be expressed in terms of a parameter k at a given temperature.

II. Let us now consider the case of a **positive extrapolation length** ($\delta_+ = \delta_- = \delta > 0$). In this case the surface polarization p_s can be written as $p_+ = p_- = p_s$.

p_s is obtained as

$$p_s^2 = p_b^2 + p_e^2 - \sqrt{(p_b^2 + p_e^2)^2 + (p_c^4 - 2p_b^2 p_c^2)}. \tag{11}$$

Here, p_e is a parameter defined by

$$p_e^2 = K/(\beta \delta^2) \tag{12}$$

As before

$$p(z) = p_c \sin \theta(z) \tag{13}$$

Here, θ is given by

$$\sqrt{-\frac{\alpha}{2K}} \int_{-L/2}^{z} dz = \sqrt{\frac{1+k^2}{2}} \int_{\theta_s}^{\theta} \frac{d\theta}{\sqrt{1 - k^2 \sin^2 \theta}} \tag{14a}$$

and θ_s is defined by $p_s = p_c \sin \theta_s$.

θ_s can also be expressed as

$$\theta_s = \sin^{-1} \frac{p_b}{\sqrt{p_b^2 + p_e^2}} \tag{14b}$$

The critical thickness L_c below which ferroelectricity is lost is

$$L_c = 2\sqrt{\frac{-K}{\alpha}} \left(\frac{\pi}{2} - \theta_s\right) \tag{15}$$

The polarization profiles $p = p(z)$ for $\delta = 1$ are shown in Fig. 9.9. The value of z where the $p(z)$ curve terminates corresponds to $L/2$.

The critical thickness becomes thinner with increasing extrapolation length for a given temperature. The thickness transition is of second order.

The case of a **negative extrapolation length** may be only of academic interest as here the surface polarizations and the transition temperatures are higher than for the 'bulk' transitions.

III. So far, we have considered symmetric films where the extrapolation length at both surfaces is the same. For practical applications **asymmetric films** are more important. We can distinguish three different cases:

224 Ferroelectric thin films

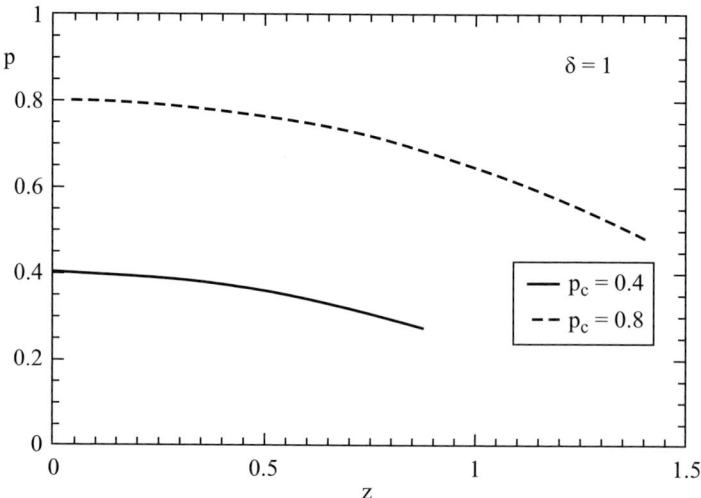

Fig. 9.9 Polarization profiles for $\delta = 1$.

a. the positive–positive case ($\delta_+ > 0, \delta_- > 0$);
b. the negative–negative case ($\delta_+ < 0, \delta_- < 0$);
c. the mixed case ($\delta_+ > 0, \delta_- < 0$).

The average free energy is for asymmetric films given by [9.24]

$$f = \frac{1}{L}\left\{\int K\left(\frac{\partial p}{\partial z}\right)^2 dz + c + \frac{K}{2}\left(\frac{p_+^2}{\delta_+} + \frac{p_-^2}{\delta_-}\right)\right\} \quad (16)$$

where c is

$$c = \frac{\alpha}{2}p_m^2 + \frac{\beta}{4}p_m^4 \quad (17)$$

p_m is the extreme value in the polarization profile located at $z = z_m$, where $\frac{dp}{dz} = 0$.
The free energy can be expanded into a power series of p_m:

$$F = \frac{A}{2}p_m^2 + \frac{B}{4}p_m^4 + \cdots \quad (18)$$

where p_m acts as an effective order parameter. The value $A = 0$ gives the critical point.

9.1.1 The positive–positive case

For $\delta_+ > 0, \delta_- > 0$ one gets:

$$A = \alpha + \frac{\sqrt{-\alpha K}}{L}\left(\pi - \sin^{-1}\frac{p_b^2}{\sqrt{p_b^2 + p_{e+}^2}} - \sin^{-1}\frac{p_b^2}{\sqrt{p_b^2 + p_{e-}^2}}\right) \quad (19)$$

where

$$p_{e+}^2 = \frac{K}{\beta \cdot \delta_+^2} \tag{20}$$

$$p_{e-}^2 = \frac{K}{\beta \cdot \delta_-^2} \tag{21}$$

The transition temperature is always lower than in the bulk. It decreases with decreasing thickness of the film.

9.1.2 The negative–negative case

For $\delta_+ < 0, \delta_- < 0$ one finds:

$$A = \alpha - \frac{\sqrt{\alpha K}}{L}\left(\tanh^{-1}\sqrt{\frac{K}{\alpha\delta_-^2}} + \tanh^{-1}\sqrt{\frac{K}{\alpha\delta_+^2}}\right) \tag{22}$$

Here, the transition temperature increases with decreasing film thickness.

9.1.3 The mixed case

For $|\delta_-| < |\delta_+|$ the theoretical minimum value of the polarization is located outside the film at $z > L/2$. The transition temperature increases with decreasing film thickness [9.11, 9.24].

For $|\delta_-| > |\delta_+|$ the theoretical minimum value of the polarization is located outside the film [9.11, 9.24] at $z < L/2$. The transition temperature decreases with decreasing film thickness.

The main conclusions can be thus summarized as follows:

a) For the $(\delta_+ > 0, \delta_- > 0)$ case the transition temperature is always lower than in the 'bulk' and decreases with decreasing thickness of the film.
b) For the $(\delta_+ < 0, \delta_- < 0)$ case the transition temperature increases with decreasing film thickness.
c) In the mixed case the transition temperature increases with decreasing film thickness for $|\delta_-| < |\delta_+|$. For $|\delta_-| > |\delta_+|$, on the other hand, the transition temperature decreases with decreasing film thickness.
d) The dielectric susceptibility obeys the Curie–Weiss law in the paraelectric phase.

It should be mentioned that finite-size effects such as depressions of T_c and reductions in P_S have been indeed observed. One should also mention that the local polarization P_S as a function of the depth z may actually increase as the surface is approached. This leads to an increase in T_c as the thickness of the film decreases. Both effects have been observed by now, in, e.g., PZT. This can be described by the fact that the extrapolation length can have either sign leading to an increase or decrease of P_S at the surface. Surface strains and inhomogeneity effects seem to be important.

Nanophase ferroelectric cells for G bit non-volatile memories thus seem to be a real possibility.

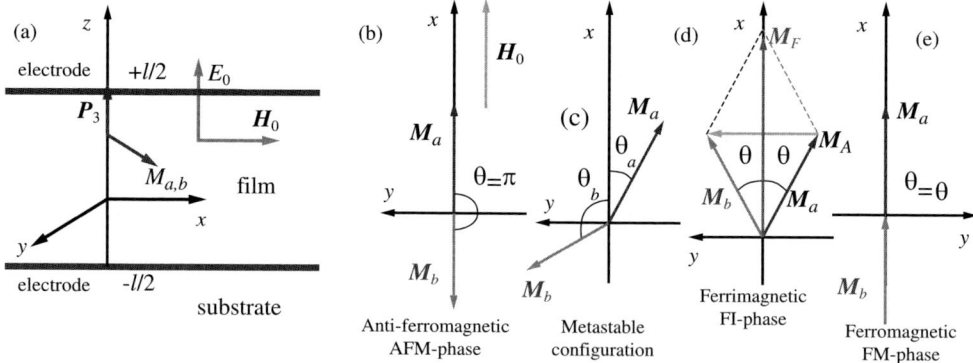

Fig. 9.10 (a) Geometry of the film. x is the weak magnetic anisotropy axis; z is the polar ferroelectric axis. The external electric field E_0 is directed along polar axes, the magnetic field H_0 is directed along the x-axes. $M_{a,b}$ are sublattice magnetization vectors. Possible stable magnetic phases: (b) anti-ferromagnetic phase (AFM), (d) ferrimagnetic phase (FI) and (e) ferromagnetic phase (FM). 2θ is the angle between the vectors M_a and M_b. Phase (c) is unstable in the bulk, however it may be stable near the film surface [9.34]. Reproduced with permission from M. D. Glinchuk, A. N. Morozovska, E. A. Eliseev, and R. Blinc: Misfit strain induced magnetoelectric coupling in thin ferroic films, J. Appl. Phys. 105, 084108 (2009).

9.2 Misfit-strain-induced magnetoelectric coupling in thin films

Recently, Wang *et al.* [9.25] and Tian *et al.* [9.26] reported about the dramatically higher quadratic magnetoelectric (ME) coefficients and spontaneous polarization values in heteroepitaxially strained thin films of $BiFeO_3$ in comparison with the bulk material. Similar effects are also found in thin polycrystalline films [9.27, 9.28].

Here we show – following the work of Glinchuk *et al.* [9.34] – that epitaxial misfit strain due to lattice mismatch at the film/substrate interface may significantly change the ME coupling coefficients, the surface energy parameters, and the polar and magnetic phase diagrams of anti-ferromagnetic (AFM)–ferroelectric (FE) films. Thus, it allows for tailoring the electric and magnetic properties of ferroic films opening the way to advanced applications.

Let us consider an AFM–FE uniaxial insulating film of thickness $l(l/2 \leq z \leq l/2)$ epitaxially grown on a thick rigid substrate. The film is in perfect electric contact with thin planar conducting electrodes (Fig.9.10(a)). For the sake of simplicity we suppose that piezomagnetism is absent, whereas magnetostriction exists in the bulk of the film.

In the Landau-Ginsburg phenomenological theory the free energy is

$$\Delta G = \frac{1}{l} \int_{-l/2}^{l/2} dz \; g_V(z) + G_S\left(\frac{l}{2}\right) + G_S\left(-\frac{l}{2}\right) \qquad (23)$$

where g_V and G_S describe the order parameter dependent contribution of the bulk and surface of the film. For the description of the phase transitions in FE-AFM films consisting of two magnetic sublattices with magnetization vectors $\vec{M}_a$ and $\vec{M}_b$, we suppose that the polarization P_3 and the electric field E_0 are directed along the polar axis z. The axis x is assumed to be the weak magnetic anisotropy axis. When we study size-induced phase transitions in thin films the dependence of polarization P_3 and magnetization of two sublattices $\vec{M}_{a,b}$ on depth z should be considered [9.29, 9.30]. The expansion of the Gibbs energy density g_V in terms of the order parameters P_3 and $\vec{M}_{a,b}$ has the form

$$g_V = \begin{pmatrix} a_1 P_3^2 + a_{11} P_3^4 + a_{111} P_3^6 + \gamma \left(\frac{dP_3}{dz}\right)^2 - Q_{ij33}\sigma_{ij} P_3^2 - \frac{A_{ij}}{2}\sigma_{ij}^2 P_3^2 - P_3\left(E_0 + \frac{E_d}{2}\right) \\ -\frac{s_{ijkl}}{2}\sigma_{ij}\sigma_{kl} + b(\vec{M}_a^2 + \vec{M}_b^2) + c\,\vec{M}_a\vec{M}_b + d(\vec{M}_a^4 + \vec{M}_b^4) + k\,\vec{M}_a^2\vec{M}_b^2 \\ +\delta\left(\frac{d\vec{M}_a}{dz}\right)^2 + \delta\left(\frac{d\vec{M}_b}{dz}\right)^2 - (M_{a1} + M_{b1})H_0 + b_1(M_{a1}^2 + M_{b1}^2) \\ +c_1 M_{a1} M_{b1} - Z_{ij11}\sigma_{ij}(M_{a1}^2 + M_{b1}^2) - W_{ij11}\sigma_{ij} M_{a1} M_{b1} \end{pmatrix}.$$
(24)

Subscripts 1, 2, and 3 denote Cartesian coordinates x, y, and z and summation rules are used. We assume that the bulk material has cubic symmetry in the paraphase. The bulk energy, the correlation energy, the interaction with the external field E_0, the striction terms, the elastic energy, and the depolarization field E_d are included in Eq. (24). The coefficients $a_1(T) = \alpha_P(T - T_C^b)$ and $b(T) = \alpha_M(T - T_N^b)$ explicitly depend on temperature T, whereas all other expansion coefficients are assumed to be temperature independent. Here, T_C^b and T_N^b are the Curie and Néel transition temperatures. σ_{ij} is the elastic stress, Q_{ijkl}, Z_{ijkl} and W_{ijkl} are the electro- and magnetostriction coefficients, respectively, and s_{ijkl} are components of the elastic compliance tensor. Note that the demagnetization field is absent when $M_{a,b3} = 0$. Typically $|b| \gg |c| \gg |b_1| + |c_1|$. Thus, we deliberately neglect the striction contribution into the highest terms b and c, while we consider its influence on the weak anisotropy terms. For the AFM phase with weak axis x to be stable in the bulk sample the inequalities $c > 0$ and $2b_1 - c_1 < 0$ should be valid. The case $2b_1 - c_1 > 0$ corresponds to the weak plane.

For the case of a single-domain insulator film with ideal electrodes the depolarizing field E_d has the form $E_d = 4\pi[\bar{P}_3 - P_3(z)]$ [9.30], where the bar designates spatial averaging over the film thickness, i.e. $\bar{P}_3 \equiv \int_{-l/2}^{l/2} P_3(z)dz/l$.

In the vicinity of the surface inversion symmetry breaking takes place and the surface piezoelectric effect g_{ijk}^e has to be taken into account in the surface free energy [9.31, 9.32],

$$G_S\left(\pm\frac{l}{2}\right) = \frac{1}{l}\left(\frac{\delta}{\lambda_M}(\vec{M}_a^2 + \vec{M}_b^2) + \frac{\delta}{\lambda_{MA}}(M_{a1}^2 + M_{b1}^2) + \frac{\gamma}{\lambda_P}P_3^2 - g_{3jk}^e\sigma_{jk}P_3\right)\bigg|_{z=\pm\frac{l}{2}}$$
(25)

where λ_P, λ_M and λ_{MA} are FE [9.30, 9.32] and magnetic [9.29] extrapolation lengths, respectively, and $\lambda_{MA} \gg \lambda_M$ allowing for weak magnetic anisotropy.

We introduce ferromagnetic (FM) $\vec{M}_F = \vec{M}_a + \vec{M}_b$ and AFM $\vec{M}_A = \vec{M}_a - \vec{M}_b$ order parameters and $\vec{M}_a^2 = \vec{M}_b^2 = M^2$.

The averaged magnetization $\overline{M}$ depends on polarization $\overline{P}_3$ via the ME coupling $\tilde{f}^{\text{DP}}$ in the following way:

$$\overline{M}^2 = -\left\{\alpha_M\left[T - T_{\text{cr}}^{\text{DF}}(l)\right] + 2\tilde{f}^{\text{DF}}\overline{P}_3^2\right\}/16\tilde{d} \tag{26}$$

So, coupling-induced phase transitions could appear. At zero resulting field, $E_m + E_0 = 0$, each of the phases could be either paraelectric (PE) with $P_3 = 0$ or FE with $P_3 \neq 0$. Estimation of material parameters shows that size effects and misfit strains substantially renormalize the free-energy coefficients. The misfit strain may significantly increase the values of the quadratic ME coupling coefficients $\tilde{f}^{\text{AFM,FM}}(l)$ in comparison with bulk values $f_\pm$.

The dependence of the normalized ME coupling coefficients $\tilde{f}^{\text{AFM}}/f_-$ and $\tilde{f}^{\text{FM}}/f_+$ on the film thickness for different misfit strains is illustrated in Fig. 9.11(a). Because $\tilde{f}^{\text{AFM,FM}}$ can be positive or negative, they lead to an increase or a decrease of the order parameters, as shown in Fig. 9.11(b) for polarization $P_3 \neq 0$.

It should be stressed that the order parameters $\overline{M} = \overline{M}(T, l, u_m)$ and $\overline{P}_3 = \overline{P}_3(T, l, u_m)$ can be tuned by the misfit strain u_m and film thickness l, thus leading to size – and ME coupling-induced phase transitions. The significant increase in the polarization compared to the bulk is clearly seen from Fig. 9.11(b).

The phase diagrams of strained ferroic films at zero external fields are shown in Fig. 9.12 for reasonable material parameters. The stabilization of the AFM phase

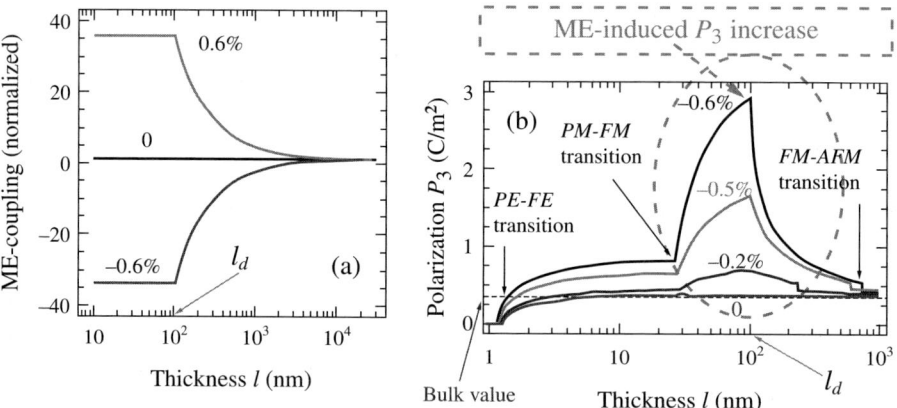

Fig. 9.11 (a) Dependence of normalized ME-coupling coefficients $\tilde{f}^{\text{AFM}}/f_- = \tilde{f}^{\text{FM}}/f_+$ and (b) polarization P_3 dependence on film thickness l for different misfit strains u_m in per cent (labels near the curves) and $l_d = 100$ nm. Reasonable material parameters in International units: $a_1(T) = (T-1103) \times 5 \times 10^5$, $a_{11} = 6.5 \times 10^8$, $b(T) = (T-642) \times 10^{-5}$, $T = 300$ K; $\gamma = 10^{-9}$, $\delta = 10^{-20}$, $c = 10^{-5}$, $b_1 = -5 \times 10^{-6}$, $c_1 = 10^{-7}$, $a_{11} = 6.5 \times 10^8$, $d = 10^{-15}$, $k = 3 \times 10^{-16}$, $s_{11} = 5.3 \times 10^{-12}$, $s_{12} = -1.85 \times 10^{-12}$, $Q_{12} = -0.005$, $Z_{11} = W_{11} = -10^{-14}$, $Z_{12} = W_{12} = 4 \times 10^{-15}$, $A_{11} = -10^{-10}$; $g_{31}^e = 0$. Lengths $\lambda_P = 4$ nm, $\lambda_M = 0.4$ nm, $\lambda_{MA} = 400$ nm [9.34]. Reproduced with permission from M. D. Glinchuk, A. N. Morozovska, E. A. Eliseev, and R. Blinc: Misfit strain induced magnetoelectric coupling in thin ferroic films, J. Appl. Phys. 105, 084108 (2009).

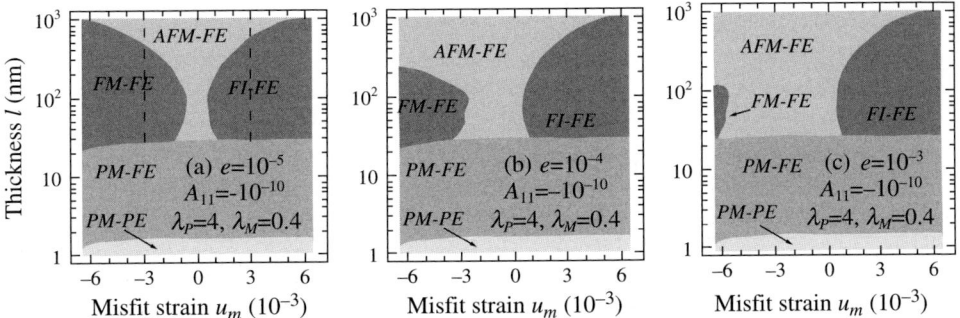

Fig. 9.12 Phase diagrams of strained ferroic films: AFM–FE designates anti-ferromagnetic and ferroelectric phases, FM–FE are ferromagnetic and ferroelectric phases (secondary ferroic phase), FI–FE is a ferrimagnetic (week plane at $\theta = \pi/2$) and ferroelectric phase, PM–FE is a paramagnetic-ferroelectric phase, PM–PE is a paramagnetic-paraelectric phase. External fields are zero. A built-in electric field is absent ($g_{31}^e = 0$). Material parameters are listed in Fig. 9.10. Different values of A_{11} and c, λ_P and λ_M (in nanometres) are listed [9.34]. Reproduced with permission from M. D. Glinchuk, A. N. Morozovska, E. A. Eliseev, and R. Blinc: Misfit strain induced magnetoelectric coupling in thin ferroic films, J. Appl. Phys. 105, 084108 (2009). See Plate 4 in the colour plate section.

with the increase in the sublattice interaction constant c is similar to what is known for bulk materials (compare Fig. 9.12(a) with Figs. 9.12(b) and 9.12(c)). It is clear that size effects and misfit strain (at a film thickness less than the critical thickness for the appearance of the misfit dislocations l_d) cause strong changes in phase diagrams. In particular, FM and ferrimagnetic (FI) phases may appear in thin films of AFM bulk materials. Small extrapolation length and depolarization field effects decrease the corresponding order parameter values and cause thickness-induced PE phase transitions in thin ferroic films (compare Figs. 9.12(a)–(c) for $\lambda_P = 4$ nm with Figs. 9.12(d) and 9.12(e) for $\lambda_P = 0.4$ nm). The decrease in λ_P, λ_M stabilizes the PE or paramagnetic (PM) phase, respectively. This is because the extrapolation length reflects the rate of polarization or magnetization profile change with film thickness so that the thinner the film the sharper is the decrease in the polarization or magnetization profile. This increases the region of $P_3 = 0$ or $M = 0$, i.e. the region of the existence of PE or PM phase.

The size-induced AFM to FM phase transition in thin films of ferroics is most probable for compressive misfit strains of more than 10^{-3}, a negative electrostriction coefficient Q_{12}, a relatively high striction coupling constant $|A_{11}|$ and a small sublattice interaction constant c. In contrast to the appearance of the FM phase transition for compressive strains, tensile misfit strains of about 10^{-3} or higher may cause an AFM spin-flop transition from the weak anisotropy axis into the perpendicular plane at zero external magnetic field, i.e. a spontaneous size-induced weak-axis–weak-plane transition [9.33].

10
Nanoferroelectrics

The search for smaller and smaller ferroelectric devices has resulted in the study of nanowires [10.1, 10.2], nanotubes [10.3–10.5] and nanodots [10.6, 10.7] with novel geometries [10.8]. Since ferroelectricity is a collective phenomenon it was initially assumed that the ferroelectric polarization would simply disappear with decreasing size of the system. However the situation turned out to be more complex. Naumov, et al. [10.9] showed that before vanishing the polarization may undergo a size-induced transition to polar geometries, which do not exist in the bulk.

The basic difference between nanoferroelectrics and bulk ferroelectrics is the increased role of the surface. This is due to the increase in the surface to volume ratio with decreasing size of the system. Therefore, the surface-energy term may become dominant. Another important point is the symmetry breaking inherently present in the vicinity of the surface. This gives rise to the appearance of spontaneous piezomagnetic, piezoelectric and magnetoelectric (ME) effects in nanomaterials even if they are absent in bulk material [10.10]. Coupled with a curvature-induced surface stress in nanoparticles like nanowires, nanotubes and nanospheres as well as lattice-mismatch strains in thin films on substrates, the surface piezomagnetic and piezoelectric effects lead to the appearance of internal – built-in – magnetic fields. These fields lead to the appearance of spontaneous magnetization and spontaneous polarization in nanosystems that are paramagnetic, respectively, paraelectric in the bulk. Such effects have been indeed observed.

The main advantage of nanomaterials is thus the possibility to vary their properties and obtain the required characteristics by the choice of their geometry and size.

10.1 Surface piezoelectric, piezomagnetic and ME tensors

The piezomagnetic tensor is a third-rank axial tensor $d_{ijk}^{(m)}$ that couples the axial magnetization vector $\vec{M}$ with the polar strain tensor $u_{ij} = u_{jk}$. In bulk materials the piezomagnetic effect can exist in 66 magnetic classes [10.11]. This was obtained as the difference between the total number of magnetic classes (90) and the number of classes that possess simultaneously the operations of time and space reversal (21) as well as the three classes of cubic symmetry $\bar{4}3m$, 432 and m3m. Since the centre of inversion is absent in the vicinity of the surface for nanomaterials, the piezomagnetic effect here should be present in the previously mentioned 21 magnetic classes. The surface piezomagnetic and piezoelectric tensors $d_{ijm}^{(Sm)}$ and $d_{ijm}^{(Se)}$ should exist in cubic lattices, too, in view of the disappearance of the centre of inversion. Thus, the piezomagnetic effect should exist in the vicinity of the surface in all 90 magnetic classes. The effect of the surface should extend over distances of several tens of nanometres from the surface.

The situation is similar with the second-rank ME tensor γ_{ij} that describes the linear coupling between the polar polarization vector $\vec{P}$ and the axial magnetization vector $\vec{M}$. In the bulk it exists in 58 magnetic classes, whereas in nanomaterials it exists in 90 magnetic classes. As an example let us treat [10.10] the bulk m3m, m'3m', m'3m and m3m' cubic classes [10.10] where the prime stands for the coupling with time reversal. These symmetry groups correspond to the bulk symmetry of the non-piezoelectric and non-piezomagnetic binary oxides MnO, FeO, CO, NiO, EuO, PrO as well as the paraphase of $BiFeO_3$. For nanomaterials the inversion centre disappears in the direction of the normal to the surface and only the symmetry axes and planes normal to the surface are conserved [10.12]. As a result, the magnetic and space symmetry groups are changed. From the above-mentioned bulk m3m, m'3m m'3m' and m3m' symmetry groups, the surface symmetry groups 4mm, 4m'm', 4'm'm and 4'mm' are obtained. Here, the symbol 4 stands for the four-fold rotation axis that is parallel to the surface normal x_3 ↑↑ 4. The corresponding bulk and surface piezomagnetic, piezoelectric and linear ME tensors are presented in Table 10.1 [10.10, 10.13].

It is seen that the cubic non-piezoelectric and non-piezomagnetic bulk materials become piezoelectric and piezomagnetic in the vicinity of the surface. Non-piezoelectric but piezomagnetic bulk materials with m3m' symmetry remain piezomagnetic and become piezoelectric. The magnetic symmetry group, however, changes.

The surface piezomagnetic and piezoelectric effects coupled with curvature-induced surface stress for nanoparticles and mismatch strains for thin films on substrates lead to built-in internal fields that can produce ferroelectricity, ferromagnetism and magnetoelectricity in nanomaterials which are only paraelectric and paramagnetic in the bulk. The values of these fields increase with decreasing nanoparticle radii R and decreasing film thickness h. The surface ME coupling γ_{ij}^S is similarly inversely proportional to the sizes in both cases. In the adopted model the part of the free energy we are interested in, namely the one that is dependent on piezo- and ME-coupling of electric (E_i) and magnetic (H_i) fields, can be written as

$$\Delta G_R = -\left(d_{ijk}^{(e)}\sigma_{jk}E_i + d_{ijk}^{(m)}\sigma_{jk}H_i + \varepsilon_0 E_i^b E_i + \mu_0 H_i^b H_i + \gamma_{ij}^R H_i E_j\right) V \qquad (1)$$

The energy includes the built-in magnetic $H_i^b = \frac{S}{\mu_0 V} d_{ijk}^{(Sm)} \sigma_{jk}$, and electric $E_i^b = \frac{S}{\varepsilon_0 V} d_{ijk}^{(Se)} \sigma_{jk}$ fields (ε_0 and μ_0 are universal dielectric and magnetic constants of vacuum, respectively). σ_{ij} is the stress tensor and the magnetoelectric energy density is $\gamma_{ij}^R H_i E_j$ with the renormalized ME coefficient $\gamma_{ij}^R \cong \gamma_{ij} + \frac{S}{V}\gamma_{ij}^S$, where γ_{ij}^S is listed in Table 10.2.

Finally, a word of caution should be added. Symmetry tells you what is allowed and what is forbidden. It does not tell you how large the allowed effects are. This depends on the material constants in each separate case.

Table 10.1 Surface and bulk piezoelectric, piezomagnetic and ME tensors [10.10, 10.13].

Symmetry group	Piezomagnetic tensor: non-trivial components	Piezoelectric tensor: non-trivial components	Linear ME tensor: non-trivial components
Bulk m3m, m̄3m	Absent in the bulk $d_{ijk}^{(m)} \equiv 0$	Absent in the bulk $d_{ijk}^{(e)} \equiv 0$	Absent $\gamma_{ij} = 0$
Bulk m'3m'	Absent in the bulk $d_{ijk}^{(m)} \equiv 0$	Absent in the bulk $d_{ijk}^{(e)} \equiv 0$	$\begin{pmatrix} \gamma_{11} & 0 & 0 \\ 0 & \gamma_{11} & 0 \\ 0 & 0 & \gamma_{11} \end{pmatrix}$
Bulk m̄3m'	$d_{123} = d_{213} = d_{312} = d_{14}^{(m)}$ $\begin{pmatrix} 0 & 0 & 0 & d_{14}^{(m)} & 0 & 0 \\ 0 & 0 & 0 & 0 & d_{14}^{(m)} & 0 \\ 0 & 0 & 0 & 0 & 0 & d_{14}^{(m)} \end{pmatrix}$	Absent in the bulk $d_{ijk}^{(e)} \equiv 0$	Absent $\gamma_{ij} = 0$
Surface 4mm (normal $x_3 \uparrow\uparrow 4$)	$d_{14}^{(Sm)} = d_{123} = -d_{213}$ $\begin{pmatrix} 0 & 0 & 0 & d_{14}^{(Sm)} & 0 & 0 \\ 0 & 0 & 0 & 0 & -d_{14}^{(Sm)} & 0 \\ 0 & 0 & 0 & 0 & 0 & 0 \end{pmatrix}$	$d_{31}^{(Se)} = d_{311} = d_{322}, d_{333} = d_{33}^{(Se)},$ $d_{131} = d_{232} = d_{15}^{(Se)}$ $\begin{pmatrix} 0 & 0 & 0 & 0 & d_{15}^{(Se)} & 0 \\ 0 & 0 & 0 & d_{15}^{(Se)} & 0 & 0 \\ d_{31}^{(Se)} & d_{31}^{(Se)} & d_{33}^{(Se)} & 0 & 0 & 0 \end{pmatrix}$	$\gamma_{12}^S = -\gamma_{21}^S$ $\begin{pmatrix} 0 & \gamma_{12}^S & 0 \\ -\gamma_{12}^S & 0 & 0 \\ 0 & 0 & 0 \end{pmatrix}$

Surface		
4m'm'	$d_{311} = d_{312} = d_{31}^{(Sm)}$, $d_{333} = d_{33}^{(Sm)}$, $d_{131} = d_{232} = d_{15}^{(Sm)}$ $\begin{pmatrix} 0 & 0 & 0 & 0 & d_{15}^{(Sm)} & 0 \\ 0 & 0 & 0 & d_{15}^{(Sm)} & 0 & 0 \\ d_{31}^{(Sm)} & d_{31}^{(Sm)} & d_{33}^{(Sm)} & 0 & 0 & 0 \end{pmatrix}$	The same as above $d_{31}^{(Se)} = d_{311} = d_{322}, d_{333} = d_{33}^{(Se)}$, $d_{131} = d_{232} = d_{15}^{(Se)}$	$\gamma_{11}^S = \gamma_{22}^S$ $\begin{pmatrix} \gamma_{11}^S & 0 & 0 \\ 0 & \gamma_{11}^S & 0 \\ 0 & 0 & \gamma_{33}^S \end{pmatrix}$
(normal $x_3 \uparrow\uparrow 4$)		
4'mm'	$d_{123} = d_{213} = d_{14}^{(Sm)}, d_{312} = d_{36}^{(Sm)}$ $\begin{pmatrix} 0 & 0 & 0 & d_{14}^{(Sm)} & 0 & 0 \\ 0 & 0 & 0 & 0 & d_{14}^{(Sm)} & 0 \\ 0 & 0 & 0 & 0 & 0 & d_{36}^{(Sm)} \end{pmatrix}$	The same as above $d_{31}^{(Se)} = d_{311} = d_{322}, d_{333} = d_{33}^{(Se)}$, $d_{131} = d_{232} = d_{15}^{(Se)}$	$\gamma_{12}^S = \gamma_{21}^S$ $\begin{pmatrix} 0 & \gamma_{12}^S & 0 \\ \gamma_{12}^S & 0 & 0 \\ 0 & 0 & 0 \end{pmatrix}$
(normal $x_3 \uparrow\uparrow 4$, $x_{1,2} \perp$ m planes)*		
Surface		
4'm'm	$d_{113} = -d_{223} = d_{15}^{(Sm)}$, $d_{322} = -d_{311} = d_{31}^{(Sm)}$, $\begin{pmatrix} 0 & 0 & 0 & 0 & d_{15}^{(Sm)} & 0 \\ 0 & 0 & 0 & -d_{15}^{(Sm)} & 0 & 0 \\ d_{31}^{(Sm)} & -d_{31}^{(Sm)} & 0 & 0 & 0 & 0 \end{pmatrix}$	The same as above $d_{31}^{(Se)} = d_{311} = d_{322}, d_{333} = d_{33}^{(Se)}$, $d_{131} = d_{232} = d_{15}^{(Se)}$	$\gamma_{11}^S = -\gamma_{22}^S$ $\begin{pmatrix} \gamma_{11}^S & 0 & 0 \\ 0 & -\gamma_{11}^S & 0 \\ 0 & 0 & 0 \end{pmatrix}$
(normal $x_3 \uparrow\uparrow 4$, $x_{1,2} \perp$ m' planes)*		

* Groups 4'mm' and 4'm'm are equivalent within the rotation of the coordinate system. m stands for magnetic, e stands for electric, S stands for surface.

Table 10.2 Surface built-in fields and ME coupling coefficients induced by the surface piezoelectric and piezomagnetic effects [10.10].

Nanosystem	Built-in magnetic field normal component H	Built-in electric field normal component E	Linear ME coupling $\gamma_{ij}^{R} = \gamma_{ij} + \frac{\gamma_{ij}^{S}}{\eta}$, η is the characteristic size
Thin layer of thickness h, surface normal $\uparrow\uparrow x_3$; σ_{ij} are the surface stress tensor components	$H_1^b = \frac{d_{113}^{(Sm)}\sigma_{13} + d_{123}^{(Sm)}\sigma_{23}}{\mu_0 h}$ $H_2^b = \frac{d_{223}^{(m)}\sigma_{23} + d_{213}^{(m)}\sigma_{13}}{\mu_0 h}$ $H_3^b = \frac{1}{\mu_0 h}\begin{pmatrix} d_{311}^{(m)}\sigma_{11} + d_{322}^{(m)}\sigma_{22} \\ + d_{333}^{(m)}\sigma_{33} + d_{312}^{(m)}\sigma_{12} \end{pmatrix}$	$E_1^b = \frac{1}{\varepsilon_0 h}d_{113}^{(Se)}\sigma_{13}$ $E_2^b = \frac{1}{\varepsilon_0 h}d_{223}^{(Se)}\sigma_{23}$ $E_3^b = \frac{1}{\varepsilon_0 h}\begin{pmatrix} d_{311}^{(Se)}\sigma_{11} + d_{322}^{(Se)}\sigma_{22} \\ + d_{333}^{(Se)}\sigma_{33} \end{pmatrix}$	Size $\eta = h$
Nanowire of radius R, wire axes $\uparrow\uparrow x_1$, local normal $\boldsymbol{n} \uparrow\uparrow x_3$	$H_\rho^b = -\frac{2\tau}{\mu_0 R^2}\begin{pmatrix} d_{322}^{(Sm)} + d_{333}^{(Sm)} \\ -2d_{311}^{(Sm)}\frac{s_{12}}{s_{11}} \end{pmatrix}$ $\sigma_{33} = \sigma_{22} = -\frac{\tau}{R}$, $\sigma_{11} = \frac{s_{12}}{s_{11}}\frac{2\tau}{R}$, $\sigma_{12} = \sigma_{13} = \sigma_{23} = 0$ τ is the intrinsic surface stress tensor coefficient	$E_\rho^b = -\frac{2\tau}{\varepsilon_0 R^2}\begin{pmatrix} d_{322}^{(Se)} + d_{333}^{(Se)} \\ -2d_{311}^{(Se)}\frac{s_{12}}{s_{11}} \end{pmatrix}$	Size $\eta = R/2$
Nanosphere of radius R, local normal $\boldsymbol{n} \uparrow\uparrow x_3$	$H_r^b = -\frac{3\tau}{\mu_0 R^2}\begin{pmatrix} d_{322}^{(Sm)} + d_{333}^{(Sm)} \\ + d_{311}^{(Sm)} \end{pmatrix}$ $\sigma_{33} = \sigma_{22} = \sigma_{11} = -\frac{\tau}{R}$, $\sigma_{12} = \sigma_{13} = \sigma_{23} = 0$	$E_r^b = -\frac{3\tau}{\varepsilon_0 R^2}\begin{pmatrix} d_{322}^{(Se)} + d_{333}^{(Se)} \\ + d_{311}^{(Se)} \end{pmatrix}$	Size $\eta = R/3$

Here, S_{ijkl} is the elastic compliances tensor, u_{ij} and σ_{ij} are strain and stress tensor components, χ^{e}_{ij} and χ^{m}_{ij} are electric and magnetic susceptibilities. S is the systems surface and V is the volume. γ_{ij} and γ^{S}_{ij} are the bulk and surface ME coefficients, τ is the intrinsic surface stress-tensor coefficient. The resulting fields may well exceed the critical values for spin-flop transitions in anti-ferromagnets and change anti-ferroelectrics into ferroelectrics. They may also lead to the appearance of ferroelectricity in paraelectric materials and ferromagnetism in paramagnetics (see also [9.33]).

10.2 Size-induced ferroelectricity in non-ferroelectric insulators

It has been recently shown that a ferroelectric phase can be induced in otherwise non-ferroelectric insulators by a strong enough surface stress present under the curved surface of nanowires [9.33, 10.14]. Small enough nanowires of BaO with a radius 1–10 nm can be ferroelectric even at room temperatures [10.14]. The spatial confinement with a curved geometry may, in the general case, lead to a strong enhancement of the ferroelectric properties up to a critical size. Such a behaviour has been indeed observed in Rochelle salt nanorods [10.15, 10.16] and was predicted by first-principles calculations [10.17] and phenomenological theory [10.18, 10.19].

The possible atomic structure of the MO wire is shown in Fig. 10.1. The surface stress leads to strains and a corresponding bond length contraction in the radial direction ρ and an elongation along the wire axes z. The surface stress induces out-of-plane displacements of the light O^{2-} anions that lead to the unit-cell tetragonal distortion and thus can cause a macroscopic dipole moment to appear in the longitudinal direction z (see Fig. 10.1).

Including the surface energy term, the Landau–Ginzburg (L–G) free energy F depends on the component P_3 of the spontaneous polarization and mechanical strains u_{ij} as [10.19–10.21]:

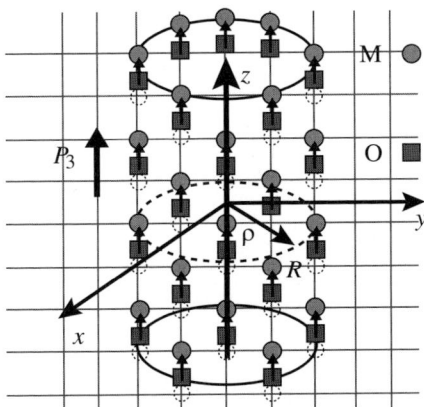

Fig. 10.1 Possible atomic structure of the MO wire. Metal M is for Ba, Zn, Mg, Eu, etc [10.14]. Reproduced with permission from Anna N. Morozovska, E. A. Eliseev, Maya D. Glinchuk, and Robert Blinc: Analytical prediction of size-induced ferroelectricity in BaO nanowires under stress, Phys. Rev. B 81, 092101 (2010). http://prb.aps.org/abstract/PRB/v81/i9/e092101

$$F = \begin{pmatrix} \int_V d^3r \left(\dfrac{\alpha}{2} P_3^2 + \dfrac{\beta}{4} P_3^4 + \dfrac{\gamma}{6} P_3^6 + \dfrac{g}{2}(\nabla P_3)^2 - P_3 \left(E_e + \dfrac{E_3^d}{2} \right) - q_{ij33} u_{ij} P_3^2 \right. \\ \left. + \dfrac{c_{ijkl}}{2} u_{ij} u_{kl} \right) \\ + \int_S d^2r \left(\dfrac{\alpha_S}{2} P_3^2 + \mu_{ij}^S u_{ij} \right) \end{pmatrix} \quad (2)$$

The surface energy coefficient α_S is positive, isotropic and weakly temperature dependent, so that higher terms can be neglected in the surface-energy expansion. The integration is performed over the systems surface S and volume V, respectively. μ_{ij}^S is the intrinsic surface stress tensor that exists under the curved surface of a solid body and determines the excess pressure on the surface [10.19]. The elastic stiffness tensor is c_{ijkl} and q_{ijkl} stands for electrostriction stress tensor. E_e is the external electric field.

The expansion coefficient γ as well as the gradient term g are positive. The coefficient $\alpha(T) = \alpha_T(T - T_c^*)$ is T dependent, where T is the absolute temperature. T_c^*, which is normally positive for conventional ferroelectrics, is here negative as bulk MO is non-ferroelectric. Its value can be determined from the critical strain at zero temperature for thick layers with a flat surface.

For long cylindrical nanoparticles with the spontaneous polarization directed along the cylinder axes the depolarization field E_d can be neglected. At the same time, a strong depolarization field exists in spherical nanoparticles [10.18, 10.20].

For cubic symmetry the non-zero components of the strain tensor inside the cylindrical wire of radius R have the following form [10.20]: radial strain $u_{\rho\rho} = u_{11} = u_{22} = -(s_{11} + s_{12})(\mu/R) + Q_{12} P_3^2$ and longitudinal strain $u_{33} = -s_{12}(\mu/R) + Q_{11} P_3^2$, where Q_{ij} are the components of electrostriction strain tensor, s_{ij} are the elastic compliances, and the isotropic approximation $\mu_{ii}^S = \mu$ has been used. Note that the radial strain $u_{\rho\rho}$ is negative, while the longitudinal strain u_{33} is positive at $P_3 = 0$, since $(s_{11} + s_{12}) > 0$ and $s_{12} < 0$ for the considered cubic symmetry. Thus, the surface stress induces a bond-length contraction in radial directions $\{x, y\}$ and an elongation in the z-direction. The corresponding tetragonality of the unit cell acquires the form:

$$\dfrac{c}{a} = \dfrac{1 + u_{33}}{1 + u_{22}} = \dfrac{1 - s_{12}(\mu/R) + Q_{11}\bar{P}^2}{1 - (s_{11} + s_{12}) \cdot (\mu/R) + Q_{12}\bar{P}^2} \approx 1 + s_{11}\dfrac{\mu}{R} + (Q_{11} - Q_{12})\bar{P}^2 \quad (3)$$

Using the expressions for strains and the direct variational method, the transition temperature into the ferroelectric phase and the spontaneous polarization $\vec{P}_3(R)$ averaged over the wire radius R can be derived similarly to [10.18, 10.20] as:

$$T_{cr}(R) \approx T_c^* - Q_{12} \dfrac{4\mu}{\alpha_T R} - \dfrac{2}{\alpha_T R} \left(\dfrac{g}{(g/\alpha_S) + 2R/k_0^2} \right) \quad (4)$$

$$\bar{P}_3(R) \approx \sqrt{\dfrac{2\alpha_T (T_{cr}(R) - T)}{\beta + \sqrt{\beta^2 + 4\gamma\alpha_T(T_{cr}(R) - T)}}} \quad (5)$$

The constant $k_0 = 2.408\ldots$ is the smallest root of the Bessel function $J_0(k)$.

Table 10.3 L–G free energy expansion coefficients for BaO bulk material.

α_T, m/(F K)	β, m^5/ (C^{2}F)	γ, m^9/ (C^{4}F)	T_c^*, K	Q_{11}, m^4/C^2	Q_{12}, m^4/C^2	s_{11}, m^2/N	s_{12}, m^2/N
6.4×10^6	1.0×10^9	4.4×10^{11}	-226	0.50 ± 0.05	-0.22 ± 0.3	10.23×10^{-12}	-2.91×10^{-12}

The first term in Eq. (4) is negative and size-independent for nonferroelectric oxide layers with a flat surface. The second one is the contribution of the intrinsic surface stress $\sim\mu/R$, whereas the third negative term $\sim g$ originates from correlation effects. The third term decreases the possible transition temperature T_{cr} at positive α_S, while the second term increases T_{cr} under the condition $Q_{12}\mu < 0$. By changing the wire radius one can tune the transition temperature and polarization value in a wide range.

The L–G free-energy expansion coefficients for BaO bulk material have been extracted from the results of the first-principles calculation [10.19] and experimental data [10.22]. All parameters are listed in Table 10.3.

In analogy to perovskites Q_{12} is negative. The surface curvature can thus induce ferroelectricity in BaO nanowires. The phase diagrams of BaO wires in the temperature–wire-radius plane are shown in Fig. 10.2. The dependence of the spontaneous polarization of BaO nanowires on temperature and radius are shown in Fig. 10.3. The figures were calculated from Eqs. (3–5), with the parameters from Table 10.3 and the gradient coefficient $g = 2\times10^{-9}$ m^3/F. The surface tension coefficient μ varies in the typical range of 5-50 N/m depending on the nanowire ambient.

The two limiting cases $|\alpha_S| \ll 2\mu|Q_{12}|$ and $|\alpha_S| \gg 2\mu|Q_{12}|$ correspond to large and small contributions of the intrinsic surface stress.

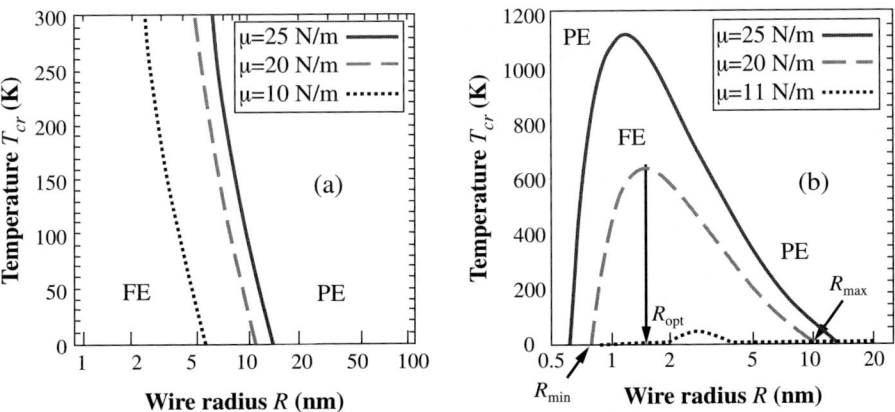

Fig. 10.2 Temperature–radius phase diagrams of BaO nanowires for (a) $|\alpha_S| \ll 2\mu|Q_{12}|$ and (b) $|\alpha_S| \gg 2\mu|Q_{12}|$ [10.14]. Reproduced with permission from Anna N. Morozovska, E. A. Eliseev, Maya D. Glinchuk, and Robert Blinc: Analytical prediction of size-induced ferroelectricity in BaO nanowires under stress, Phys. Rev. B 81, 092101 (2010). http://prb.aps.org/abstract/PRB/v81/i9/e092101

It is seen from Figs. 10.2 and 10.3 that BaO nanowires of radius $\sim(1-10)$ nm can be ferroelectric at room and even at higher temperatures (with spontaneous polarization values up to $0.5\,\text{C/m}^2$) for surface stress coefficients $\mu\sim(10-50)\,\text{N/m}$. The spontaneous polarization increases with the increase of the surface-tension coefficient.

Let us underline that the dependence of the transition temperature T_{cr} on the wire radius R is monotonic for the case $|\alpha_S| \ll 2\mu|Q_{12}|$: T_{cr} decreases with increasing wire radius R (Fig. 10.2(a)). Actually, for the case $T_{\text{cr}}(R) \approx T_c^* - Q_{12}\frac{4\mu}{\alpha_T R}$ the critical radius for the appearance of the ferroelectricity is $R_{\text{cr}}(T) \approx \frac{4\mu Q_{12}}{\alpha_T(T_c^*-T)}$ at a given temperature T. For the limiting case the influence of the correlation effects is negligibly small in comparison with the surface-stress contribution. The spontaneous polarization in this case monotonically decreases with the increase of the wire radius and disappears at $R = R_{\text{cr}}(0)$(see Fig. 10.3(b)).

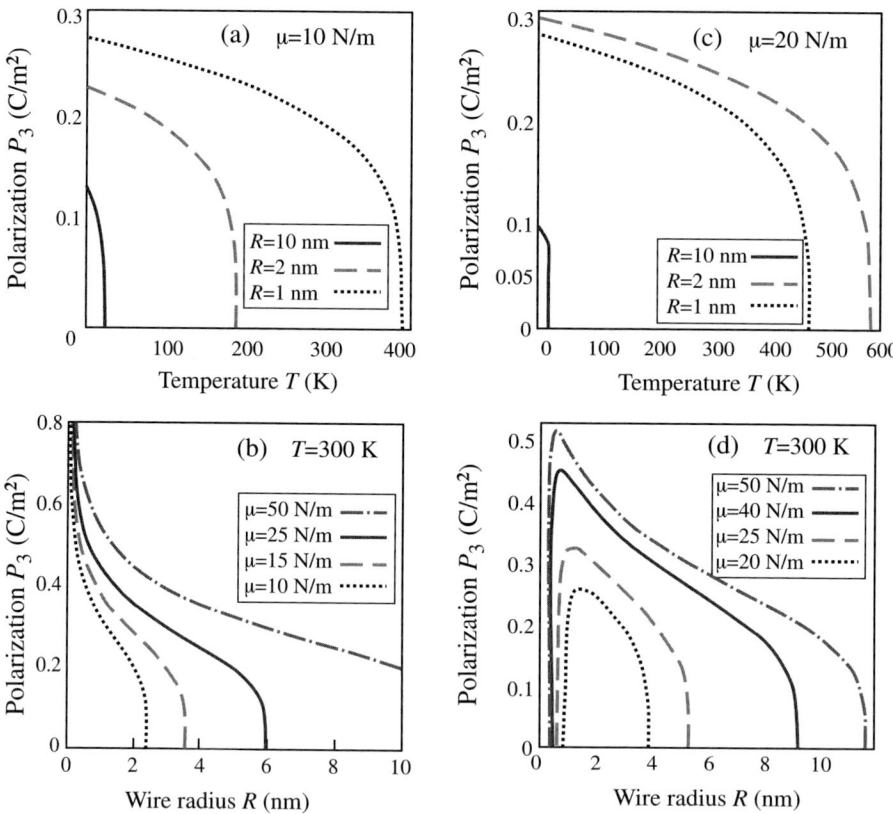

Fig. 10.3 Spontaneous polarization of BaO nanowires vs. temperature (a,c) and radius at $T = 300\,\text{K}$ (b,d). Plots (a,b) correspond to $|\alpha_S| \ll 2\mu|Q_{12}|$, plots (c,d) correspond to $|\alpha_S| \gg 2\mu|Q_{12}|$ [10.14]. Reproduced with permission from Anna N. Morozovska, E. A. Eliseev, Maya D. Glinchuk, and Robert Blinc: Analytical prediction of size-induced ferroelectricity in BaO nanowires under stress, Phys. Rev. B 81, 092101 (2010). http://prb.aps.org/abstract/PRB/v81/i9/e092101

In the other limiting case $|\alpha_S| \gg 2\mu|Q_{12}|$ the region of the appearance of ferroelectricity is smaller (compare Figs. 10.2(a) and 10.2(b)). Here, the dependence of the transition temperature T_{cr} on the wire radius R is not monotonic (see Fig. 10.3(d)).

10.3 Spontaneous flexoelectric effect in nanoferroics

The most general definition of the direct flexoeffect is the appearance of the polarization or magnetization in response to an inhomogeneous mechanical impact (elastic stress or strain gradient). The converse flexoeffect corresponds to the appearance of mechanical stress or strain in response to the gradient of the order parameter. A typical example is the flexoelectric effect [10.23, 10.24, 10.25] due to the coupling of the polarization gradient with the elastic strain and polarization with elastic-strain gradient. It is worth underlining that the flexomagnetic effect should exist in materials without an inversion symmetry of time (e.g., in some anti-ferromagnetics [10.26]), so that the symmetry consideration for the flexomagnetic effect will be similar to the well-studied flexoelectric phenomena [10.62].

The flexoelectric effect was predicted by Mashkevich and [10.27]. Later, a theoretical study of the flexoelectric effect in bulk crystals was performed by Tagantsev [10.24, 10.25]. Experimental measurements of flexoelectric tensor components were carried out by Ma and Cross [10.28, 10.29, 10.30] and Zubko et al. [10.31]. A new development of the theoretical description for the flexoelectric response of different nanostructures starts from the papers of Catalan and coworkers [10.32, 10.33], while recent achievements are presented in the papers of Majdoub et al. [10.34] and Kalinin and Meunier [10.35]. However, in these papers the flexoelectric effect was considered as a coupling between intrinsic polar properties (e.g., polarization) and extrinsic factors like the misfit-strain relaxation [10.32, 10.33] or the system's bending by external forces [10.34, 10.35], while the coupling between intrinsic parameters (i.e. spontaneous polarization gradient and strain) was not considered. The crucial role of the surface in all physical properties of nanosystems including the strong order-parameter gradients in ferroic nanostructures [10.36] inevitably leads to a noticeable spontaneous flexocoupling. It is almost negligible for bulk materials, where the order parameters are usually homogeneous.

It is worth underlining that the macroscopic phenomenological approach needs an estimation of spatial scales where it is valid. For ferroelectrics the correlation radii determine the minimum possible scale of order-parameter changes, so the phenomenological approach can be valid if the correlation radii are higher than the lattice constant. The critical sizes for the existence of long-range order in ferroics derived on the basis of microscopic [10.37] and phenomenological calculations [10.38] are in reasonable agreement both with each other and with the experimental observation of a ferromagnetic critical size of 7–30 nm [10.39] and the ferroelectric critical size of 1–10 nm [10.40, 10.41, 10.42]. The critical sizes are thus in the region from several lattice constants to several nanometres. Majdoub et al. [10.34, 10.43] considered the response of a cantilever-shaped beam to an external stress gradient or the dielectric properties of a thin film taking into account the polarization gradient and allowing for the flexoelectric effect. They apply phenomenological macroscopic models to nano-sized systems and compare the obtained results with microscopic modeling (*ab initio* and molecular-dynamics simulations). Majdoub et al. [10.34, 10.43] found that the results of the microscopic modeling are qualitatively (and in some cases quantitatively) reproduced by the phenomenological model taking into account flexoelectricity and the polarization gradient.

10.3.1 Basic equations for the flexoeffect in ferroic nanoparticles

For the description of flexocoupling in ferroic nanoparticles we shall use the Landau–Ginsburg (L–G) phenomenological approach [10.44, 10.45] including the surface energy, gradient energy, depolarization or demagnetization fields, mechanical stress and flexoeffect. We need to minimize the Helmholtz free energy F [10.21, 10.46, 10.47, 10.48, 10.49]. For ferroics with second-order phase transitions the corresponding L–G expansion of the bulk (F_V) and surface (F_S) parts of the Helmholtz free energy F in terms of the order parameter η and strain tensor components u_{ij} is:

$$F_V = \int_V d^3r \left[\frac{a_{ij}(T)}{2}\eta_i\eta_j + \frac{a_{ijkl}}{4}\eta_i\eta_j\eta_k\eta_l - \eta_i\left(E_{0i} + \frac{E_i^d}{2}\right) + \frac{g_{ijkl}}{2}\left(\frac{\partial \eta_i}{\partial x_j}\frac{\partial \eta_k}{\partial x_l}\right) \right.$$
$$\left. - \frac{f_{ijkl}}{2}\left(\eta_k \frac{\partial u_{ij}}{\partial x_l} - u_{ij}\frac{\partial \eta_k}{\partial x_l}\right) - q_{ijkl}u_{ij}\eta_k\eta_l + \frac{c_{ijkl}}{2}u_{ij}u_{kl} + \frac{v_{ijklmn}}{2}\left(\frac{\partial u_{ij}}{\partial x_m}\frac{\partial u_{kl}}{\partial x_n}\right) \right]$$

$$F_S = \int_S d^2r \left(\frac{a_{ij}^S}{2}\eta_i\eta_j + \frac{a_{ijkl}^S}{4}\eta_i\eta_j\eta_k\eta_l + \mu_{\alpha\beta}^S u_{\alpha\beta} + d_{ijk}^S u_{ij}\eta_k + \frac{w_{ijkl}^S}{2}u_{ij}u_{kl} + \cdots \right)$$
(6)

The coefficients $a_{ij}(T)$ explicitly depend on temperature T. The coefficients a_{ij}^S, a_{ijkl}, and a_{ijkl}^S are supposed to be temperature independent; the constants g_{ijkl} and v_{ijklmn} determine the magnitude of the gradient energy. The tensors g_{ijkl}, a_{ijkl} and a_{ijkl}^S are positively defined. The tensor w_{ijkl}^S represents the surface excess elastic moduli, $\mu_{\alpha\beta}^S$ is the surface stress tensor [10.50, 10.51], and d_{ijk}^S is the surface piezoelectric tensor [10.24, 10.52]. q_{ijkl} are the bulk striction coefficients and c_{ijkl} are components of the elastic stiffness tensor [10.53].

The tensor f_{ijkl} is the flexocoupling coefficient tensor [10.31, 10.32]. In fact, only the Lifshitz invariant $\frac{f_{ijkl}}{2}(\eta_k \frac{\partial u_{ij}}{\partial x_l} - u_{ij}\frac{\partial \eta_k}{\partial x_l})$ is relevant for the bulk contribution. The gradient terms like $v_{ijklmn}(\partial u_{ij}/\partial x_k)(\partial u_{lm}/\partial x_n)$ are responsible for the stable smooth distribution of the order parameter at non-zero strain gradients since the presence of the Lifshitz invariant essentially changes the stability conditions [10.54]. In the scalar case, the inequality $f^2 < gc$ should be valid for the stability of the order-parameter distribution. In the considered tensorial case the terms $v_{ijklmn}(\partial u_{ij}/\partial x_k)(\partial u_{lm}\partial x_n)$ can be neglected under the condition $f_{klmn}^2 < g_{ijkl}c_{ijmn}$.

$\vec{E}_0$ in Eq. (6) is the external field coupled with the order parameter $\vec{\eta} \cdot \vec{E}^d$ is the depolarization or demagnetization field that appears due to the non-zero divergence [div$(\vec{\eta}) \neq 0$] of the order parameter $\vec{\eta}$ in a confined system [10.55, 10.56].

10.3.2 Thin pills

Here, we set $a_1(T) = \alpha_T(T - T_C)$ and use for screened ferroelectric thin pills the depolarization field $E_3^d = (\langle P_3 \rangle - P_3)/(\varepsilon_0 \varepsilon_b)$. P_3 is the polarization directed along the pill symmetry axes (see inset in Fig. 10.4), where ε_b is the dielectric permittivity of the background [10.57] or reference state [10.58] unrelated with the ferroelectric soft mode (typically $\varepsilon_b < 10$). The averaged value of susceptibility in the paraelectric phase is:

$$\langle \chi_{33} \rangle \approx \left[1 - \frac{2R_z^2}{(R_z + \lambda^*)h}\right] \left[a_1 + \frac{2g_{11}^*}{(R_z + \lambda^*)h} - \frac{24f^2 R_z}{(R_z + \lambda^*)h^2}\right]^{-1} \quad (7)$$

Here we introduced the characteristic length $R_z^2 = g_{11}^* \varepsilon_0 \varepsilon_b$. Equation (7) was derived for typical conditions $h \gg R_z$. Using the divergence of susceptibility (Eq. (7)), one could find the critical temperature of the transition between paraelectric and ferroelectric phases:

$$T_{\mathrm{cr}}(h, f_{11}) \approx T_C - \frac{1}{\alpha_T} \left[\frac{2g_{11}^*}{(R_z + \lambda^*)h} - \frac{24f^2 R_z}{(R_z + \lambda^*)h^2}\right]. \quad (8)$$

The first term in Eq. (8) is the bulk transition temperature, the second term is mainly determined by the influence of surface effects and depolarization field renormalized by the flexoeffect. The third term originates from the influence of the flexoterm $\sim \frac{f^2}{h} \langle \frac{\partial P_3}{\partial z} \frac{z}{h} \rangle$.

The dependence of the ferroelectric transition temperature on the thickness h and the flexoelectric coefficient f_{11} is shown in Fig. 10.4.

The effect of the increase of the transition temperature with flexoelectric coefficient f_{11} is demonstrated in Figs. 10.4(c) and 10.4(d) for several fixed thicknesses h. The smaller the thickness h, the higher the slope of $T_{\mathrm{cr}}(f_{11})$. For ultrathin pills with thickness $h \leq h_{\min}$ non-monotonic effects appeared.

10.3.3 Nanowires

The renormalization of the gradient coefficient and extrapolation length strongly affects all the properties and in particular the transition temperature shift and correlation radius in single-domain ferroelectric nanowires with axial symmetry of the polarization P_3 (see Fig. 10.5). For short-circuit boundary conditions

$$E_3^d(\rho, z) \approx \left[1 + (h/2R)^2\right]^{-1} \left[\overline{P}_3 - P_3(\rho, z)\right] / (\varepsilon_0 \varepsilon_b) \quad (9)$$

[10.18], while

$$E_3^d(\rho, z) \approx -\left[1 + (h/2R)^2\right]^{-1} P_3(\rho, z)/(\varepsilon_0 \varepsilon_b) \quad (10)$$

for the open-circuit ones. So, one can neglect the small depolarization field $E_d \sim (R/h)^2$ for the case $h \gg R$ typical for nanowires.

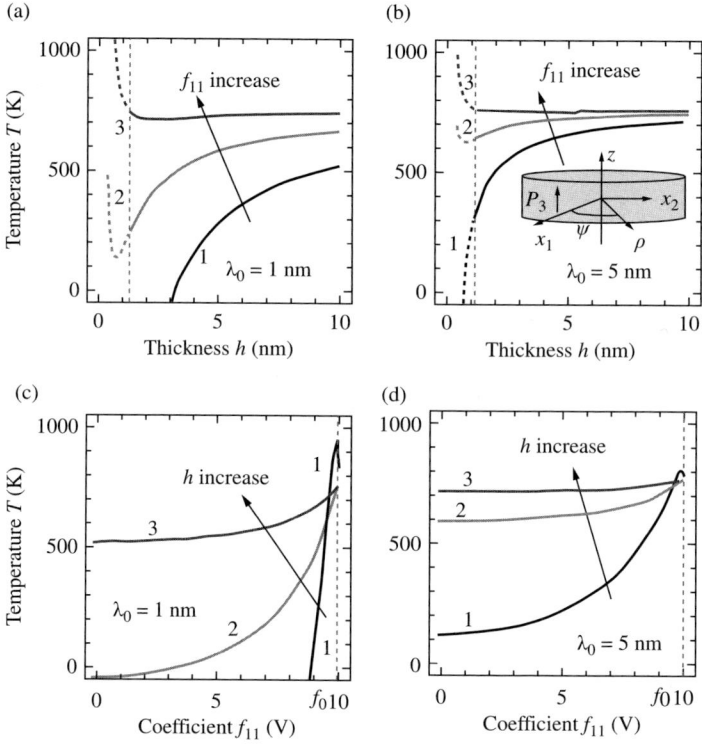

Fig. 10.4 (a) and (b) The dependence of the transition temperature T_{cr} on pill thickness h for different values of flexoelectric coefficient $f_{11} = 0, 11, 12$, and $13\,\text{V}$ (curves 1–4). (c) and (d) The dependence of the transition temperature T_{cr} on the flexoelectric coefficient for different values of thickness $h = 0.8, 2, 3$, and $10\,\text{nm}$ (curves 1–4). Material parameters correspond to PbTiO$_3$: $g_{11} = 10^{-9}\,\text{Jm}^3/\text{C}^2$, $\varepsilon_b = 1$, $T_C = 765\,\text{K}$, $\alpha_T = 7.53 \times 10^6\,\text{Jm}/\text{C}^2\,\text{K}$, $c_{11} = 1.75 \times 10^{11}\,\text{J/m}^3$, and $c_{12} = 0.79 \times 10^{11}\,\text{J/m}^3$ [10.62].

The approximate expression for the ferroelectric to the paraelectric phase-transition temperature $T_{cr}(R)$ for nanowires can be rewritten as

$$T_{cr}(R, f_{44}) \approx \begin{cases} T_C - \dfrac{2}{\alpha_T}\left[\dfrac{g_{12}^*}{R\lambda^*(R) + 2R^2/k_{01}^2}\right], & \lambda^* > 0 \\ T_C - \dfrac{2}{\alpha_T}\left[g_{12}^*\dfrac{2\lambda^*(R) - R}{2R\lambda^{*2}}\right], & \lambda^* < 0 \end{cases} \quad (11)$$

where $k_{01} = 2.408\ldots$ is the smallest positive root of equation $J_0(k) = 0$. The dependences of the renormalized transition temperature T_{cr} on nanowire radius and flexoelectric coefficients f_{44} are shown in Fig. 10.5 for dimensionless variables and in Figs. 10.6(a) and 10.6(b) for PbTiO$_3$ material parameters. It is clear from the plots

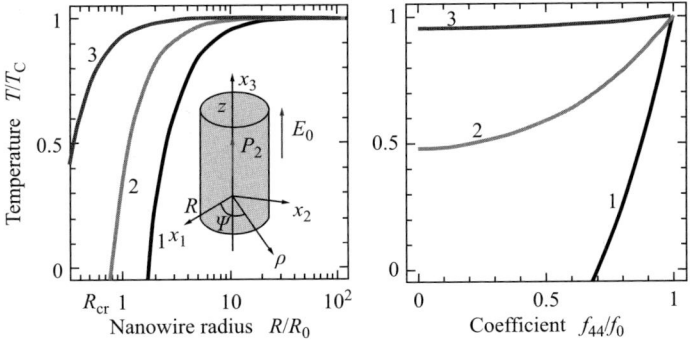

Fig. 10.5 Dependence of the phase-transition temperature T_{cr} on the nanowire radius at fixed flexoelectric coefficient $f_{44}/f_0 = 0, 0.9, 0.95$, and 0.99 (curves 1–3) and T_{cr} dependence on the flexoelectric coefficient at fixed radius $R/R_c = 1.25, 2.5, 5$, and 10 (curves 1–3). Parameter $g_{12}/(a_1^S R_c) = 1$ PbTiO$_3$ material parameters are $T_c = 479°C, \alpha_T = 3.8 \times 10^5$ J m/C^2 K, $c_{44} = 1.1 \times 10^{11}$ J/m^3, $g_{12} = 10^{-9}$ Jm^3/C^2, and extrapolation length $\lambda_0 = 1$ nm [10.62].

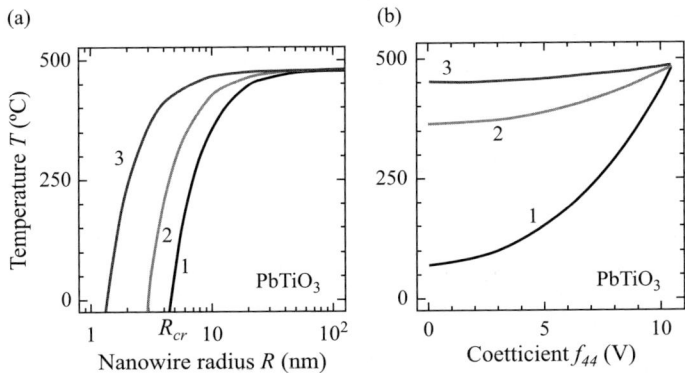

Fig. 10.6 T_{cr} dependence on radius (a) at fixed values of flexoelectric coefficient $f_{44} = 0, 8$, and 10 V (curves 1–3) and dependence on flexoelectric coefficient (b) at fixed values of radius $R = 5, 10$, and 20 nm (curves 1–3). PbTiO$_3$ material parameters are $T_c = 479°C, \alpha_T = 3.8 \times 10^5$ J m/C^{2}K, $c_{44} = 1.1 \times 10^{11}$ J/m^3, $g_{12} = 10^{-9}$ J m^3/C^2, and extrapolation length $\lambda_0 = 1$ nm [10.62].

that the higher the f_{44} value, the higher the transition temperature T_{cr} and the smaller the critical radius R_{cr} that corresponds to $T = T_{cr}$ (see also [9.33]).

10.4 Ferroelectric vortex states: Phase transitions in zero-dimensional nanoferroelectrics

Bulk ferroelectrics are characterized by multiple degenerate states with a differently oriented spontaneous polarization $\vec{P}$, i.e. ferroelectric domains, which can be switched by an external electric field. Previous studies [10.2, 10.8, 10.59] have suggested that phase transitions leading from a paraelectric to a spontaneously polarized ferroelectric state smoothly disappear with decreasing size of the system. Since ferroelectricity is a collective phenomenon, it is clear that below a critical size the spontaneous polarization $\vec{P}$ must vanish. It has been, however, recently shown [10.9] that before the ferroelectric state vanishes, the system may first undergo a size-induced phase transition to a ferroelectric vortex state [10.9, 10.60]. This is a polar configuration where dipoles form a ferroelectric vortex that does not exist in the bulk (Fig. 10.7). This state persists down to very small diameters where it finally vanishes and ferroelectricity disappears.

The point is that the order parameter of the phase transition here is not the total net polarization

$$\vec{P} = N^{-1} \sum_i \vec{p}_i = 0 \tag{12}$$

Fig. 10.7 Local dipoles on the central x (the left panel) and y cross-section (the right panel) of the [10.7, 10.28] rod at 64 K. Local vortices are schematically shown as circles. Such a state has been observed for instance in BiFeO$_3$ nanodots.

where $\vec{p}_i$ is the local dipole moment of the cell i located at $\vec{R}_i$ but the toroid moment $\vec{G}$ of the polarization

$$\vec{G} = (2N)^{-1} \sum_i \vec{R}_i \times \vec{p}_i \tag{13}$$

Thus, a very specific ordering of the local dipoles occurs in nanoparticles, i.e. in zero-dimensional systems like small disks and rods, where the polarization vanishes, $\vec{P} = 0$, but the toroid moment is non-zero, $\vec{G} \neq 0$, below a certain critical temperature T_c [10.9, 10.61]. This is called the 'A' phase. Here, the z-components of the toroid moments G_z spontaneously order along the cylindrical z-axis of the nanoparticles below T_c (Fig. 10.8(a)). G_x and G_y, on the other hand, are close to zero.

Investigations have been performed [10.9] on nanoparticles of a cylindrical shape with a diameter D and height h. Particles with $D < h$ and $D > h$ are referred to as disks and rods, respectively. At fixed height h the 'A' phase was found to be stable at low temperatures only about a critical diameter D called $D_{c,A}$. The toroid moment G_z increases with increasing D and vanishes below $D_{c,A}$. The vanishing of $\vec{G}$ is, however, not direct. Below $D_{c,A}$ a phase B appears (Fig. 10.11(b)). Unlike the A phase the toroid moments of the B phase have no z-component, $G_z = 0$, but have non-zero x

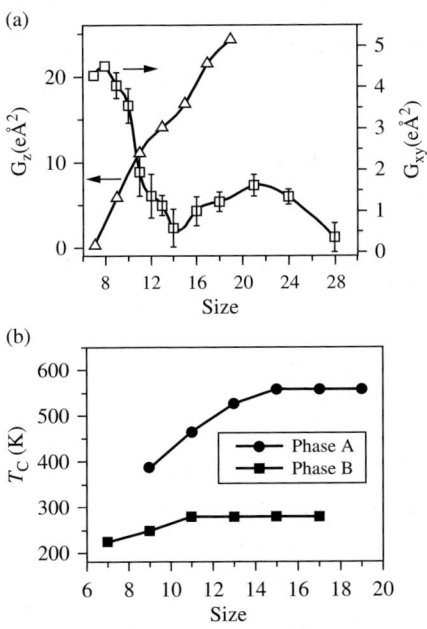

Fig. 10.8 (a) Size (D) dependence of toroid moment G_z at 64 K in phase A. (b) Size (D) dependence of the critical temperatures T_c for the A and B phases at fixed h. One unit is the bulk lattice constant of PZT $a = 4$ Å. h is taken as $h = 14$ [10.9].

and y components $G_{x,y} \neq 0$. The total toroid moment in the B phase is thus relatively small. Below a second critical diameter $D_{c,B}$ the phase is non-toroidal and non-polar.

It should be noted that the physical mechanism inducing the vortex state is the depolarization field that is minimized by this domain structure. The vortex structure of the A-phase is bistable as the toroid moment G_z can be parallel or anti-parallel to the z-axis.

States with equivalent but differently oriented toroid moments G_z can be switched from one orientation to another by applying a time-dependent magnetic field. This generates a curling electric field:

$$\text{rot } \vec{E} = -\frac{\partial \vec{B}}{\partial t} \tag{14}$$

The interaction energy is

$$U_{\text{int}} = \vec{G} \cdot \frac{\partial \vec{B}}{\partial t} \tag{15}$$

Storing data with a switchable macroscopic toroid moment may be superior to storing with a switchable spontaneous polarization. This is so because no direct electrode contact is required and cross-talk is eliminated. In addition, the minimum size of disks that display structural vortex bistability is here estimated as 3.2 nm allowing for non-volatile ferroelectric random access memories (NFERAM) with an information density of 60 Tbits inch^{-2}, i.e. 60×10^{12} bits per square inch. This is five orders of magnitude larger than the current NFERAM capabilities of 0.2 Gbit inch^{-2} and also surpasses the information density of magnetic recording devices ($\sim 1 G$bit inch^{-2}).

Appendix

Creation of Polar Nanoclusters in Relaxors

A theoretical model for the creation of polar nanoclusters in relaxors has been developed. It is shown that under certain conditions a second solution exists for the loss of stability of the paraelectric phase in addition to the one normally obtained for bulk ferroelectrics. This transition takes place at temperatures much higher than the ones in bulk ferroelectrics. New families of relaxors with properties tailored for specific applications can be thus synthetized.

Relaxors [1, 2] are characterized by a strong frequency dispersion, a high dielectric constant and the appearance of polar nanoclusters. They are important for many applications, such as memory devices and actuators, as well as in robotics and spintronics. In contrast to bulk ferroelectrics, relaxors retain their macroscopic symmetry down to the lowest temperatures studied. The question why nanoclusters are spontaneously polarized at temperatures much above the ferroelectric transition temperature T_C in bulk ferroelectrics is still not understood. This is so in spite of numerous investigations in recent years. The nature of relaxors and polar nanoclusters represents one of the most basic open problems in modern solid state physics and material science [3, 4, 5].

Here we present a theoretical solution describing the creation, existence and properties of polar nanoclusters in relaxors.

We first investigate the phase transition of a sphere in a cubic ferroelectric surrounded by a dielectric medium. In contrast to other studies we take explicitly into account the realistic effects of the depolarizing field [6].

We use the standard Ginzburg-Landau free energy in spherical coordinates r, θ, φ with gradient terms and appropriate boundary conditions [7, 8]. We are looking for the highest temperature where the stability of the paraelectric phase is lost and a homogeneous nanocluster polarization $P_\varphi \neq 0$ appears. The explicit equations are rather cumbersome. Therefore we restrict ourselves to the essentials. It is convenient to expand $P_\varphi(r, \theta)$ in a Fourier series

$$P_\varphi(r,\theta) = \sum P_{\varphi m}(r) y_m(\theta) \tag{1}$$

where

$$\frac{d^2 P_{\varphi m}}{dr^2} - \left(\frac{A}{g} + \frac{\lambda_m}{r^2}\right) P_{\varphi m} = 0$$

and

$$\lambda_m = 1/R_m \tag{2}$$

Here we have $A = -g(\mu_n/R)^2$. μ_n is the n-th root of Eq. 3. $(g\varepsilon_0)^{1/2}$ is of the order of an interatomic distance. g is the coefficient of the gradient term in the Landau expansion and ε_0 is the dielectric constant of the vacuum.

For A > 0 these equations have no solutions, satisfying the boundary conditions, whereas for A < 0 such solutions do exist.

The boundary conditions are satisfied and a loss of stability of the paraelectric phase is possible if

$$2^{-1}J\upsilon_m(X) + xJ'\nu_m(X) = 0. \tag{3}$$

Here $J\upsilon_m(X)$ and $J'\nu_m(X)$ are Bessel functions.

If the argument is negative, a solution of the above equation does not exist and no nanoclusters can be formed.

The earliest loss of stability takes place for the minimum value μ_n^m corresponding to the first root of Eq. 3 with the smallest ν_m. Numerically we find that $\mu_1^0 = 2,39$.

The condition for the loss of stability is thus

$$A = -[\varepsilon_0(\varepsilon_b + 2\varepsilon_p)]^{-1}. \tag{4}$$

Here ε_p is the dielectric constant of the medium surrounding the sphere and ε_b is the "base dielectric constant" [8]. For the critical radius we obtain

$$R_c = \mu_1^0 (g\varepsilon_0)^{1/2}(\varepsilon_b + 2\varepsilon_p)^{1/2} \tag{5}$$

Locally polarized nanoregions thus appear at high temperatures before the phase transition in the bulk if A is larger than $[\varepsilon_0(\varepsilon_b + 2\varepsilon_p)]^{-1}$ and the radii of the nanoclusters are smaller than Rc. This means that a second solution exists for the loss of stability of the paraelectric phase under certain conditions at temperatures much higher than the one known to occur for bulk ferroelectrics.

As the dielectric constant of relaxors is rather high, polarized nanoclusters do indeed appear. It should be mentioned that they do not appear at a single well-defined temperature, but rather in a temperature interval. If the sphere is replaced by an ellipsoid, the temperature where polarized nanoclusters appear, is changed. However, the situation is qualitatively the same.

These results are in good agreement with the data obtained for PMN and PMN-PT.

In conclusion, we have developed a theoretical model for the creation of polarized nanoclusters in relaxors. The results allow for a synthesis of new families of relaxors, tailored for specific applications. The appearance of so many ferroelectrics in nature may be thus understood.

References

[1] V. A. Isupov, *Zh. Tekh. Fiz.* **26**, 1912 (1956).
[2] E. Cross, *Ferroelectrics* **76**, 241 (1987) and references therein.
[3] G. Burns and F. H. Dacol, *Phys. Rev. B* **28**, 2527 (1983).
[4] R. Blinc, V. V. Laguta, B. Zalar and J. Banys, *J. Mat. Science* **41**, 27 (2006).

[5] H. Schmidt and F. Schwabl, *Phys. Lett. A* **61**, 476 (1971).
[6] A. M. Bratkovsky and A. P. Levanyuk, *J. Comp. Theor. Nanoscience* **6**, 1 (2009).
[7] I. I. Naumov and H. Fu, *Phys. Rev. Lett.* **98**, 077603 (2007).
[8] A. Tagantsev, *Ferroelectrics* **19**, 375 (2008).
[9] J. F. Scott, R. Blinc, *J. Phys.: Condens. Matter* **23**, 113202 (2011).

Derivation of the Equations

The free energy is:

$$F = \frac{A}{2}P^2 + \frac{g}{2}\left[\left(\frac{\delta P_\rho}{\delta \rho}\right)^2 + \left(\frac{\delta P_\rho}{\delta \rho}\right)^2\right] + \frac{1}{g^2}\left[\left(\frac{\delta P_\varphi}{\delta \rho} - P_\varphi\right)^2 + \left(\frac{\delta P_\varphi}{\delta \rho} + P_\varphi\right)^2\right]$$

$$+ \textit{ depolarization field terms}$$

From this:

$$AP_\varphi - g\left[\frac{\delta^2 P_\varphi}{\delta r^2} + \frac{1}{r^2}\frac{\delta^2 P_\varphi}{\delta \theta^2} - \frac{1}{r^2}\frac{P_\varphi}{sin^2\theta}\right] = 0$$

$$(g\varepsilon_0)^{-1} \cdots \textit{interatomic distance}$$

$$\varepsilon_p \cdots \textit{dielectric constant of the medium}$$

$$E = -\varepsilon_0(\varepsilon_b + 2\varepsilon_p)^{-1} \cdot P$$

$$P_\varphi(r,\theta) = \sum P_{\varphi m}(r) \cdot y_m(\theta)$$

$y_m(\theta)$ = Eigen functions of the Sturm-Liouville problem.

$$\frac{d^2 P_{\varphi m}}{dr^2} - \left(\frac{A}{g} + \frac{\lambda_m}{r^2}\right) P_{\varphi m} = 0$$

$$P_{\varphi m} = C r^{1/2} J_{\nu m}\left(\sqrt{\frac{-A}{g}} \cdot r\right)$$

$$\nu_m = \frac{1}{2}\sqrt{1 + 4\lambda_m}$$

if

$$A = -g\left(\mu_\nu^m / R\right)^2$$

μ_ν^m is the n-th root of the equation

$$\frac{1}{2}J_{\nu m}(x) + x \cdot J'_{\nu m}(x) = 0$$

Loss of stability occurs for a minimal value of λ_m:

$$\lambda_0 = \frac{1}{2} + \frac{\sqrt{5}}{2};$$

$$y_0(\theta) = (sin\theta)^{\lambda_0};$$

$$\mu_1^0 = 2,39$$

Critical radius:

$$R_c = \mu_1^0 (g\varepsilon_0)^{1/2} \cdot (\varepsilon_b + 2\varepsilon_p)^{1/2}$$

Nanoclusters exist for

$$R < R_c$$

and

$$A < 0.$$

References

Chapter 1

[1.1] M. E. Lines and A. Glass, in *Principles and Applications of Ferroelectrics and Related Materials*, Oxford University Press, New York (1977).

[1.2] M. Dawber, M. Rabe, and J. F. Scott, *Rev. Mod. Phys.* **77**, 1083 (2005).

[1.3] S. Horiuchi and Y. Tokura, *Nature Mater.* **7**, 357 (2008).

[1.4] B. Zalar, V. Laguta, and R. Blinc, *Phys. Rev. Lett.* **90**, 037601 (2003); B. Zalar, A. Lebar, J. Seliger, and R. Blinc, *Phys. Rev.* **B71**, 064107 (2005).

[1.5] N. S. Dalal, A. Klymachyov, and A. Bussmann-Holder, *Phys. Rev. Lett.* **81**, 5924 (1998).

[1.6] A. Bussmann-Holder, N. S. Dalal, R. Fu, and R. Migoni, *J. Phys. Condens. Matter* **13**, L231 (2001); A. Bussmann-Holder and N. S. Dalal in *Structure and Bonding (Springer-Verlag), 'Order/disorder versus or with displacive dynamics in ferroelectric systems'*, **124**, 1 (2007); N. S. Dalal, A. Klymachyov, and A. Bussmann-Holder, *Phys. Rev. Lett.* **81**, 5924 (1998).

[1.7] G. J. Goldsmith and J. G. White, *J. Chem. Phys.* **31**, 1175–87 (1959).

[1.8] D. Bordeaux, J. Bornarel, A. Capiomont, and J. Lajzerowicz-Bonneteau, *Phys. Rev. Lett.* **31**, 314 (1973).

[1.9] J. Kroupa, P. Vaněk, R. Krupkova, and Z. Zikmund, *Ferroelectrics* **202**, 229 (1997).

[1.10] M. Hiraoka, T. Hasegawa, T. Yamada, Y. Takahashi, S. Horiuchi, and Y. Tokura, *Adv. Mater.* **19**, 3248 (2007).

[1.11] Y. Kamishima, Y. Akishige, and M. Hashimoto, *J. Phys. Soc. Jpn.* **60**, 2147 (1991).

[1.12] Y. Akishige and Y. Kamishima, *J. Phys. Soc. Jpn.* **70**, 3124 (2001).

[1.13] H. Schultes, P. Strohriegl, and E. Dormann, *Ferroelectrics* **70**, 161 (1986).

[1.14] P. Gruner-Bauer and E. Dormann, *J. Phys. Condens. Matter* **4**, 5599 (1992).

[1.15] P. Szklarz and G. Bator, *J. Phys. Chem. Solids* **66**, 121 (2005).

[1.16] K. Noda, K. Ishida, A. Kubono, T. Horiuchi, H. Yamada, and K. Matsushige, *J. Appl. Phys.* **93**, 2866 (2003).

[1.17] S. Horiuchi, Y. Tokunaga, G. Giovannetti, S. Picozzi, H. Itoh, R. Shimano, R. Kumai, and Y. Tokura, *Nature* **463**, 789 (2010).

[1.18] H. Okamoto, H. Fujii, C. Inoue, and E. Tatsumoto, *Phys. Rev. B* **43**, 8224 (1991); S. Horiuchi, Y. Okimoto, R. Kumai, and Y. Tokura, *J. Phys. Soc. Jpn.* **69**, 1302 (2000).

[1.19] Y. Tokura, S. Koshihara, Y. Iwasa, H. Okamoto, T. Komatsu, T. Koda, N. Iwasawa, and G. Saito, *Phys. Rev. Lett.* **63**, 2405 (1989).

[1.20] S. Horiuchi, F. Ishii, R. Kumai, Y. Okimoto, H. Tachibana, N. Nagaosa, and Y. Tokura, *Nature Mater.* **4**, 163 (2005); A. Almeida, M. L. Dossantos, M. R. Chaves, M. H. Amaral, J. C. Toledano, A. Perigaud, and H. Savary, *Ferroelectrics* **79**, 253 (1988).

[1.21] J. Wang, et al., *Science* **299**, 1719 (2003).

[1.22] S. Horiuchi, R. Kumai, and Y. Tokura, *Angew. Chem. Int. Ed.* **46**, 3497–501 (2007).

[1.23] E. Murakami, M. Komukae, T. Osaka, and Y. Makita, *J. Phys. Soc. Jpn* **59**, 1147 (1990).

[1.24] T. Furukawa, *Phase Transit.* **18**, 143 (1989).

[1.25] Y. Shiozaki, E. Nakamura, and T. Mitsui (eds), *Crystal and Solid State Physics*, Vols. **16** (1982), **28** (1990) and **36** (2006) (Springer, Berlin).

[1.26] M. Szafrański, A. Katrusiak, and G. J. McIntyre, *Phys. Rev. Lett.* **89**, 215507 (2002).

[1.27] S. Hoshino, T. Mitsui, F. Jona, and R. Pepinsky, *Phys. Rev.* **107**, 1255 (1957).

[1.28] J. Van Suchtelen, *Philips Res. Rep.* **27**, 28 (1972).

[1.29] W. Eerenstein, N. D. Mathur, and J. F. Scott, *Nature* **442**, 759 (2006).

[1.30] A. M. J. G. van Run, D. R. Terrel, and J. H. Scholing, *J. Mater. Sci.* **9**, 1710 (1974).

[1.31] C.-W. Nan, L. Liu, N. Cai, J. Zhai, Y. Ye, and Y. H. Lin, *Appl. Phys. Lett.* **81**, 3831 (2002).

[1.32] N. Cai, C.-W. Nan, J. Zhai, and Y. Lin. *Appl. Phys. Lett.* **84**, 3516 (2004).

[1.33] I. Valasek, *Phys. Rev.* **17**, 475 (1921).

[1.34] S. C. Abrahams, *Acta Crystallogr.* **B45**, 228 (1989); J. Ravez, S. C. Abrahams, R. Depape, *J. Appl. Phys.* **65**, 3987 (1989); R. Blinc et al., *J. Appl. Phys.* **103**, 074114 (2008) and references therein.

[1.35] Y. Ishibashi and M. Iwata, *Jpn. J. Appl. Phys.* **37**, L985 (1998).

[1.36] E. Castel, M. Josse, D. Michau, and M. Maglione, *J. Phys. Condens. Matter* **21**, 452201 (2009).

[1.37] W. Cochran, *Phys. Rev. Lett.* **3**, 412–414 (1959).

[1.38] J. C. Slater, *Phys. Rev.* **78**, 748 (1950).

[1.39] R. Blinc and H. Žekš, in *Soft Modes in Ferroelectrics and Antiferroelectrics*, North Holland, Amsterdam (1974); R. Blinc, *J. Phys. Chem. Solids* **13**, 204 (1960).

[1.40] K. A. Müller and H. Burkard, *Phys. Rev.* **B19**, 3593 (1979).

[1.41] S. E. Rowley, L. J. Spalek, R. P. Smith, M. P. M. Dean, G. G. Lonzarich, J. F. Scott, and S. S. Saxena, *http://arxiv.org/abs/0903.1445* (2009).

[1.42] A. Bussmann-Holder and A. R. Bishop, *Phys. Rev. B* **78**, 104117 (2008); A. Bussmann-Holder, H. Büttner, and A. R. Bishop, *Phys. Rev. Lett.* **99**, 167603 (2007).

[1.43] J. G. Bednorz and K. A. Müller, *Phys. Rev. Lett.* **52**, 2289 (1984).

[1.44] J. H. Haeni et al., *Nature* **430**, 758 (2004).

[1.45] M. Itoh et al., *Phys. Rev. Lett.* **82**, 3540 (1999).

[1.46] T. Schneider, H. Beck, and E. Stoll, *Phsy. Rev.* **B13**, 1123 (1976); S. S. Saxena et al. *Nature* **406**, 587 (2000).

[1.47] R. Rousseu and A. J. Millis, *Phsy. Rev.* **B67**, 014105 (2003).

[1.48] L. Palova, P. Chandra, and P. Coleman, *Phsy. Rev.* **B79**, 075101 (2009).

[1.49] L. J. Spalek, M. Shimuta, S. E. Rowley, T. Katsufuji, O. Pitrenko, S. S. Saxena, and C. Paragopoulos, *Proc. SCES'08*.

[1.50] W. Cochran, *Adv. Phys.* **9**, 387 (1960).

[1.51] J. D. Axe, J. Harade, and G. Shirane, *Phys. Rev.* **B1**, 1227 (1970); S. M. Shapiro, G. Shirane, T. Riste, and J. D. Axe, *ibid.* **6**, 4332 (1972).

[1.52] R. Migoni, D. Bäverle, and H. Bilz, *Phys. Rev. Lett.* **37**, 1155 (1976).

[1.53] J. F. Scott, *Rev. Mod. Phys.* **46**, 83 (1974); K. A. Müller and H. Burkard, *Phys. Rev.* **B19**, 3539 (1979).

[1.54] K. A. Müller and W. Berlinger, *Phys. Rev.* **B34**, 6130 (1986).

[1.55] R. Comes, M. Lambert, and A. Guinier, *Solid State Commun.* **6**, 715 (1968).

[1.56] Y. Jakoby and E. A. Stern, *Comments Condens. Matter Phys.* **18**, 1 (1996).

[1.57] A. S. Chaves, F. C. S. Barreto, R. A. Nogueiru, and B. Žekš, *Phys. Rev.* **B13**, 207 (1976).

[1.58] R. Kind, P. M. Cereghetti, C. A. Jeitziner, B. Zalar, J. Dolinšek and R. Blinc, *Phys. Rev. Lett.* **88**, 195501 (2002) and references therein. See also Y. Feng, C. Ancona-Torres, T. F. Rosenbaum, G. F. Reiter, D. L. Price, and E. Courtens, *Phys. Rev. Lett.* **97**, 145501 (2006).

[1.59] J. C. Slater, *J. Chem. Phys.* **9**, 16 (1941); L. Pauling, in *The Nature of the Chemical Bond*, University Press, Ithaca, New York (1945).

[1.60] G. Reiter, J. Meyers, and P. Platzman, *Phys. Rev. Lett.* **89**, 135505 (2002).

[1.61] V. H. Schmidt and E. A. Uehling, *Phys. Rev.* **126**, 447 (1962); J. Dolinšek, D. Arčon, B. Zalar, R. Pirc, R. Blinc, and R. Kind, *Phys. Rev.* **B54**, R6811 (1996).

[1.62] S. Koval, J. Kohanoff, R. L. Migoni, and E. Tossati, *Phys. Rev. Lett.* **89**, 187602 (2002); S. Koval *et al.*, *Phys. Rev.* **B71**, 184102 (2005).

[1.63] E Mathsushita and T. Matsubara, Prog. Theor. Phys. **68**, 1811 (1982); *J. Phys. Soc. Jpn.* **56**, 200 (1987).

[1.64] Y. Ishibashi and M. Iwata, *J. Phys. Soc. Jpn.* **78**, 014704 (2009).

Chapter 2

[2.1] 'Incommensurate Phases in Dielectrics', R. Blinc and A. Levanyuk, editors, North Holland, Amsterdam, Vol. I and II (1986).

[2.2] I. E. Dzaloshinskii, *Sov. Phys. JETP* **19**, 960 (1964).

[2.3] G. A. Toombs, *Phys. Reports* **40C**, 181 (1978).

[2.4] J. M. Cowley, *Diffraction Physics* (North-Holland Publishing Co., Amsterdam), p. 370 (1975).

[2.5] W. Cochran, *Adv. Phys.* **18**, 157 (1969); See also A. P. Levanyuk, D. G. Sannikov, *Sov. Phys.-Solid State* **18**, 1122 (1976); Y. Ishibashi, V. Dvorak, *J. Phys. Soc. Jpn.* **44**, 32 (1978).

[2.6] Y. Yamada, I. Shibuya, and S. J. Hoshino, *J. Phys. Soc. Jap.* **18**, 1594 (1963).

[2.7] M. J. Iizumi, D. Axe, G. Shirane, and K. Shimaoka, *Phys. Rev. B* **15**, 4392 (1977).

[2.8] C. de Pater, *Ph.D. Thesis*, University of Delft (1978).
[2.9] P. M. de Wolf, *Acta Crystallogr. A* **30**, 777 (1974).
[2.10] A. Janner and Jansen, T., *Phys. Rev. B* **15**, 643 (1977).
[2.11] D. E. Moncton, I. D. Axe, and F. J. DiSalvo, *Phys. Rev. Lett.* **34**, 734 (1975).
[2.12] W. L. McMillan, *Phys. Rev. B* **12**, 1187 (1975).
[2.13] W. L. McMillan, *Phys. Rev. B* **14**, 1496 (1976).
[2.14] S. Aubry, *Solitons and Condensed Matter Physics*, eds. A. R. Bishop, and T. Schneider, (Springer-Verlag, Berlin), p. 264 (1978).
[2.15] P. Bak and J. von Boehm, *Phys. Rev. Lett.* **42**, 122 (1978).
[2.16] W. A. Overhauser, *Phys. Rev. B* **3**, 3173 (1971).
[2.17] A. P. Levanyuk and D. G. Sannikov, *Usp. Fiz. Nauk* **112**, 561 (1974).
[2.18] E. Fatuzzo and W. V. Merz, *Ferroelectricity* (North Holland, Amsterdam), 1967.
[2.19] I. Hatta, *J. Phys. Soc. Jpn* **28**, 1266 (1970).
[2.20] G. J. Goldsmith and J. G. White, *J. Chem. Phys.* **31**, 1175 (1959).
[2.21] K. Aizu, *J. Phys. Soc. Jpn* **46**, 1232 (1979).
[2.22] M. Iizumi and K. Gesi, *J. Phys. Soc. Jpn* **45**, 711 (1978).
[2.23] Y. Ishibashi and M. Shiba, *J. Phy. Soc. Jpn* **45**, 409 (1978).
[2.24] Y. Ishibashi, *Ferroelectrics* **24**, 119 (1980).
[2.25] K. Aiki, K. Hukuda, H. Koga, and T. Kobayashi, *J. Phys. Soc. Jpn* **28**, 389 (1970).
[2.26] I. Hatta, *J. Phys. Soc. Jpn* **28**, 1266 (1970); R. Osredkar, S. Južnič, V. Rutar, J. Seliger, and R. Blinc, *Ferroelectrics* **24**, 147 (1980); See also C. Berthier, D. Jerome, P. Molinie, and J. Rouxel, *Solid State Commun.* **19**, 131 (1976) where the first NMR study of an incommensurate charge-density wave system is reported.
[2.27] B. A. Strukov, V. M. Arutyunova, and Y. Uesu, *Fiz. Tverd. Tela* **24**, 3061 (1982). R. Blinc, V. Rutar, J. Seliger, S. Zumer, Th. Rasing, and I.P. Aleksandrova, *Solid State Commun.* **34**, 895 (1980).
[2.28] C. J. de Pater, *Phys. Status Solidi B* **48**, 503 (1978).
[2.29] K. Hamano, Y. Ikeda, T. Fujimoto, K. Ema, and S. Hirotsu, *J. Phys. Soc. Jpn* **49**, 2278 (1980).
[2.30] H. G. Unruh and J. Strömich, *Solid State Commun.* **39**, 737 (1981); R. Blinc, S. Južnič, V. Rutar, J. Seliger, and S. Žumer, *Phys. Rev. Lett.* **44**, 609 (1980).
[2.31] P. Prelovšek, A. Levstik, and C. Filipič, *Phys. Rev. B* **28**, 6610 (1983).
[2.32] P. Bak and V. J. Emery, *Phys. Rev. Lett.* **36**, 978 (1976).
[2.33] K. Hamano, Y. Ikeda, T. Fujimoto, K. Ema, and S. Hirotsu, *J. Phys. Soc. Jpn* **49**, Suppl. B, 10 (1980).
[2.34] K. Hamano, T. Hishinuma, and K. Ema, *J. Phys. Soc. Jpn* **50**, 2666 (1981).
[2.35] H. Mashiyama and H. G. Unruh, *J. Phys (GB) C* **16**, 5009 (1983); S. Pleško, R. Kind, and H. Arend, *Phys. Status Solidi A* **61**, 87 (1980).
[2.36] A. Levstik, P. Prelovšek, C. Filipič, and B. Žekš, *Phys. Rev. B* **25**, 3416 (1982).
[2.37] C. J. de Pater and C. van Dijk, *Phys Rev. B* **18**, 1281 (1978).
[2.38] L. Bernard, R. Currat, P. Delamoye, C. M. E. Zeyen, S. Hubert, and R. de Kouchkovsky, *J. Phys. (GB) C* **16**, 433 (1983).
[2.39] J. D. Axe, *Phys. Rev. B* **21**, 4181 (1980).

[2.40] C. F. Majkrzak, J. F. Axe, and A. D. Bruce, *Phys. Rev. B* **22**, 5278 (1980).
[2.41] H. Mashiyama, *J. Phys. Soc. Jpn* **50**, 2655 (1981).
[2.42] R. A. Cowley, *Adv. Phys.* **29**, 1 (1980).
[2.43] R. Blinc, V. Rutar, B. Topič, F. Milia, I. P. Aleksandrova, A. S. Chaves, and R. Gazzinelli, *Phys. Rev. Lett.* **46**, 1406 (1981).
[2.44] S. R. Andrews and H. Mashiyama, *J. Phys. C* **16**, 4985 (1983).
[2.45] H. Mashiyama, *J. Phys. Soc. Jpn* **50**, 2655 (1981); F. Borsa, and A. Rigamonti, *Magnetic Resonance of Phase Transitions*, eds. C. P. Poole Jr. and H. A. Farach (Academic, New York, 1979) pp. 79–169 and references therein.
[2.46] H. Terauchi, H. Takenaka, and K. Shimaoka, *J. Phys. Soc. Jpn* **39**, 435 (1975).
[2.47] J. D. Axe, M. Izumi, and G. Shirane, *Phys. Rev.* **B22**, 3408 (1980); S. Žumer, and R. Blinc, *J. Phys. C: Solid State Phys.* **14**, 465 (1981); R. Blinc, *Phys Rep.* **79**, 331 (1981).
[2.48] R. Osredkar, S. Južnič, V. Rutar, J. Seliger, and R. Blinc, *Ferroelectrics* **24**, 147 (1980); See also C. Berthier, D. Jerome, P. Molinie, and J. Rouxel, *Solid State Commun.* **19**, 131 (1976) where the first NMR study of an incommensurate charge-density wave system is reported.
[2.49] R. Blinc, V. Rutar, J. Seliger, S. Zumer, Th. Rasing, and I. P. Aleksandrova, *Solid State Commun.* **34**, 895 (1980).
[2.50] R. Blinc, S. Južnič, V. Rutar, J. Seliger, and S. Žumer, *Phys. Rev. Lett.* **44**, 609 (1980).
[2.51] S. Pleško, R. Kind, and H. Arend, *Phys. Status Solidi A* **61**, 87 (1980).
[2.52] S. Žumer and R. Blinc, *J. Phys. C: Solid State Phys.* **14**, 465 (1981); R. Blinc, *Phys Rep.* **79**, 331 (1981).
[2.53] A. D. Bruce and R. A. Cowley, *J. Phys. C: Solid State Phys.* **11**, 3609 (1978).
[2.54] A. D. Bruce, *J. Phys. C: Solid State Phys.* **13**, 4615 (1980).
[2.55] L. N. Bulayevsky and D. I. Kloomsky, *ZhETF* **74**, 1863 (1978).
[2.56] V. L. Pokrovskii and A. L. Talapov, *ZhETF* **75**, 1151 (1978).
[2.57] R. Blinc, P. Prelovšek, V. Rutar, J. Seliger, and S. Žumer, *Incommensurate Phases in Dielectrics*, p. 143, Vol. I, (1986) in [2.1]; R. Blinc and T. Apih, *Prog. Nucl. Magn. Reson. Spectrosc.* **41**, 49 (2002).
[2.58] P. Bak, *Rep. Prog. Phys.*, **45**, 587 (1982) and references therein.
[2.59] H. Z. Cummins, *Phys. Rep.*, **185**, 211 (1990). An updated database of incommensurate phases, based on reference [2.59], is maintained by R. Caracas (*http://www.mapr.ucl.ac.be/~crystal*).
[2.60] See, e.g., C. Janot, *Quasicrystals* (Clarendon, Oxford, 1994).
[2.61] *Charge-Density Waves in Solids*, eds Gy. Hutiray and J. Solyom, *Lecture Notes in Physics*, Vol. 217 (Springer-Verlag, Berlin, 1985); Nuclear Spectroscopy on Charge Density Wave Systems, eds T. Butz, *Physics and Chemistry of Materials with Low-Dimensional Structures*, Vol. 15 (Kluwer Academic Publishers, Dodrecht, 1992).
[2.62] J.-C. Tolédano and P. Tolédano, *The Landau Theory of Phase Transitions* (World Scientific, Singapore, 1987).
[2.63] P. Bak, D. Mukamel, J. Villain, and K. Wentowska, *Phys. Rev. B* **19**, 1610 (1979).

[2.64] G. Grüner, *Density Waves in Solids* (Addison Wesley, Menlo Park, CA, 1994).
[2.65] Y. Gao, P. Lee, P. Coppens, M. A. Subramanian, and A. W. Sleight, *Science* **241**, 954 (1988).
[2.66] S. Žumer and J. Seliger, *Ferroelectrics*, **36**, 301 (1981).
[2.67] A. Gölzhäuser, B. Topič, H. Zimmermann, U. Haeberlen, R. Blinc, and S. Žumer (unpublished). See also H. Cailleau, p. 71, Vol. II in Ref. [2.1].
[2.68] J. C. Toledano, J. Schneck, and G. Errandonea, p. 233, Vol. II in Ref. [2.1].
[2.69] D. E. Cox, S. M. Shapiro, R. J. Nelmes, T. W. Ryan, H. Bleif, M. Eibschutz, and H. J. Guggenheim, *Phys. Rev. B,* **28**, 1640 (1983).
[2.70] G. Dolino, J. P. Bachheimer, and C. M. E. Zeyen, *Solid State Commun.*, **45**, 295 (1983).
[2.71] P. Mischo, F. Decker, U. Häcker, K.-P. Holzer, J. Petersson, and D. Michel, *Phys. Rev. Lett.*, **78**, 2152 (1997); K.–P. Holzer, J. Petersson, D. Schüssler, R. Walisch, U. Häcker, and D. Michel, *Phys. Rev. Lett.*, **71**, 89 (1993); J. M. Perez–Mato, R. Walisch, and J. Petersson, *Phys. Rev. B*, **35**, 6529–6537 (1987).
[2.72] F. Borsa and A. Rigamonti, *Magnetic Resonance of Phase Transitions*, eds. C. P. Poole Jr. and H. A. Farach (Academic, New York, 1979) pp. 79–169 and references therein.
[2.73] R. Blinc and S. Žumer, *Phys. Rev. B*, **41**, 11314 (1990).
[2.74] T. Apih, U. Mikac, A. V. Kityk, and R. Blinc, *Phys. Rev. B*, **55**, 2693 (1997).
[2.75] For a review, see G. Dolino, p.205, Vol. II in Ref. [2.1].
[2.76] P. Saint Gregoire and I. Luk'yanchuk, *Ferroelectrics*, **191**, 267 (1991); M. Vallade, V. Dvorak, and J. Lajzerovitz, *J. Phys. (Paris)*, **48**, 1171 (1987).
[2.77] T. Apih, U. Mikac, J. Dolinšek, J. Seliger, and R. Blinc, *Phys. Rev. B*, **61**, 1003 (2000).
[2.78] R. Blinc and J. Seliger, *Solid State Commun.*, **56**, 295 (1985).
[2.79] T. Apih, U. Mikac, J. Seliger, J. Dolinšek, and R. Blinc, *Phys. Rev. Lett.*, **80**, 2225 (1998).
[2.80] D. Harker, *J. Chem. Phys.*, **4**, 381 (1936).
[2.81] S. S. Khasanov and V. Sh. Shekhman, *Ferroelectrics*, **67**, 55 (1986).
[2.82] D. Baisa, A. Bondar, A. Gordon, and S. Maltsev, *Phys. Status Solidi B*, **93**, 805 (1979).
[2.83] P. J. S. Ewen, W. Taylor, and G. L. Paul, *J. Phys. C*, **16**, 6475 (1983).
[2.84] R. J. Nelmes, C. J. Howard, T. W. Ryan, W. I. F. David, A. J. Schultz, and P. C. W. Leung, *J. Phys. C*, **17**, L861 (1984).
[2.85] W. Ryan, A. Gibaud, and R. J. Nelmes, *J. Phys. C*, **18**, 5279 (1985).
[2.86] L. Pokrovsky and L. P. Pryadko, *Fiz. Tverd. Tela (Leningrad)*, **29**, 1492 (1987); *Sov. Phys. Solid State*, **29**, 853 (1987).
[2.87] S. Allen, *Phase Transit.* **6**, 1 (1985).
[2.88] S. R. Yang and K. N. R. Taylor, *Phase Transit.*, **36**, 233 (1991).

Chapter 3

[3.1] R. B. Meyer, L. Liebert, L. Strzelecki, and P. Keller. *J. Phys. Lett. (Paris)* **36**, 69 (1975); S. Garrof and R. B. Meyer, *Phys. Rev. A***19**, 338 (1979), *Phys. Rev. Lett.* **38**, 848 (1977).

[3.2] R. Blinc, *Phys. Status Solidi B* **70**, K 29 (1975).
[3.3] P. G. De Gennes, *The Physics of Liquid Crystals*, (Oxford University Press, Oxford, 1974) and references therein.
[3.4] I. Muševič, R. Blinc, and B. Žekš, *The Physics of Ferroelectric and Antiferroelectric Liquid Crystals*, (World Scientific, Singapore, 2000) and references therein.
[3.5] R. Blinc, *Ferroelectrics* **14**, 603 (1976).
[3.6] V. I. Indenbom, S.A. Pikin, and E.B. Loginov, *Kristalografija* **21**, 632 (1976).
[3.7] R. Blinc and B. Žekš, *Phys. Rev. A* **18**, 740 (1978).
[3.8] I. Muševič, B. Žekš, Th. Rasing, and P. Wyder, *Phys. Rev. Lett.* **48**, 192 (1982).
[3.9] A. Levstik, T. Carlsson, C. Filipič, and B. Žekš, *Phys. Rev. A* **35**, 3527 (1987).
[3.10] R.M. Hornreich, M. Luban, and S. Shtrikman, *Phys. Lett. A* **36**, 39 (1975).
[3.11] A. Michelson, *Phys. Rev. Lett.* **39**, 464 (1977).
[3.12] A. Michelson, L. Benguigui, and D. Cabib, *Phys. Rev. A* **16**, 394 (1977); A. Michelson, D. Cabib, and L. Benguigui, *J. Phys. (Paris)* **38**, 961 (1977).
[3.13] I. Muševič, R. Blinc, B. Žekš, C. Filipič, M. Čopič, A. Seppen, P. Wyder, and A. Levanyuk, *Phys. Rev. Lett.* **60**, 1530 (1988).
[3.14] A. Michelson and D. Cabib, *J. Phys. Lett.* **38**, L-321 (1977).
[3.15] N. A. Clark and S. T. Lagerwall, *Appl. Phys. Lett.* **36**, 899 (1980).
[3.16] S. Heinekamp, R. A. Pelcovits, E. Fontes, E. Yi. Chen, R. Pindak, and R. B. Meyer, *Phys. Rev. Lett.* **52**, 1027 (1984).
[3.17] E. I. Demikhov, E. Hoffmann, H. Stegemeyer, S. A. Pikin, and A. Strigazzi, *Phys. Rev. E* **51**, 5954 (1995); E. I. Demikhov, *Phys. Rev. E* **51**, R12 (1995).

Chapter 4

[4.1] K. Binder and A. P. Young, *Rev. Mod. Phys.*, **58**, 801 (1986); R. Pirc, B. Tadić, and R. Blinc, *Physica A*, **185**, 322 (1992).
[4.2] R. Blinc and B. Žekš, '*Soft Modes in Ferroelectrics and Antiferroelectrics*', (North Holland, Amsterdam, 1974).
[4.3] R. Pirc, B. Tadić, and R. Blinc, *Phys. Rev. B*, **36**, 8607 (1987); G. Parisi, *Phys. Rev. Lett.* **73A**, 203 (1979).
[4.4] R. Blinc, J. Dolinšek, R. Pirc, B. Tadić, B. Zalar, R. Kind, and O. Liechti, *Phys. Rev. Lett.*, **63**, 2248 (1989); J. Dolinšek, *J. Magn. Reson.*, **92**, 312 (1991).
[4.5] A. Levstik, C. Filipič, Z. Kutnjak, I. Levstik, R. Pirc, B. Tadić, and R. Blinc, *Phys. Rev. Lett.*, **66**, 2368 (1991).
[4.6] J. L. Bjorkstam, *Adv. Magn. Reson.*, **7**, 1 (1974); J. L. Bjorkstam and E. A. Uehling, *Phys. Rev.*, **114**, 961 (1959); V. H. Schmidt and E. A. Uehling, *Phys. Rev.*, **126**, 447 (1962).
[4.7] R. Blinc, in '*Magnetic Resonance of Phase Transitions*', eds F. J. Owens, C. P. Poole, and H. A. Farach, (Academic Press, London, 1979) p. 247.
[4.8] E. Matsushita and T. Matsubara, *Prog. Theor. Phys.*, **67**, 1 (1982).
[4.9] J. Seliger, V. Žagar, R. Blinc, and V. H. Schmidt, *Phys. Rev. B*, **42**, 3881 (1990); S. G. P. Brosnan and D. T. Edmonds, *J. Magn. Reson.*, **38**, 47 (1980); D. T. Edmonds, *Phys. Rep.*, **29**, 233 (1977).
[4.10] R. Blinc, *Ferroelectrics,* **16**, 33 (1977).

[4.11] J. S. Waugh, L. Huber, and U. Haeberlen, *Phys. Rev. Lett.*, **20**, 180 (1968).
[4.12] B. Schröter, H. Rosenberger, and D. Hadži, *J. Mol. Struct.*, **96**, 301 (1983).
[4.13] R. Blinc, I. Zupančič, G. Lahajnar, J. Slak, V. Rutar, M. Verbec, and S. Žumer, *J. Chem. Phys.*, **72**, 3626 (1980).
[4.14] F. Ermark, B. Topič, U. Haeberlen, and R. Blinc, *J. Phys.: Condens. Matter*, **1**, 5489 (1989); A. Bussmann-Holder and N. S. Dalal in *Structure and Bonding (Springer-Verlag), 'Order/disorder versus or with displacive dynamics in ferroelectric systems'*, **124**, 1 (2007); N. S. Dalal, A. Klymachyov, and A. Bussmann-Holder, *Phys. Rev. Lett.* **81**, 5942 (1998).
[4.15] R. Blinc and J. L. Bjorkstam, *Phys. Rev. Lett.*, **23**, 788 (1969); J. L. Bjorkstam, *J. Phys. Soc. Jpn. Suppl.*, **28**, S101 (1970).
[4.16] J. C. Slater, *J. Chem. Phys.*, **9**, 16 (1941); *Phys. Rev.*, **78**, 748 (1950); T. Nagamiya, *Prog. Theor. Phys.*, **7**, 275 (1952).
[4.17] R. Blinc and S. Žumer, *J. Chem. Phys.*, **62**, 3118 (1975); R. Blinc, J. Dolinšek, and S. Žumer, *J. Non-Crystalline Solids*, **131–133**, 125 (1991); R. Blinc in *'Nuclear Magnetic Resonance in Solids'*, Ed. L. Van Gerven, (Plenum, New York, 1977) p. 303.
[4.18] K. A. Müller (ed.), *'Local Properties at Phase Transitions'*, (North Holland, Amsterdam, 1976); R. M. Cotts and N. D. Knight, *Phys. Rev.*, **96**, 1285 (1954); R. R. Hewitt, *Phys. Rev.*, **121**, 45 (1961); G. F. Bonera, F. Borsa, and A. Rigamonti, *Rev. Nuovo Cim.*, **2**, 325 (1972).
[4.19] E. Courtens, *J. Phys. (Paris) Lett.*, **43**, 199 (1982).
[4.20] J. Dolinšek, B. Zalar, and R. Blinc, *Europhys. Lett.*, **23**, 289 (1993); J. Dolinšek, B. Zalar, and R. Blinc, *Phys. Rev.* **B 50**, 805 (1994); R. Blinc, J. Dolinšek, D. Arčon, B. Zalar, *J. Phys. Chem. Solids*, **57**, 1479 (1996).
[4.21] K. Schmidt-Rohr and H. W. Spiess, *Phys. Rev. Lett.*, **66**, 3020 (1991).
[4.22] G. Papantopoulos, G. Papavassiliou, F. Milia, V. H. Schmidt, J. E. Drumheller, N. J. Pinto, R. Blinc, and B. Zalar, *Phys. Rev. Lett.*, **73**, 276 (1994).
[4.23] R. Kind, P. M. Cereghetti, C. A. Jeitziner, B. Zalar, J. Dolinšek, and R. Blinc, *Phys. Rev. Lett.*, **88**, 195501 (2002).
[4.24] D. Sherrington and S. Kirkpatrick, *Phys. Rev. Lett.* **35**, 1792 (1975).
[4.25] J. R. L. de Almeida and D. J. Thouless, *J. Phys.* **A 11**, 983 (1987).
[4.26] G. Parisi, *J. Phys.* **A 13**, 1887 (1980); *Phys. Rev. Lett.* **50**, 1946 (1983). See also I. Kondor, *Sci. Prog. (Oxford)* **71**, 145 (1987).
[4.27] V. V. Shvartsman, S. Bedanta, P. Borisov, W. Kleemann, A. Tkach, and P. M. Vilarinho, *Phys. Rev. Lett.* **101**, 165704 (2008).

Chapter 5

[5.1] P. Curie, *J. Physique* **3**, 393 (1894).
[5.2] I.E. Dzyaloshinskii, *Zh. Eksp. Teor. Fiz.* **32**, 1547 (1957); [I. E. Dzyaloshinskii, *Sov. Phys. – JETP* **5**, 1259 (1957)].

[5.3] D. N. Astrov, *Zh. Eksp. Teor. Fiz.* **38**, 984 (1960); [D. N. Astrov, *Sov. Phys. - JETP* **11**, 708 (1960)].

[5.4] G. T. Rado, *J. Appl. Phys.* **33**, 1126 (1962); J. V. Folen, G. T. Rado, and E. W. Stalder, *Phys. Rev. Lett.* **6**, 607 (1961).

[5.5] M. Fiebig, *J. Phys. D* **38**, R123 (2005).

[5.6] N. A. Spaldin, and M. Fiebig, *Science* **309**, 391 (2005).

[5.7] W. Eerenstein, N. D. Mathur, and J. F. Scott, *Nature* **442**, 759 (2006).

[5.8] M. Fiebig, Th. Lottermoser, D. Fröhlich, A.V. Goltsev, and R.V. Pisarev, *Nature* **419**, 818 (2002).

[5.9] Th. Lottermoser, Th. Lonkai, U. Amman, D. Hohlwein, J. Ihringer, and M. Fiebig, *Nature* **430**, 541 (2004).

[5.10] C.W. Nan, G. Liu, Y. Liu, and H. Chen, *Phys. Rev. Lett.* **94**, 197203 (2005).

[5.11] E. Ascher, H. Rieder, H. Schmid, and H. Stössel, *J. Appl. Phys.* **37**, 1404 (1966).

[5.12] S. Shastry, G. Srinivasan, M.I. Bichurin, V.M. Petrov, and A.S. Tatarenko, *Phys. Rev. B* **70**, 064416 (2004).

[5.13] H. Zheng, J. Wang, S. E. Lofland, Z. Ma, L. Mohaddes-Ardabili, T. Zhao, L. Salamanca-Riba, S. R. Shinde, S. B. Ogale, F. Bai, D. Viehland, Y. Jia, D. G. Schlom, M. Wuttig, A. Roytburd, and R. Ramesh, *Science* **303**, 661 (2004).

[5.14] N. A. Hill, *J. Phys. Chem.* **B104**, 6694 (2000).

[5.15] J. Wang, J. B. Neaton, H. Zheng, V. Nagarajan, S. B. Ogale, B. Liu, D. Viehland, V. Vaithyanathan, D. G. Schlom, U. V. Waghmare, N. A. Spaldin, K. M. Rabe, M. Wuttig, and R. Ramesh, *Science* **299**, 1719 (2003).

[5.16] C. Ederer and N.A. Spaldin, *Phys. Rev. B* **71**, 060401(R) (2005).

[5.17] C. Ederer and N. Spaldin, *Phys. Rev. Lett.* **95**, 257601 (2005).

[5.18] R. Blinc, M. Kosec, J. Holc, Z. Trontelj, Z. Jaglicic, and N. Dalal, *Ferroelectrics* **349**, 16–20 (2007); A. Levstik *et al.*, *Appl. Phys. Lett.* **91**, 012905 (2007).

[5.19] C. Michel, J.-M. Moreau, G.D. Achenbach, R. Gerson, and W.J. James, *Solid State Commun.* **7**, 701 (1969).

[5.20] N. Ikeda *et al.*, *Nature* **436**, 1136 (2005).

[5.21] D. Klomskii, *Bull. Am. Phys. Soc.* **A13** 00006 (2007); *J. Phys.: Condens. Matter* **20**, 434217 (2008).

[5.22] R. Seshadri and N. A. Hill, *Chem. Mater.* **13**, 2892 (2001).

[5.23] B. B. Van Aken, T. T. M. Palstra, A. Filippetti *et al.*, *Nature Mater.* **3**, 164 (2004).

[5.24] L. C. Chapon, P. G. Radaelli, G. R. Blake *et al.*, *Phys. Rev. Lett.* **96**, 097601 (2006); S. Samarin, O. M. Artamonov, A. D. Sergeant *et al.*, *Phys. Rev. Lett.* **97**, 096402 (2006).

[5.25] D. V. Efremov, J. Van den Brink, and D. I. Khomskii, *Nature Mater.* **3**, 853 (2004); N. Ikeda, H. Ohsumi, K. Ohwada *et al.*, *Nature* **436**, 1136 (2005); Y. Tokunaga, T. Lottermoser, Y. Lee *et al.*, *Nature Mater.* **5**, 937 (2006).

[5.26] H. Katsura, N. Nagaosa, and A. V. Balatsky, *Phys. Rev. Lett.* **95**, 057205 (2005); M. Mostovoy, *Phys. Rev. Lett.* **96**, 067601 (2006).

[5.27] M. Pregelj *et al.*, *Phys. Rev. Lett.* **103**, 147202 (2009).

[5.28] W. Eerenstein, N.D. Mathur, and J.F. Scott, *Nature* **442/17**, 759 (2006).

[5.29] N. A. Spaldin and R. Ramesh, *MRS Bull.* **33**, 1047 (2008).

[5.30] T. Kimura et al., *Nature (London)* **426**, 55 (2003).
[5.31] G. Lawes et al., *Phys. Rev. Lett.* **95**, 087205 (2005).
[5.32] Y. Yamasaki et al., *Phys. Rev. Lett.* **96**, 207204 (2006).
[5.33] Y. J. Choi et al., *Phys. Rev. Lett.* **100**, 047601 (2008).
[5.34] K. Kimura et al., *Phys. Rev. Lett.* **103**, 107201 (2009).
[5.35] K. Kimura et al., *Phys. Rev.* **B78**, 140401 (2008).
[5.36] A. Kumar, G. L. Sharma, R. S. Katiyar, R. Pirc, R. Blinc, and J. F. Scott, *J. Phys.: Condens. Matter* **21**, 382204 (2009).
[5.37] R. Pirc and R. Blinc, *Ferroelectrics* **400**, 387 (2010).
[5.38] H. Schmid, *J. Phys.: Condens. Matter* **20**, 434201 (2008).
[5.39] J. Zhai, J. Li, D. Viehland, and M. I. Bichurin, *J. Appl. Phys.* **101**, 014102 (2007).
[5.40] S. A. Ivanov, S.-G. Eriksson, R. Tellgren, and H. Rundlöf, *Mater. Res. Bull.* **39**, 2317 (2004).
[5.41] R. Blinc, P. Cevc, A. Zorko, J. Holc, M. Kosec, Z. Trontelj, J. Pirnat, N. Dalal, V. Ramachandran, and J. Krzystek, *J. Appl. Phys.* **101**, 033901 (2007); A. Levstik, V. Bobnar, C. Filipič, J. Holc, M. Kosec, R. Blinc, Z. Trontelj, and Z. Jagličič, *Appl. Phys. Lett.* **91**, 012905 (2007).
[5.42] G. Catalan and J. F. Scott, *Nature (London)* **448**, E4 (2007).
[5.43] V. V. Shvartsmann, S. Bedanta, P. Borisov, W. Kleemann, A. Tkach, and P. M. Vilarinho, *Phys. Rev. Lett.* **101**, 165704 (2008).
[5.44] G.A. Samara, *Solid State Phys.* **56**, 239 (2001).
[5.45] Y. Okimoto, Y. Ogimoto, M. Matsubara, Y. Tomioka, T. Kageyama, T. Hasegawa, H. Koinuma, M. Kawasaki, and Y. Tokura, *Appl. Phys. Lett.* **80**, 1031 (2002).
[5.46] J. Dho, W. S. Kim, and N. H. Hur, *Phys. Rev. Lett.* **89**, 027202 (2002).
[5.47] R. Pirc and R. Blinc, *Phys. Rev.* **B 76**, 020101(R) (2007).
[5.48] R. Pirc and R. Blinc, *Phys. Rev.* **B 60**, 13472 (1998).
[5.49] R. Pirc, R. Blinc, V. Bobnar, and A. Gregorovič, *Phys. Rev.* **B 72**, 014202 (2005).
[5.50] K. Uchino, S. Nomura, L. E. Cross, S. J. Jang, and R. E. Newnham, *J. Appl. Phys.* **51**, 1142 (1980).
[5.51] R. Pirc, R. Blinc, V. Bobnar, and A. Gregorovič, *Ferroelectrics* **347**, 7 (2007).
[5.52] R. Pirc, R. Blinc, and V. Bobnar, *Phys. Rev.* **B 63**, 054203 (2001).
[5.53] J. Macutkevic, S. Kamba, J. Banys, A. Brilingas, A. Pashkin, J. Petzelt, K. Bormanis, and A. Sternberg, *Phys. Rev.* **B 74**, 104106 (2006).
[5.54] M. B. Isichenko, *Rev. Mod. Phys.* **64**, 961 (1992).
[5.55] R. Pirc, R. Blinc, and J. F. Scott, *Phys. Rev.* **B 79**, 214114 (2009).

Chapter 6

[6.1] G. A. Smolensky, *J. Phys. Soc. Jpn.* **28**, 26 (1970) and references therein.
[6.2] E. Cross, *Ferroelectrics* **76**, 241 (1987).
[6.3] G. Burns and F. H. Dacol, *Solid State Commun.* **48**, 853 (1983).

[6.4] G. Samara, *Solid State Physics*, vol. **56**, 239 (Academic Press, New York, 2001) and references therein.
[6.5] Z. G. Ye, *Ferroelectrics* **184**, 193 (1996).
[6.6] N. deMathan, E. Husson, G. Calvarin, J. R. Gavarri, A. W. Hewat, and A. Morell, *J. Phys. Condens. Matter* **3**, 8159 (1991) and references therein.
[6.7] A. Levstik, Z. Kutnjak, C. Filipič, and R. Pirc, *Phys. Rev. B* **57**, 11204 (1998).
[6.8] R. Pirc and R. Blinc, *Phys. Rev.* **B 60**, 13470 (1999).
[6.9] V. Bobnar, Z. Kutnjak, R. Pirc, and A. Levstik, *Phys. Rev. B* **60**, 6420 (1999).
[6.10] M. V. Gorev, V. S. Bondarev, and K. S. Aleksandrov, *Ferroelectrics* **360**, 37 (2007).
[6.11] Y. Moriya, H. Kawaji, T. Tojo, and T. Atake, *Phys. Rev. Lett.* **90**, 205901 (2003)
[6.12] V. Westphal, W. Kleemann, and M. D. Glinchuk, *Phys. Rev. Lett.* **68**, 847 (1992).
[6.13] Y. Uesu, H. Tazawa, K. Fujishiro, and Y. Yamada, *J. Korean Phys. Soc.* **29**, S703 (1996); M. Tachibana and E. Takayama-Muromachi, *Phys Rev B* **79**, 100104(R) (2009).
[6.14] R. Pirc, R. Blinc, V. Bobnar, and A. Gregorovič, *Phys Rev B* **72**, 014202 (2005).
[6.15] R. Pirc, R. Blinc, and V. Bobnar, *Phys. Rev. B* **63**, 054203 (2001).
[6.16] R. Blinc and R. Pirc, *AIP Conf. Proc.* **535**, 30 (2000).
[6.17] V. Bobnar, Z. Kutnjak, R. Pirc, R. Blinc, and A. Levstik, *Phys. Rev. Lett.* **84**, 5829 (2000).
[6.18] Z. Kutnjak, R. Pirc, A. Levstik, I. Levstik, C. Filipič, R. Blinc, and R. Kind, *Phys. Rev.* **B50**, 12421 (1994).
[6.19] Z. Kutnjak, R. Pirc, and R. Blinc, *Appl. Phys. Lett.* **80**, 3162 (2002) and references therein.
[6.20] R. Pirc, R. Blinc, and Z. Kutnjak, *Phys. Rev. B* **65**, 214101 (2002).
[6.21] R. Pirc and R. Blinc, *Phys. Rev. B* **76**, 020101 (2007).
[6.22] Z. Kutnjak, C. W. Garland, C. G. Schatz, P. J.. Collings, C. J. Booth, and J. W. Goodby. *Phys. Rev. E* **53**, 4955–63 (1996).
[6.23] P. M. Gehring, S. E. Park, and G. Shirane, *Phys. Rev. Lett.* **84**, 5216 (2000).
[6.24] S. Wakimoto, C. Stock, R. J. Birgeneau, Z.-G. Ye, W. Chen, W. J. L. Buyers, P. M. Gehring, and G. Shirane, *Phys. Rev.* **B65**, 172105 (2002).
[6.25] I. P. Swainson, C. Stock, P. M. Gehring, G. Xu, K. Hirota, Y. Qiu, H. Luo, X. Zhao, J. F. Li, and D. Viehland, *Phys. Rev.* **B79**, 224301 (2009)
[6.26] M. Iwata, N. Tomisato, H. Orihara, H. Ohwa, N. Yasuda, Y. Ishibashi, *Ferroelectrics* **261**, 83 (2001).
[6.27] G. A. Samara and L. A. Boatner, *Phys. Rev. B* **61**, 3889 (2000); G. A. Samara, *Solid State Phys.* **56**, 240 (2001).
[6.28] G. A. Samara, *Physica B* **150**, 179 (1988).
[6.29] R. Blinc, V. Laguta, and B. Zalar, *Phys. Rev. Lett.* **91**, 247601 (2003); see also C. K. D. Stock, 'Interplay between Static and Dynamic Polar Nanoregions in PbMg$_{1/3}$Nb$_{2/3}$O$_3$', *Aspen Center for Physics Winter Conference (31 January – 5 February 2010), Book of Abstracts*, p. 157 (2010).
[6.30] J. J. Van der Klink, S. Rod, and A. Chaterlain, *Phys. Rev. B* **33**, 2084 (1986).

[6.31] U. T. Hochli, K. Knorr, and A. Loidl, *Adv Phys.* **39**, 405 (1990).
[6.32] M. Maglione, S. Rod, and U. T. Hochli, *Europhys. Lett.* **4**, 631 (1987).
[6.33] K. B. Lyons, P. A. Fleury, and D. Rytz, *Phys. Rev. Lett.* **57**, 2207 (1986).
[6.34] R. Blinc, J. Dolinšek, A. Gregorovič, B. Zalar, C. Filipič, Z. Kutnjak, A. Levstik, and R. Pirc, *Phys. Rev. Lett.* **83**, 424 (1999).
[6.35] R. Blinc, J. Dolinšek, D. Arčon, B. Zalar, *J. Phys. Chem. Solids*, **57**, 1479 (1996).
[6.36] Z. Kutnjak, J. Petzelt, and R. Blinc, *Nature* **441**, 956 (2006).
[6.37] Z. Kutnjak, Blinc, and Y. Ishibashi, *Phys. Rev. B* **76**, 104102 (2007).
[6.38] M. Iwata, Z. Kutnjak, Y. Ishibashi, and R. Blinc, *J. Phys. Soc. Jpn.* **77**, 034703 (2008); see also M. Iwata, Z. Kutnjak, Y. Ishibashi, and R. Blinc, *J. Phys. Soc. Jpn.* **77**, 065003 (2008).
[6.39] I. E. Cross, *Ferroelectric Ceramics* (Birkhauser, Berlin, 1993).
[6.40] Z. Kutnjak, C. Filipič, R. Pirc, A. Levstik, R. Farhi, and M. El Marssi, *Phys. Rev. B* **59**, 294 (1999).
[6.41] G. Schmidt, H. Arndt, J. von Cieminski, T. Petzsche, H. J. Voigt, and N. N. Krainik, *Krist. Tech.* **15**, 1415 (1980).
[6.42] U. Bottger, A. Biermann, and A. Arlt, *Ferroelectrics* **134**, 253 (1992).
[6.43] E. V. Colla, E. Y. Koroleva, N. M. Okuneva, and S. B. Vakhrushev, *Phys. Rev. Lett.* **74**, 1681 (1995).
[6.44] D. Viehland, J. F. Li, S. J. Jang, L. E. Cross, and M. Wuttig, *Phys. Rev. B* **46**, 8013 (1992).
[6.45] V. E. Colla, N. K. Yushin, and D. Viehland, *J. Appl. Phys.* **83**, 3298 (1998).
[6.46] G. Xu, D. Viehland, J. F. Li, P. M. Gehring, and G. Shirane, *Phys. Rev. B* **68**, 212410 (2003).
[6.47] F. Bai, N. Wang, J. F. Li, D. Viehland, P. M. Gehring, G. Xu, and G. Shirane, *J. Appl. Phys.* **96**, 1620 (2004).
[6.48] D. Viehland and J. Powers, *J. Appl. Phys.* **89**, 1820 (2001).
[6.49] M. Davis, D. Damjanović, and N. Setter, *J. Appl. Phys.* **97**, 064101 (2005).
[6.50] M. Davis, D. Damjanović, and N. Setter, *Phys. Rev. B* **73**, 014115 (2006).
[6.51] C.-S. Tu, I.-C. Shih, V. H. Schmidt, and R. Chien, *Appl. Phys. Lett.* **83**, 1833 (2003).
[6.52] R. R. Chien, V. H. Schmidt, C.-S. Tu, L.-W. Hung, and H. Luo, *Phys. Rev. B* **69**, 172101 (2004).
[6.53] Z. Wu and R. E. Cohen, *Phys. Rev. Lett.* **95**, 037601 (2005).
[6.54] H. Fu and R. E. Cohen, *Nature (London)* **403**, 281 (2000).
[6.55] V. A. Shuvaeva, A. M. Glaser, and D. Zekria, *J. Phys.: Condens. Matter* **17**, 5709 (2005).
[6.56] S. Wada, S. Suzuki, T. Noma, T. Suzuki, M. Osada, M. Kakihana, S.-E. Park, L. E. Cross, and T. R. Shrout, *Jpn. J. Appl. Phys.*, Part 1 **38**, 5505 (1999).
[6.57] V. Bobnar, Z. Kutnjak, and A. Levstik, *Jpn. J. Appl. Phys.*, Part 1 **37**, 5634 (1998).
[6.58] X. Zhao, W. Qu, X. Tan, A. A. Bokov, and Z. G. Ye, *Phys. Rev. B* **75**, 104106 (2007).
[6.59] H. Yao, K. Ema, and C. W. Garland, *Rev. Sci. Instrum.* **69**, 172 (1998).

[6.60] Y. Lu, D.-Y. Jeong, Z.-Y. Cheng, Q. M. Zhang, H.-S. Luo, Z.-W. Yin, and D. Viehland, *Appl. Phys. Lett.* **78**, 3109 (2001).
[6.61] Z. Kutnjak, S. Kralj, G. Lahajnar, and S. Žumer, *Phys. Rev. E* **68**, 021705 (2003).
[6.62] M. Sulc and M. Pokorny, *J. Phys. IV* **126**, 77 (2005).
[6.63] M. Porta, T. Lookman, and A. Saxena, *J. Phys.: Condens. Matter* **22**, 345902 (2010).

Chapter 7

[7.1] For a review see: G. A. Samara, *Solid State Phys.* **56**, 1 (2001).
[7.2] R. Pirc, R. Blinc, V. Bobnar, and A. Gregorovič, *Phys Rev B* **72**, 014202 (2005).
[7.3] A. J. Lovinger, *Science* **220**, 1115 (1983).
[7.4] H. Kawai, *Jpn. J. Appl. Phys.* **8**, 975 (1969).
[7.5] T. Furukawa, *Phase Transit.* **18**, 143 (1989) and references therein.
[7.6] R. Hasegawa, Y. Takahashi, Y. Chatani, and H. Tadokoro, *Polym. J.* 3, 600 (1972).
[7.7] G. A. Samara and F. Bauer, *Ferroelectrics* **135**, 385 (1992), **171**, 299 (1995).
[7.8] A. I. Lovinger, *Developments in Crystalline Polymers*, J. D. C. Bassett, ed. (Applied Science, London, 1982 p. 195 and references therein); *Jap. J. Appl. Phys.*, Supplement 2, **24**, 18 (1985).
[7.9] J. W. Lee, Y. Takasa, B. A. Newman, and J. I. Sheimbeim, *J. Polym. Sci. B. Polym. Phys.* **29**, 279 (1991) and references therein.
[7.10] J. F. Legrand, B. Daudin, and E. Bellet-Amalric, *Nucl Instr. Methods Phys. Res. B* **105**, 225 (1995).
[7.11] Q. M. Zhang, V. Bharti, and X. Zhao, *Science* **280**, 2101 (1998).
[7.12] A. V. Bune, V. M. Fridkin, S. Ducharme, L. M. Blinov, S. P. Palto, A. Sorokin, S. G. Yudin, and A. Zlatkin, *Nature* **391**, 874 (1998) and references therein.
[7.13] S. Ducharme, S. P. Palto, L. M. Blinov, and V. M. Fridkin, Proc. of the 2000 Aspen Center of Physics Meeting on Fundamental Physics of Ferroelectrics, *AIP Conf. Proc.* **535**, p. 354 (2000).
[7.14] Z.-Y. Cheng, D. Olson, H. Xu, F. Xia, J. S. Hundal, Q. M. Zhang, F. B. Bateman, G. J. Kavarnos, and T. Ramotowski, *Macromolecules* **35**, 664 (2002).
[7.15] H. Xu, Z.-Y. Cheng, D. Olson, and Q. M. Zhang, *Appl. Phys. Lett.* **78**, 2360 (2001).
[7.16] J. M. Kosterlitz, D. J. Thouless, and R. C. Jones, *Phys. Rev. Lett.* **36**, 1217 (1976).
[7.17] K. H. Fischer and J. A. Hertz, *Spin Glasses* (Cambridge University Press, Cambridge, UK, 1991), p. 131.
[7.18] A. Gregorovič, *Ph.D. thesis*, University of Ljubljana, 2002; See also A. Levstik, Z. Kutnjak, C. Filipič, and R. Pirc, *Phys. Rev. B* **57**, 11204 (1998); V. Bobnar, Z. Kutnjak, R. Pirc, and A. Levstik, *Phys. Rev. B* **60**, 6420 (1999).
[7.19] R. Pirc and R. Blinc, *Phys. Rev.* **B 60**, 13470 (1999).

[7.20] V. Bobnar, B. Vodopivec, A. Levstik, Z.-Y. Cheng, and Q. M. Zhang, *Phys. Rev. B* **67**, 094205 (2003).

[7.21] V. Bobnar, B. Vodopivec, A. Levstik, M. Kosec, B. Hilczer, and Q. M. Zhang, *Macromolecules* **36**, 4436 (2003).

[7.22] M. Gabay and G. Toulouse, *Phys. Rev. Lett.* **47**, 201 (1981).

[7.23] A. E. Glazounov and A. K. Tagantsev, *Phys. Rev. Lett.* **85**, 2192 (2000).

[7.24] Q. M. Zhang, H. Li, M. Poh, F. Xia, Z.-Y. Cheng, H. Xu, and C. Huang, *Nature (London)* **419**, 284 (2002).

[7.25] Q. M. Zhang, V. Bharti, and X. Zhao, *Science* **280**, 2101 (1998).

[7.26] J. Y. Li, *Phys. Rev. Lett.* **90**, 217601 (2003).

[7.27] V. Bobnar, A. Levstik, C. Huang, and Q. M. Zhang, *Phys. Rev. Lett.* **92**, 047604 (2004).

[7.28] Z.-Y. Cheng, T.-B. Xu, V. Bharti, S. Wang, and Q. M. Zhang, *Appl. Phys. Lett.* **74**, 1901 (1999).

[7.29] R. Pirc, R. Blinc, and V. S. Vikhnin, *Phys. Rev. B* **69**, 212105 (2004).

[7.30] V. Sundar and R. E. Newnham, *Ferroelectrics* **135**, 431 (1992).

[7.31] Q. M. Zhang, Z.-Y. Cheng, V. Bharti, T.-B. Xu, H. Xu, T. X. Mai, and S. J. Gross, in *Smart Structures and Materials 2000: Electroactive Polymer Actuators and Devices (EAPAD)*, eds Yoseph Bar-Cohen, *Proc. SPIE* **3987**, 34 (2000).

[7.32] V. Bharti, H. S. Xu, G. Shanti, and Q. M. Zhang, *J. Appl. Phys.* **87**, 452 (2000).

Chapter 8

[8.1] M. Lines and A. Glass, *Principles and Applications of Ferroelectric and Related Materials*, (Clarendon Press, Oxford, 1977) p. 148.

[8.2] F. Jona and G. Shirane, *Ferroelectric Crystals*, (Dover, New York, 1992) p. 134.

[8.3] B. A. Tuttle and D. A. Payne, *Ferroelectrics* **37**, 603 (1981).

[8.4] A. S. Mischenko, Q. Zhang, J. F. Scott, R. W. Whatmore, and N. D. Mathur, *Science* **311**, 1270 (2006).

[8.5] B. Neese, B. Chu, S. G. Lu, Y. Wang, E. Furman, and Q. M. Zhang, *Science* **321**, 821 (2008).

[8.6] A. S. Mischenko, Q. Zhang, R. W. Whatmore, J. F. Scott, and N. D. Mathur, *Appl. Phys. Lett.* **89**, 242912 (2006).

[8.7] Z. Kutnjak, J. Petzelt, and R. Blinc, *Nature* **441**, 956 (2006).

[8.8] Sheng-Guo Lu, B. Rožič, Q. M. Zhang, Z. Kutnjak, X. Li, E. Furman, L. J. Gorny, M. Lin, B. Malič, M. Kosec, R. Blinc, and R. Pirc, *Appl. Phys. Lett.* **97**, 162904–1 (2010).

[8.9] G. Akcay, S. P. Alpay, G. A. Rossetti Jr., and J. F. Scott, *J. Appl. Phys.* **103**, 024104 (2008).

[8.10] R. Pirc, and R. Blinc, *Phys. Rev. B* **60**, 13470 (1999).

[8.11] M. Iwata, Z. Kutnjak, Y. Ishibashi, and R. Blinc, *J. Phys. Soc. Jpn.* **77**, 034703 (2008).

[8.12] R. Pirc, Z. Kutnjak, R. Blinc, and Q. M. Zhang, *Appl. Phys. Lett.* **98**, 021909 (2011).

Chapter 9

[9.1] J. F. Scott and C. A. Aranjo, *Science* **246**, 1400 (1989).
[9.2] J. F. Scott, *Ferroelectric Memories*, (Springer-Verlag, Heidelberg, Germany, 2000); *Ferroelectrics* **236**, 247 (2000). See also Y. Ishibashi, H. Orihara, and D. Tilley, *J. Phys. Soc. Jpn.* **67**, 3292 (1998); *ibidem* **71**, 1471 (2002).
[9.3] M. Alexe, C. Harnagea, W. Erfurth, D. Hesse, and V. Goesde, *Appl. Phys.* **A70**, 1, (2000).
[9.4] I. P. Batra and B. D. Silverman, *Solid State Commun.* **11**, 291 (1972).
[9.5] T. Yamamoto, *Integr. Ferroelectr.* **12**, 161 (1997).
[9.6] H. Kohlstedt *et al.*, *Phys. Rev.* **B72**, 125341 (2005).
[9.7] M. Ye. Zhuravlev *et al.*, *Phys. Rev. Lett.* **94**, 246802 (2005).
[9.8] E. Y. Tsymbal and H. Kohlstedt, *Science* **3**, 2705 (1962).
[9.9] D. J. Kim *et al.*, *Phys. Rev. Lett.* **95**, 237602 (2005).
[9.10] O. Auciello, J. F. Scott, and R. Ramesh, *Phys. Today* **51**, 22 (1998).
[9.11] M. Okuyama and Y. Ishibashi, *Ferroelectric Thin Films*, (Springer-Verlag, Heidelberg, Germany 2005).
[9.12] Th. Tybell, C. H. Ahn and I.M. Triscone, *Appl. Phys. Lett.* **75**, 856 (1999).
[9.13] J. Junquera and P. H. Ghosez, *Nature* **422**, 506 (2003).
[9.14] I. I. Ivanchik, *Sov. Phys. Solid State* **3**, 2705 (1962).
[9.15] I.P. Batra and B. D. Silverman, *Solid State Commun.* **11**, 291 (1972); I. P. Batra, P. Würfel, and B. D. Silverman, *Phys. Rev. Lett.* **30**, 384 (1973).
[9.16] N. Sai, A. M. Kolpak, and A. M. Rappe, *Phys. Rev.* **B72**, 020101(R) (2005).
[9.17] G. Gerra *et al.*, *Phys. Rev. Lett.* **96**, 107603 (2006).
[9.18] D. D. Fong *et al.*, *Phys. Rev. Lett.* **96**, 127601 (2006); L. Despont *et al.*, *Phys. Rev.* **B73**, 094110 (2006).
[9.19] A. L. Roytburd, S. Zhong, and S. P Alpay, *Appl. Phys. Lett.* **87**, 092902 (2005).
[9.20] N. A. Pertsev and H. Kohlstedt, *Phys. Rev. Lett.* **98**, 257603 (2007).
[9.21] J. Y. Jo *et al.*, *Phys. Rev. Lett.* **97**, 247602 (2006).
[9.22] R. Landauer, *J. Appl. Phys.* **28**, 227 (1957).
[9.23] D. Tilley and B. Žekš, *Solid State Commun.* **49**, 823 (1984).
[9.24] Y. Ishibashi, H. Orihara, and D. Tilley, *J. Phys. Soc. Jpn.* **71**, 1471 (2002).
[9.25] J. Wang, J.B. Neaton, H. Zheng, V. Nagarajan, S.B. Ogale, B. Liu, D. Viehland, V. Vaithyanathan, D.G. Schlom, U.V. Waghmare, N.A. Spaldin, K.M. Rabe, M. Wuttig, and R. Ramesh, *Science* **299**, 1719 (2003).
[9.26] W. Tian, V. Vaithyanathan, D.G. Schlom, Q. Zhan, S.Y. Yang, Y.H. Chu, and R. Ramesh, *Appl. Phys. Lett.* **90**, 172908 (2007).
[9.27] H. Naganuma, N. Shimura, J. Miura, H. Shima, Sh. Yasui, K. Nishida, T. Katoda, T. Iijima, H. Funakubo, and S. Okamura, *J. Appl. Phys.* **103**, 07E314 (2008).
[9.28] B. Ruette, S. Zvyagin, A.P. Pyatakov, A. Bush, J.F. Li, V.I. Belotelov, A.K. Zvezdin, and D. Viehland, *Phys. Rev.* **B 69**, 064114 (2004).
[9.29] M.I. Kaganov and A.N. Omelyanchouk, *Zh. Eksp. Teor. Fiz.* **61**, 1679 (1971); *Sov. Phys. JETP* **34**, 895 (1972).
[9.30] R. Kretschmer and K. Binder, *Phys. Rev.* **B 20**, 1065–76 (1976).

[9.31] M.D. Glinchuk and A.N. Morozovska, *J. Phys.: Condens. Matter.* **16**, 3517 (2004).

[9.32] C.-L. Jia, V. Nagarajan, J.-Q. He, L. Houben, T. Zhao, R. Ramesh, K. Urban, and R. Waser, *Nature Mater.* **6**, 64 (2007).

[9.33] M. D. Glinchuk and A. V. Ragulya: 'Nanoferroics', (Scientific Book Project, Kiev, Naukova Dumka, 2010).

[9.34] M. D. Glinchuk, A. N. Morozovska, E. A. Eliseev, and R. Blinc, *J. Appl. Phys.* **105**, 084108 (2009).

Chapter 10

[10.1] J. J. Urban, W. S. Yun, Q. Hu, and H. Park, *J. Am. Chem. Soc.* **124**, 1186 (2002).

[10.2] W. S. Yun, J. J. Urban, Q. Hu, and H. Park, *Nano. Lett.* **2**, 447 (2002).

[10.3] Y. Luo, I. Szafraniak, N. D. Zakharov, V. Nagarajan, M. Steinhart, R. B. Wehrspohn, J. H. Wendorff, R. Ramesh, and M. Alexe, *Appl. Phys. Lett.* **83**, 440 (2003).

[10.4] F. D. Morrison, L. Ramsay, and J. F. Scott, *J. Phys. Condens. Matter* **15**, L527 (2003); F. D. Morrison *et al.*, *Rev. Adv. Mater. Sci.* **4**, 114 (2003).

[10.5] Y. Mao, S. Banerjee, and S. S. Wong, *Chem. Commun.* **3**, 408–9 (2003).

[10.6] H. J. Shin, J. H. Choi, H. J. Yang, Y. D. Park, Y. Kuk, and C. J. Kang, *Appl. Phys. Lett.* **87**, 113114 (2005).

[10.7] M.-W. Chu, I. Szafraniak, R. Scholz, C. Harnagea, D. Hesse, M. Alexe, and U. Gösele, *Nature Mater.* **3**, 87–90 (2004).

[10.8] H. Fu and L. Bellaiche, *Phys. Rev. Lett.* **91**, 257601 (2003).

[10.9] I. I. Naumov, L. Bellaiche, and H. Fu, *Nature* **432**, 737 (2004).

[10.10] E. A. Eliseev, A. N. Morozovska, M. D. Glinchuk, B. Y. Zaulichnyi, V. V. Skorohod, and R. Blinc, *to be published*.

[10.11] Modern Crystallography, Vol. IV: Physical Properties of Crystals, ed. L. A. Shuvalov, (Springer-Verlag, Berlin, 1988).

[10.12] M. D. Glinchuk and M. F. Deigen, *Surf. Sci.* **3**, 243 (1965).

[10.13] J. P. Rivera, *Eur. Phys. J.* **B71**, 299 (2009).

[10.14] A. N. Morozovska, E. A. Eliseev, M. D. Glinchuk, and R. Blinc, *Phys. Rev.* **B81**, 092101 (2010).

[10.15] D. Yadlovker and S. Berger, *Phys. Rev.* **B71**, 184112 (2005).

[10.16] D. Yadlovker and S. Berger, *J. Electroceram.* **22**, 281 (2009).

[10.17] G. Geneste, E. Bousquet, J. Junquera, and P. Chosez, *Appl. Phys. Lett.* **88**, 112906 (2006).

[10.18] A. N. Morozovska, E. A. Eliseev, and M.D. Glinchuk, *Phys. Rev.* **B73**, 214106 (2006).

[10.19] E. Bousquet, N. Spaldin, and Ph. Ghosez, Strain-induced ferroelectricity in simple rocksalt binary oxides. *arXiv:0906.4235v1*.

[10.20] A. N. Morozovska, M. D. Glinchuk, and E.A. Eliseev, *Phys. Rev.* **B76**, 014102 (2007).

[10.21] Z.-G. Ban, S. P. Alpay, and J. V. Mantese, *Phys. Rev.* **B67**, 184104 (2003).
[10.22] O. Madelung, *Semiconductors: Data Handbook*, 3rd edn, (Springer Verlag, Berlin, 2004).
[10.23] Sh. M. Kogan, *Sov. Phys. Solid State* **5**, 2069 (1964).
[10.24] A. K. Tagantsev, *Phys. Rev.* **B34**, 5883 (1986).
[10.25] A. K. Tagantsev, *Phase Transit.* **35**, 119 (1991).
[10.26] R. Sabirianov and P. Lukashev, Materials Research Society Fall Meeting, Boston MA, 1–4 December 2008 (unpublished), Paper No. K5.7.
[10.27] V. S. Mashkevich and K. B. Tolpygo, *Zh. Eksp. Teor. Fiz.* **31**, 520 (1957); *Sov. Phys. JETP* **4**, 455 (1957).
[10.28] W. Ma and L. E. Cross, *Appl. Phys. Lett.* **79**, 4420 (2001).
[10.29] W. Ma and L. E. Cross, *Appl. Phys. Lett.* **81**, 3440 (2002).
[10.30] W. Ma and L. E. Cross, *Appl. Phys. Lett.* **82**, 3293 (2003).
[10.31] P. Zubko, G. Catalan, P. R. L. Welche, A. Buckley, and J. F. Scott, *Phys. Rev. Lett.* **99**, 167601 (2007).
[10.32] G. Catalan, L. J. Sinnamon, and J. M. Gregg, *J. Phys.: Condens. Matter* **16**, 2253 (2004).
[10.33] G. Catalan, B. Noheda, J. McAneney, L. J. Sinnamon, and J. M. Gregg, *Phys. Rev.* **B72**, 020102(R) (2005).
[10.34] M. S. Majdoub, P. Sharma, and T. Cagin, *Phys. Rev.* **B77**, 125424 (2008).
[10.35] S. V. Kalinin and V. Meunier, *Phys. Rev.* **B77**, 033403 (2008).
[10.36] M. D. Glinchuk, E. A. Eliseev, A. N. Morozovska, and R. Blinc, *Phys. Rev.* **B77**, 024106 (2008).
[10.37] G. Geneste, E. Bousquest, J. Junquera, and P. Chosez, *Appl. Phys. Lett.* **88**, 112906 (2006).
[10.38] C. L. Wang and S. R. P. Smith, *J. Phys.: Condens. Matter* **7**, 7163 (1995).
[10.39] A. Sundaresan, R. Bhargavi, N. Rangarajan, U. Siddesh, and C. N. R. Rao, *Phys. Rev.* **B74**, 161306(R) (2006).
[10.40] E. Erdem, H.-Ch. Semmelhack, R. Bottcher, H. Rumpf, J. Banys, A. Matthes, H.-J. Glasel, D. Hirsch, and E. Hartmann, *J. Phys.: Condens. Matter* **18**, 3861 (2006).
[10.41] D. Yadlovker and S. Berger, *Phys. Rev.* **B71**, 184112 (2005).
[10.42] D. D. Fong, G. B. Stephenson, S. K. Streiffer, J. A. Eastman, O. Auciello, P. H. Fuoss, and C. Thompson, *Science* **304**, 1650 (2004).
[10.43] M. S. Majdoub, R. Maranganti, and P. Sharma, *Phys. Rev.* **B79**, 115412 (2009).
[10.44] M. E. Lines and A. M. Glass, *Principles and Applications of Ferroelectrics and Related Phenomena* (Clarendon Press, Oxford, 1977).
[10.45] M. I. Kaganov and A. N. Omelyanchouk, *Zh. Eksp. Teor. Fiz.* **61**, 1679 (1971); *Sov. Phys. JETP* **34**, 895 (1972).
[10.46] I. Rychetsky, *J. Phys.: Condens. Matter* **9**, 4583 (1997).
[10.47] S. P. Alpay, I. B. Misirlioglu, A. Sharma, and Z.-G. Ban, *J. Appl. Phys.* **95**, 8118 (2004).
[10.48] G. Ackay, S. P. Alpay, G. A. Rossetti, and J. F. Scott, *J. Appl. Phys.* **103**, 024104 (2008).

[10.49] Q. Y. Qiu, V. Nagarajan, and S. P. Alpay, *Phys. Rev.* **B78**, 064117 (2008).
[10.50] V. I. Marchenko and A. Ya. Parshin, *Zh. Eksp. Teor. Fiz.* **79**, 257 (1980); *Sov. Phys. JETP* **52**, 129 (1980).
[10.51] V. A. Shchukin and D. Bimberg, *Rev. Mod. Phys.* **71**, 1125 (1999).
[10.52] M. D. Glinchuk, A. N. Morozovska, and E.A. Eliseev, *J. Appl. Phys.* **99**, 114102 (2006).
[10.53] L. D. Landau and E. M. Lifshitz, *Theory of Elasticity. Theoretical Physics* (Butterworth-Heinemann, Oxford, UK, 1998), Vol. 7.
[10.54] G. A. Smolenskii, V. A. Bokov, V. A. Isupov, N. N. Krainik, R. E. Pasynkov, and A. I. Sokolov, *Ferroelectrics and Related Materials* (Gordon and Breach, New York, 1984).
[10.55] L. D. Landau and E. M. Lifshitz, *Electrodynamics of Continuous Media* (Butterworth-Heinemann, Oxford, UK, 1980).
[10.56] R. Kretschmer and K. Binder, *Phys. Rev.* **B20**, 1065 (1979).
[10.57] A. K. Tagantsev and G. Gerra, *J. Appl. Phys.* **100**, 051607 (2006).
[10.58] C. H. Woo and Y. Zheng, *Appl. Phys.* **A91**, 59 (2007).
[10.59] N. D. Mermin and H. Wagner, *Phys. Rev. Lett.* **17**, 1133 (1966).
[10.60] A. A. Gorbatsevich and Yu. V. Kopaev, *Ferroelectrics* **161**, 321 (1994).
[10.61] S. P. Timoshenko and J. N. Goodier, *Theory of Elasticity* (McGraw-Hill, New York, 1970).
[10.62] E. A. Eliseev, A. N. Morozovska, M. D. Glinchuk, and R. Blinc, *http://arxiv.org/ftp/arxiv/papers/0811/0811.1031.pdf*, preprint.

Subject index

2D ferroelecricity in Langmuir–Blodgett films 191
2-q modulation in incommensurate structures 59
3-q modulation in incommensurate structures 62, 69

Almeida-Thouless line 95, 97, 98
Amplitudon soft mode 1, 30

Birelaxors 131

Chemical shift tensor 108
Chiral SmC* phase 79
Classical to quantum cross-over 12, 15, 20, 120
Commensurate phase 26, 33
Comparison between properties of normal ferroelectrics and relaxors 144–147
Compton high energy neutron scattering 19, 20
Critical end point in relaxors 170–186, 208–210

Dependence of normalized ME-coupling coefficient on film thickness for different misfit strains 228–229
Determination of a EA order parameter in the fast-motion limit 112–114
Determination of a EA order parameter in the slow-motion limit 115–117
Deuteron glasses 119
Devil's staircase 28
Dipolar glasses 95
Disorder in displacive ferroelectrics 15–19
Displacive component in order–disorder ferroelectrics 9
Distribution of relaxation times 144
DOBAMBC 80
D vs. E hysteresis loop of nylon 190
Dynamic response 137

EC cooling line 212
EC in thin films 203
EC in bulk PMN 208
  P (VDF-TrFE) 191
  polymer film 191
PLZT ceramics 154
EC in relaxors 208
Edwards–Anderson order parameter 112, 115
Effect of pressure in relaxors 163–165
Electric field induced Lifshitz point 91–93
Electric field-temperature phase diagram of relaxors 161
Electrocaloric (EC) effect: analogy with vapour-compression cooling cycle 203, 204
E–T phase diagram and the Widom line in PMN 175
Excess heat capacity at different electric fields in PMN-PT with x = 0.295 172–176

Ferrodistortive and antiferrodistortive transitions 25
Ferroelectric field effect transistors (FET) 5
Ferroelectric liquid crystals 79
Ferroelectric multiferroics 125
Ferroelectric random access memories (FERAM) 5
Ferroelectric vortex states 243–246
Ferroelectrics 1, 2
Field cooled and zero field cooled static dielectric constant 99, 144
Field-induced relaxor-ferroelectric transition 159, 160
Flexoelectric coupling between tilt and polarization 82
Free energy of a deuteron dipolar pseudo-spin glass 96
Freely suspended ferroelectric thin films 93, 94

Giant electrostriction 202
Goldstone mode, aplitudon mode and soft mode 84–88
Goldstone mode – phason 1, 78

Hysteresis loops of 1-μm and 100-nm large PZT cells 216

Incommensurate systems 2, 23
Ishibashi interaction 22

Isotope effects:
    ^{18}O in SrTiO$_3$ 13
    deuteration in KH$_2$PO$_4$ 22

Lifshitz invariant 25
Linear dielectric response of relaxors 153
Linear magnetoelectric response 123
Local and non-local case 64
Local polarization distribution
    function 96–98, 112–114, 120
Lock-in transition 28

Magnetic field induced Liftshitz point 89–91
Magnetic resonance frequency distribution
    for 1-h case 43
Magnetoelectric coefficient α_{ij} 123
Magnetoelectric systems 1
Maximum "efficiencies" $\Delta T/\Delta E$ and
    $\Delta S/\Delta E$ for different systems
    213
Meetings on ferroelectricity 2, 4
Memory retention time, thickness
    dependence 216
Misfit strain induced magnetoelectric
    coupling in thin fims 226
Morphotropic phase boundary (MPB)
    5, 7, 8
Multidimensionally modulated systems 55
Multi-soliton limit 44

Nanoferroelectrics 230
Negative-negative case in thin films
    $\delta_+ < 0, \delta_- < 0$ 225
Neutron and X-ray scattering in
    I systems 38–43
NMR lineshapes for multiple q modulated
    systems 64
    plane-wave limit 64
    multisoliton limit 64
NMR of relaxors and polar
    nanocusters 165–170
Non-linear dielectric response of
    relaxors 153–157
Non-volatile ferroelectric random-access
    memory (NVFRAM)
    architecture 218
Nylons or odd polyamides 190

Orientational elementary excitations 84, 86

Pauling ice rules 19
Phase diagram of a deuteron glass 97
Phase diagrams of strained ferroic films 229
Phason and amplitudon dispersion
    relations 52–55
Piezoelectric coefficient in PMN 171,
    177–183

Piezoelectric coupling between tilt and
    polarization 83
Polarization profiles for $\delta = 0$ 222
Polar nanocluster size 129
Polar nanoclusters 144
Polyvinylidene fluoride (PVDF) 187–189,
    191
Positive-positive case in thin films
    $\delta_+ > 0, \delta_- > 0$ 224
Proustite 70
Pseudo-spin phonon coupling in SRBRF
    150
Publications on ferroelectricity 1–3

Quadratic magnetoelectric response
    123, 126
Quantum paraelectrics 12, 13

Raman spectra of relaxors 159, 163
Random bond-random field Ising model 118,
    149
Relation between T_1^{-1} and position on the
    I lineshape 59
Relaxors 1, 144
Reorientable polar electric and magnetic
    nanoclusters in birelaxors 132
Role of the surface in small enough
    nanoparticles 230, 231

Sine-Gordon equation 29, 49
Size dependence of toroid moment 245
Size induced ferroelectricy 235–239
Slater configurations 20, 95
Soliton density in 3-q modulated
    proustite 77
Soliton density n_s in Rb$_2$ZnCl$_4$ 38
Solitons 28, 41, 56
Specific heat of relaxors 148
Spontaneous flexoelectric effect 239–243
SRBRF model of relaxor polymers 191
Static dielectric constant in C and I
    phases 36–38
Static dielectric properties under constant
    magnetic field 135–137
Surface built-in fields 234
Surface symmetry and surface piezoelectric,
    piezomagnetic and ME
    tensors 230
Symmetric case $\delta_+ = \delta_- = 0$ 221

Takagi pairs 20
T dependence of EC 210–214
    electromechanical response near the
    critical point 203
Tetrafluoroethylene (TeFE) 188
The mixed case 225
The rigid spherical
    random-bond–random-field
    (SRBRF) model of relaxors 149

The SRBRF phase diagram 151
Thickness dependence of out-of-plane
    polarization in ultrathin films and
    depolarizing field 217
Thin films 2
Tilley–Žekš (T–Z) model of phase transitions
    in ferroelectric films 219
Translational lattice periodicity 23, 24, 44

Trifluoroethylene (TrFE) 188, 191
Tunneling 100, 117

Variation of critical thickness with misfit
    strain for epitaxial BT films 217

Water fall effect in relaxors 162